Palgrave Studies in Green Criminology

Series Editors
Angus Nurse
Nottingham Trent University
Nottingham, UK

Rob White
School of Social Sciences
University of Tasmania
Hobart, Australia

Melissa Jarrell
Department of Social, Cultural, and Justice Studies
University of Tennessee
Chattanooga, USA

Criminologists have increasingly become involved and interested in environmental issues to the extent that the term Green Criminology is now recognised as a distinct subgenre of criminology. Within this unique area of scholarly activity, researchers consider not just harms to the environment, but also the links between green crimes and other forms of crime, including organised crime's movement into the illegal trade in wildlife or the links between domestic animal abuse and spousal abuse and more serious forms of offending such as serial killing. This series will provide a forum for new works and new ideas in green criminology for both academics and practitioners working in the field, with two primary aims: to provide contemporary theoretical and practice-based analysis of green criminology and environmental issues relating to the development of and enforcement of environmental laws, environmental criminality, policy relating to environmental harms and harms committed against non-human animals and situating environmental harms within the context of wider social harms; and to explore and debate new contemporary issues in green criminology including ecological, environmental and species justice concerns and the better integration of a green criminological approach within mainstream criminal justice. The series will reflect the range and depth of high-quality research and scholarship in this burgeoning area, combining contributions from established scholars wishing to explore new topics and recent entrants who are breaking new ground.

Daan van Uhm • Dina Siegel
Editors

Global Green Crime and Ecojustice

palgrave
macmillan

Editors
Daan van Uhm
Willem Pompe Institute for Criminal
Law and Criminology
Utrecht University
Utrecht, The Netherlands

Dina Siegel
Willem Pompe Institute for Criminal
Law and Criminology
Utrecht University
Utrecht, The Netherlands

ISSN 2946-269X ISSN 2946-2703 (electronic)
Palgrave Studies in Green Criminology
ISBN 978-3-031-91556-7 ISBN 978-3-031-91557-4 (eBook)
https://doi.org/10.1007/978-3-031-91557-4

This Palgrave Macmillan imprint is published by the registered company Springer Nature Switzerland AG.
The registered company address is: Gewerbestrasse 11, 6330 Cham, Switzerland

Foreword

Green criminology has now—for decades—been pushing the boundaries and expanding consideration of what constitutes harm. Harm transcends notions of legality; it includes the lawful and unlawful, the administrative violations as well as the criminal ones. For many green criminologists, harm also transcends humans; it includes people, but also all parts of nature—the non-human is central to consideration. Regardless of the species—and not only species but also ecosystems and the environment—the criteria for something to be environmentally harmful are suffering, damage, injury, injustice.

(In)justice is also a concept that green criminology has challenged understanding of. And this edited collection furthers this debate. Ecological, or ecojustice, is justice for non-human nature (Washington et al., 2018) or the justice between human and non-human species (Naess, 1973). To have a just relationship with the Earth and all the beings on it, humans should minimize—or, at best, end—suffering, damage, and injury. From a largely global perspective detailing examples from Africa, America, Asia, and Europe, the chapters in this collection explore the myriad ways in which humans act unjustly with regard to the non-human. Many of these instances involve non-human animals as victims of illegal hunting and wildlife trafficking. They are taken from their homes, kept as companion animals, forced to race, or killed for food or medicine. All these situations are forms of injustice—to the individual non-humans who become victims, to non-human animal communities who lose members, and to the larger environment due to the degradation and disruption of ecosystems.

As the collection documents, eco-injustice is not confined to non-human animals. Human-induced wildfires are decimating landscapes as are human wars. The undeterred and unsustainable use of fossil fuels has set climate change in motion; this, in turn, is already impacting nature and human communities. In terms of human responses to climate change, one consequence is humans preying on other humans by trafficking them when they are forced to move.

The scale of injustice can be overwhelming, but contributions to the collection offer ways forward. One is foregrounding justice in the international justice system by recognizing ecocide as the fifth crime against peace within the Rome Statute. This would mean severe environmental harm becoming an internationally defined criminal offence. Another is expanding the use of restorative justice systems for green crimes so that when there is an injustice a remedy is central to the punishment. As the collection concludes, however, there are challenges for the furtherance of ecojustice. On a planet largely continuing environmental exploitation to sustain often lavish human lifestyles, where is the entry to scale up ecojustice? For anyone interested in learning more about the differing forms of eco-injustice and how even small improvements can be made to achieving more justice for the nonhuman, this collection is a vital resource.

Vienna, Austria — Tanya Wyatt

References

Naess, A. (1973). The shallow and the deep: Long-range ecology movement: A summary. *Inquiry, 16*, 95–99.

Washington, H. Chapron, G., Kopnina, H., Curry, P., Gray, J., & Piccolo, J. J. (2018). Foregrounding ecojustice in conservation. *Biological Conservation, 228*, 367–374. https://doi.org/10.1016/j.biocon.2018.09.011

Contents

Notes on Contributors

Paul Allin obtained his masters in Wildlife Management at the University of Pretoria in 2016 and is currently conducting his PhD research at the University of Stellenbosch, looking at using remote sensed imagery and machine learning to improve game censuses. Paul leads the research department within Transfrontier Africa and believes that the best contribution research can bring in this field is through applied studies that increase our understanding and foster coexistence.

Andrey G. Anisimov is a criminologist from Russia, with a PhD in Criminology from Saratov State Academy of Law (2016). He served 17 years in the Ministry of Internal Affairs, focusing on crime research and prevention in the Irkutsk region. He is currently an Associate Professor of Criminal Procedure and Forensic Science at the Irkutsk Law Institute of the All-Russian State University of Justice. His scientific interests include environmental crimes, crime and globalization, and crime and culture.

Inés Arroyo-Quiroz (Mexico), Biologist (UNAM, 1996), with a PhD in Biodiversity Management (University of Kent, 2004), works as a researcher on animal use, wildlife trade, green criminology and ecoglobal history, and is currently developing the projects 'A historic perspective of the fur and leather trade from wild species involving Mexico centuries' and ‚Representation of wild animals in Mexican cinema, from the perspective of green cultural criminology'. Inés also acts as co-chair of the IUCN Green Criminology Specialist Group.

Mark Brown is a criminologist at the School of Law of the University of Sheffield (UK) and the School of Social and Political Sciences, University of Melbourne (Australia), with an interest in transnational organized crime. Some of the research upon which his chapter is based was undertaken while working as an advisor exploring the intersection of Myanmar's illicit economy and the long-running civil war and periodic insurgencies in the country's borderlands.

Manuel Cabezas Vicente holds a PhD with cum laude distinction from the University of Salamanca, where he also earned a Law degree, a Master's in Access to the Legal Profession, and a Master's in Criminal Law. A member of CIDH-Diversitas, he has contributed to various European, national, and regional R+D+i projects. He teaches at the University of Salamanca, collaborates on postgraduate programs, has published internationally, and completed research stays at the University of Montreal, McGill University, and the University of Granada.

Ruishan Chen is Professor and Vice Dean at the School of Design, where he works on biodiversity conservation. Chen is the lead author of the IPBES Global Land Degradation and Restoration Assessment Report and the Transformation Change Report.

Nick van Doormaal completed his PhD in Criminology at the Netherlands Institute for the Study of Crime and Law Enforcement (NSCR) and Utrecht University (UU) in 2020. Throughout his studies, Nick collaborated multiple times with Transfrontier Africa, both in conducting studies on wildlife poaching within Olifants West Nature Reserve and in setting up research experiments on wildlife security. Nick's research interests include wildlife crime, environmental criminology, and human–wildlife conflicts by using quantitative methods.

Orion L. B. Goodman is a PhD student at John Jay College of Criminal Justice, where he intends to continue his research on trade in wildlife and other natural resources.

Janine Heim is a PhD researcher at Maastricht University, researching the exotic pet trade between Germany and South Africa. With a background in communication and security studies, her interests also include the trade structures and networks of other niche sectors.

Alison Hutchinson is a green criminologist with research interests including wildlife and environmental harms and crimes, global biodiversity governance, and social-ecological justice. She has researched the intersections of legal and illegal trade in European species through the ESRC-funded *Beastly Business* project and, prior to this, developed a global legislation dashboard on wildlife offences. She earned her PhD from Northumbria University (UK), where her thesis examined the harms and crimes against commercially exploited marine species.

Susanne C. Knittel is Associate Professor of Comparative Literature at Utrecht University in the Netherlands. Her research centres on the question of how societies remember atrocities—specifically how they deal with the uncomfortable issues of guilt and responsibility—and what role literature, art, film, and other cultural representations play in this process. Currently, she is the principal investigator of the ERC-funded EcoViolence project, which explores how contemporary culture represents and remembers the genocide–ecocide nexus.

Marijke van Kuijk is an Associate Professor at Utrecht University, the Netherlands. She is a forest ecologist committed to global forest conservation, in particular in tropical areas. Her research focuses on plant–animal interactions across diverse forest conditions, including intact, hunted, logged, fragmented, and recovering ecosystems. She leverages her expertise to guide forest management, conservation, and restoration efforts.

Jennifer Maher is Associate Professor of Criminology at the University of South Wales, UK. Her main research interests lie in green criminology and interspecies and interpersonal violence and harms, including wildlife crime, illegal pet trades and dangerous dogs. Her research has attracting funding from the European Commission, the European Parliament, UK governments, South Wales Police and the RSPCA, and her expertise is sought internationally, including completing the UN ICCWC Toolkit review of the UK response to wildlife and forest crime. She is currently reviewing the Welsh Government's Animal Health and Welfare Framework and evaluating alternative approaches to effective dog control for the RSPCA. She is co-Chair of the GREEN Working Group of the European Society of Criminology (ESC).

Anna Markovska is an Associate Professor in Criminology at a Deputy Director of the International Policing and Public Protection Research Institute (IPPPRI) at Anglia Ruskin University. Her research interests encompass policing and organized crime within the post-Soviet context. She has contributed to several research projects focused on Ukrainian society during the war.

Angus Nurse is Professor of Law and Environmental Justice, Director, Centre for Access to Justice and Inclusion (CAJI), Anglia Ruskin University, Cambridge, UK. Angus researches and teaches critical criminal justice and redress for social harms, with specific interests in animal law, corporate environmental responsibility, environmental law and human rights.

Mònica Pons-Hernández is an MSCA SEAS Postdoctoral Research Fellow at the University of Bergen, specializing in green criminology. She holds a BS in Criminology from the University of Barcelona, an MS in Transnational Crime from the University of Glasgow, and a PhD in Law from the University Rovira i Virgili. Her research on glass eel trafficking has been given an award by the Spanish Society of Criminological Research and the American Society of Criminology.

Andrew L. Rhyne is a professor of marine biology at Roger Williams University, where he has invested his career in understanding the complexities of marine ornamental aquaria.

Jessica Bell Rizzolo is a conservation social scientist who holds a joint PhD in Sociology and Environmental Science and Policy from Michigan State University. She has published work on wildlife crime prevention, the wildlife trade, and wildlife consumption in journals such as *EcoHealth*, *Frontiers in Conservation Science*, *Global Ecology and Conservation*, and *Nature Conservation*.

Cedric Ryngaert (PhD Leuven 2007) is Professor of Public International Law at Utrecht University (Netherlands). His research interests include the law of jurisdiction and extraterritoriality, immunities, sanctions, international security law, international criminal law, non-state actors, the role of international law before domestic courts, international responsibility, and international organizations. He is the chair of the Dutch Advisory Council on International Law (CAVV) and the editor-in-chief of the *Netherlands International Law Review* and the *Utrecht Law Review*.

Nieves Sanz Mulas is a Professor of Criminal Law, with three years' experience as a Prosecutor. An expert in Criminal Policy, Professor Sanz has given more than 300 lectures at national and international level, and has been awarded an honorary doctorate on four occasions. She is the author of more than 100 publications and is currently the director of the Centre for Research in Human Rights and Public Policy (*CIDH-Diversitas*) at the University of Salamanca, Spain.

Oleksii Serdiuk is an associate professor and Head of the Research Laboratory for Social and Psychological Support and Rehabilitation of MoIA Employees at Kharkiv National University of Internal Affairs, Ukraine. He is also a National Contact Point (NCP) for Horizon Europe Cluster 3 and holds the rank of Lieutenant Colonel of Police. In 2015, he was a visiting scholar and senior research fellow at the Addiction Center, University of Michigan, USA. With over 25 years of experience in fostering cross-disciplinary and international research collaborations, he is the author of more than 190 academic and professional publications.

Dina Siegel is Professor of Criminology at the Willem Pompe Institute for Criminal Law and Criminology at Utrecht University. Dina Siegel studied Social Anthropology and Sociology at Tel Aviv University and obtained her PhD in 1996 from the Vrije Universiteit Amsterdam. Between 2001 and 2008, she worked as an assistant professor at the Faculty of Law at the Vrije Universiteit Amsterdam, and since 2009, she has been a professor of Criminology at Utrecht University. Dina Siegel conducts research on various forms of organized crime. She has published on topics such as the Russian mafia, human trafficking, drug smuggling, terrorism, caviar trade, underground banking, dog fighting, crime in the diamond industry, mobile banditry, and the intersection of cultural criminology and green criminology.

Iryna Soldatenko is the head of the department of Applied Sociology and Social Communications at V. N. Karazin Kharkiv National University, where she has held the title of associate professor since 2006, having earned her PhD in Sociology in 2004. She is currently a British Academy Research Fellow at the International Policing and Public Protection Research Institute (IPPRI), Anglia Ruskin University, UK, and has previously held a visiting researcher position at the Leibniz Institute for Media Research | Hans-Bredow-Institut (HBI) in Hamburg, Germany (July–December 2022). Her academic expertise spans communication, sociology, public relations, and anticorruption methodologies.

Ragnhild Sollund is Professor of Criminology at the University of Oslo. She has mostly carried out research in the areas of migration, racial profiling, and green criminology. From 2010, she has concentrated her research on human–animal relations, particularly wildlife crimes and harms. Sollund was partner to the EFFACE project, funded by the European Union, and project leader of Criminal Justice, Wildlife Conservation and Animal Rights in the Anthropocene (CRIMEANTHROP) (2019–2024) funded by the Norwegian Research Council. Sollund has a wide range of publications, including *The Crimes of Wildlife Trafficking: Issues of Justice, Legality and Morality* (Routledge, 2019) and *Wildlife Trade and Animal Victimization: Parallel Harms and Crimes* (Routledge, 2025).

Nigel South is Emeritus Professor of Sociology and Criminology at the University of Essex, UK. He has made various contributions to the development of 'Green Criminology'. In 2022, he received the 'Outstanding Achievement Award' from the British Society of Criminology and in 2013 received a 'Lifetime Achievement Award' from the American Society of Criminology, Division on Critical Criminology and Social Justice.

Matilde Souto Mantecón who has a PhD in History from El Colegio de México, specializes in 18th-century economic, political, and maritime history. Her most recent publications include, co-authored with Fausta Gantús, "One City, Two Cartographies: Campeche, 18th and 19th

Centuries" (2024), and "The Mexican Press and Historiography in the Face of Spanish Conflict. The Use of Resentment in the Case of José María Lobato's Pronouncement" (2024).

Craig Spencer is the founder and executive director of Transfrontier Africa, as well as the warden of Olifants West Nature Reserve, which is located in the Balule region of the Greater Kruger National Park in South Africa. Through his NGO, Craig has developed novel solutions, including the Black Mamba wildlife security program which focuses on young women from the neighbouring tribal villages.

Michael F. Tlusty is an associate professor of sustainability and food solutions in the School for the Environment at the University of Massachusetts at Boston.

Daan van Uhm is Professor of Environmental Crime and Associate Professor of Criminology (Open University/Utrecht University). Daan van Uhm has conducted research on various forms of environmental crime, including illegal mining in Latin America, wildlife trafficking in Southeast Asia, deforestation in Central Africa, and the criminalization of ecocide. He obtained his PhD in Criminology at Utrecht University in 2016 (now published by Springer as *The Illegal Wildlife Trade: Inside the World of Poachers, Smugglers and Traders*). In 2018, he received the Veni grant of the Dutch Organisation for Scientific Research (NWO) for his research project 'The Diversification of Organized Crime into the Illegal Trade in Natural Resources' and in 2022 the ERC Starting Grant (European Research Council) for the research project 'Green Crimes and Joint Crime Ventures: Laundering Natural Resources'. Van Uhm primarily focuses on research in the context of environmental crime and green criminology.

Rebecca W. Y. Wong is an Associate Professor in the Department of Social and Behavioural Sciences at the City University of Hong Kong. Rebecca's research interests concern the intersections of green criminology, illegal trade in protected and endangered wildlife, animal abuse, criminal networks and issues of trust in the underworld. Rebecca holds a DPhil. in Sociology from the University of Oxford. She is also an Associate Fellow of the Extra Legal Governance Institute at the University of Oxford. In collaboration with the Netherlands Institute for the Study of Crime and Law Enforcement, she was awarded a Visiting Professorship by the Royal Netherlands Academy of Arts and Sciences.

Annah Lake Zhu is Assistant Professor of Environmental Policy at Wageningen University in the Netherlands. She received her PhD in Society and Environment from the University of California, Berkeley and her Master's in environmental management from Duke University. Annah's research investigates China as an emerging global environmental actor from a critical social science perspective. She is the author of the book *Rosewood: Endangered Species Conservation and the Rise of Global China*, published by Harvard University Press.

List of Figures

List of Images

Wildlife Trafficking in Spain: Causes, Consequences and Future Considerations

Photovoice for Ecojustice: Following the European Eel from the Sargasso Sea to the River Severn

List of Tables

A Source of Disease, Health, or Both? Wildlife Consumption in China after the Emergence of COVID-19

Environmental Crimes in East Siberia and Activities of Chinese Mafia

Global Green Crime and Ecojustice

Daan van Uhm and Dina Siegel

Over the past decades, green crimes and ecojustice has gained significant international attention. Characterized by profound impacts such as deforestation, mining, climate change, and species extinction, 'green' issues have become global in nature, with far-reaching environmental and social consequences (Dirzo et al., 2014; Johnson et al., 2017; Steffen et al., 2015). Scientists have warned of a new, so-called Anthropocene epoch, which is marked by significant changes to the Earth and atmosphere as the result of human activities. The collapse of pre-industrial institutions and the development of new free trade zones and technologies have radically transformed the world and its ecosystems (Beck, 1992; Crutzen & Stoermer, 2000).

Understanding the interconnectedness of all life forms and their contributions to ecosystem stability and resilience is crucial for comprehending the evolving landscape of environmental protection and ecojustice (Benton, 1998; White, 2011). This emphasizes the necessity

D. van Uhm (✉) • D. Siegel
Willem Pompe Institute for Criminal Law and Criminology
Utrecht University, Utrecht, The Netherlands
e-mail: d.p.vanuhm@uu.nl; d.siegel@uu.nl

D. van Uhm, D. Siegel (eds.), *Global Green Crime and Ecojustice*, Palgrave Studies in Green Criminology, https://doi.org/10.1007/978-3-031-91557-4_1

of preserving the environment not merely for its utilitarian benefits to humans but also for its inherent worth (Wilson, 1993; Crook et al., 2018). As a result, green crimes have emerged as significant global issues, necessitating their anticipation and integration into both scientific and social discussions.

Green Crimes and Criminogenic Implications

In recent years, green crime has emerged as one of the most significant black markets worldwide (UNEP & INTERPOL, 2016). The financial incentives for engaging in green crimes are substantial, with illegal activities such as deforestation, wildlife trafficking, illegal fishing, mining and waste dumping generating billions of dollars annually (Moreto & Pires, 2018; Petrossian, 2019; Siegel, 2020; Sollund, 2019; van Uhm, 2023; Wyatt, 2013; Wong, 2019). These activities are often considered as less risky compared to other forms of crime, as enforcement and penalties remain relatively weak. Therefore, crime groups are increasingly diversifying into the lucrative markets of natural resources, alongside other criminal activities (van Uhm & Nijman, 2022).

The criminogenic nature of global environmental issues perpetuates unequal power relations across the world. The increasing scarcity of natural resources, rising deforestation, and global pollution are all significant predictors of crime, conflict, and violence (Le Billon, 2012; Siegel, 2020). These environmental stressors exacerbate social inequalities and contribute to the marginalization of vulnerable communities, who are often the most affected by environmental degradation (Duffy, 2010; Ross, 2003). The competition for dwindling resources can lead to disputes, often escalating into broader conflicts. The exploitation of natural resources by crime groups further destabilizes these regions, creating a vicious cycle of environmental harm and social unrest (Welsch, 2008; van Uhm, 2024).

Social Constructions and Global Asymmetries

Forms of green crime and harm are numerous, ranging from everyday practices of abuse and exploitation to deliberate actions by corporations, states, armed groups, or organized crime (Brisman & South, 2020; van Uhm & Siegel, 2021). However, many harmful environmental activities have not been criminalized (yet). One significant reason for this is the close collaboration between governments and industry, which preserves the social and economic structures underlying nature exploitation (Michalowski & Kramer, 2006). The unbridled economic growth of powerful industrialized states is particularly implicated in the ongoing damage to the environment worldwide (Stretesky et al., 2013). These economic interests reproduce inequalities globally and have long hindered the criminalization processes of environmental harm (Passas & Goodwin, 2005; Whyte, 2020).

This is further complicated by the varying levels of environmental protection and enforcement across different countries, leading to a patchwork of regulations that are often insufficient to address the global nature of these crimes (Elliott & Schaedla, 2016; Spapens et al., 2018). These disparities also reflect differing societal and cultural values, historical relationships with nature, and priorities in environmental governance (Zhu, 2022). The asymmetries in international regulations and enforcement mechanisms exacerbates the problem, allowing perpetrators to exploit legal loopholes and jurisdictional gaps (White & Pink, 2016; Smith & Sahramaki, 2024).

Green Criminology and Ecojustice

With the emergence of green criminology, the focus is shifting towards examining the symbiotic relationship between humans and nature through a broadly conceived harm principle (Beirne & South, 2007; Sollund, 2008). This field of study expands the traditional scope of criminology, which typically centres on human victims and perpetrators, to include the environment and non-human species as potential victims. By adopting an ecocentric perspective, green criminology emphasizes the interconnectedness of all life forms and the importance of protecting ecosystems for the benefit of both human and non-human communities (Benton, 1998; Halsey & White, 1998). This green criminological approach emphasizes the role of power dynamics behind environmental exploitation, noting how marginalized and vulnerable populations often suffer the most from environmental degradation (Goyes, 2019; Ruggiero & South, 2013).

Recognizing ecosystems, rivers, and non-human animals as potential rightsholders challenges traditional legal frameworks and underscores the need to acknowledge their intrinsic value and reconsider human-centred approaches (Hall, 2013; Nurse, 2015). The discrimination against non-human animals based on perceived 'inferior' characteristics, which is socially legitimized through abusive or exploitative acts, highlights the artificial boundaries constructed between human and non-human entities (Beirne, 1999; Maher et al., 2017; Sollund, 2019). An ecojustice perspective recognizes the intrinsic value of ecosystems and non-human species and calls for a re-evaluation of our ethical and legal responsibilities towards the environment and its inhabitants.

The socio-economic impacts of green crimes are profound, affecting not only the environment but also the livelihoods and health of vulnerable communities, particularly in developing regions (Gore, 2017; Zabyelina & van Uhm, 2020). For example, illegal logging and deforestation can lead to loss of biodiversity, the disruption of water cycles, and increased greenhouse gas emissions, which contribute to climate change. These activities often displace Indigenous communities and destroy their traditional ways of life, reflecting the intersection of environmental and social justice issues (Bedford et al., 2020; Duffy, 2010). Therefore, interdisciplinary and transdisciplinary collaborations are crucial to the development of comprehensive strategies that reflect both human and ecological well-being. By fostering a deeper understanding, we can better anticipate and mitigate the impacts of global green crime.

This Book

In 2023, a two-day international conference on global green crime and ecojustice was organized by the editors of this book at Utrecht University. Academics and practitioners from different scientific fields including green criminology, environmental sociology, environmental law, political ecology, and conservation studies, among others, gathered in Utrecht to discuss the urgent challenges presented by environmental crimes. These challenges are not just environmental, but also social, economic and political in nature, and they have serious implications for human rights, public health, and social justice. Green crimes not only harm the environment but also affect the lives of millions of people, particularly those living in poverty and marginalized communities. They also undermine the rule of law, democratic institutions and governance, and foster corruption and organized crime.

During the conference leading scholars and practitioners shared their insights and experiences on a wide range of topics related to green crime and ecojustice, exploring the causes and consequences of green crime, discussing the legal frameworks and policy measures and considering ways to improve them to ensure that they respect the rights of all stakeholders, including human and non-human entities. This conference was therefore an opportunity for critical dialogue and the exchange of ideas, plans and projects on how to address the challenges posed by environmental crime and pursue eco justice.

The participants also tried to explore the intersectionality of green crimes with other forms of crime and social justice issues, including the smuggling of natural resources, human trafficking, labor exploitation, gender-based violence, and discrimination. We discussed the theoretical and methodological issues of research into green crimes. The most important purpose of the conference, and consequently of this volume, is to be aware of the fact that behind the statistics and policies there are real people with real hopes and dreams, who are often forced to bear the burden of environmental crime and degradation. Equally important is the recognition that non-human life—animals, plants, ecosystems—suffers deeply from these harms, often without voice or representation, yet with consequences that ripple across the entire web of life. This book includes a selection of contributions to the conference and raises important questions about social constructions, perpetrators, cultures, traditions, victims, markets, conflicts, as well as responsibility, liability and duty of care. It highlights the importance of interdisciplinary academic research and dialogue on global green crime and ecojustice as foundations for policy and law-making in the near future.

This book consists of four parts:

Part I Green crime and ecojustice
Part II Flora and fauna markets
Part III Culture, politics, and social constructions
Part IV Conflict, organized crime, and crime convergence

Part I Green Crime and Ecojustice

Part I provides a comprehensive introduction to green crime and ecojustice, encompassing discussions on conceptual approaches and critical reflections on the proliferation of green crimes and the responses to them. This section delves into the theoretical foundations of green criminology and ecojustice by integrating diverse disciplinary perspectives for a deeper understanding of the complexities and challenges in addressing green crime and ecojustice.

In the chapter, "Planetary Health and Eco-Justice: Contributions from Green, Southern, Cultural and Decolonial Criminologies", Nigel South examines the theme of planetary health and ecojustice through the lens of green, southern, cultural, and decolonial criminologies. He argues that a holistic approach should recognize the interdependence of all species on the planet and include considerations for the health of both human and non-human species, as well as the natural environment. South concludes that integrating these perspectives is crucial for fostering a sustainable and just future.

In the next chapter, "Greening Justice: Ecojustice and Remedying Environmental Harm", Angus Nurse argues that effective ecojustice necessitates an approach focused on remedying environmental harms. This approach should consider the needs of the environment and non-human nature by integrating punitive measures with restorative and preventative elements, centring the harm caused to nature.

Chapter "Living, Dying, and Sharing a Planet", by Ragnhild Sollund, analyses the dynamics of sharing space with other animals. She examines scenarios where animals enter spaces humans consider their own, where humans encroach upon spaces regarded by animals as theirs, and where humans and animals coexist in shared spaces. Sollund discusses how humans transgress the boundaries of other animals, contributing to the continuous destruction and encroachment of the steadily shrinking habitats of nonhuman species.

Chapter "Conceptualizing Ecocide – Interdisciplinary Perspectives" by Daan van Uhm, Cedric Ryngaert, Susanne Knittel and Marijke van Kuijk, is based on an interdisciplinary project focused on conceptualizing ecocide. It explores how perspectives from ecology, criminology, international criminal law, and the humanities can contribute to a holistic understanding of ecocide.

Part II Flora and Fauna Markets

Part II presents several case studies on flora and fauna markets, examining their significant impact on the environment and non-human animals. By analyzing these markets, the authors explore various aspects, including exploitation and illegal trade.

Chapter "The European Illegal Puppy Trade & Organised Crime", by Jennifer Maher, investigates the extensive illicit trade in puppies across Europe. This chapter addresses irresponsible and illegal breeding practices, the often-illegal movement of puppies, and the involvement of organized crime groups. Through the analysis of hundreds of online advertisements, expert interviews, a stakeholder survey, and dog owner focus groups across Great Britain, the data reveals an organized illicit market operating parallel to the legal trade, underscoring the need for enhanced legislation, traceability, and enforcement.

Chapter "Wildlife Trafficking in Spain: Causes, Consequences and Future Considerations", by Mònica Pons-Hernández, provides insights into wildlife trafficking in Spain. This chapter is based on a content analysis of data from press releases of the Nature Protection Service (SEPRONA), the Wildlife Trade Portal, governmental and non-governmental documents, academic articles, and journalistic reports. The findings show that species are trafficked in Spain for various reasons and highlight Spain's pivotal role as a country of origin, transit, and destination for illegally traded fauna and flora.

The next chapter, "Not My Snake, Not My Circus: Customary Illegality and Moral Economies in the Bitis Trade", by Janine Heim, delves into the concept of 'customary illegality' in the African viper trade. The research employs the 'zooming-in, zooming-out' technique, a method predominantly used in organizational research, to trace the formation of the trade by observing practices and connections. This includes digital platforms like Facebook, Instagram, and TikTok, as well as physical events like exotic pet expos. The concept of customary illegality is influenced by a moral economy shaped by desires, experiences, and attitudes.

In Chapter "Poaching Trends and Patterns During the COVID-19 Pandemic: Insights from a Private Nature Reserve in South Africa", Nick van Doormaal, Paul Allin, and Craig Spencer retrospectively investigate the impact of COVID-19 lockdown restrictions on rhino and bushmeat poaching in South Africa. The study found that rhino poaching did not increase during or after the lockdown, although more fresh snares for bushmeat were discovered post-lockdown. Law enforcement rangers, deemed essential services, continued patrols, and the heightened security likely increased the risk for poachers. However, the lockdown's severe impact on local communities bordering the reserve may have led to a rise in bushmeat poaching after the restrictions were lifted.

Chapter "Understanding and Navigating International Wildlife Trade Data: Avoiding Pitfalls Caused by Systemic Detail Obfuscation", by Orion Goodman, Michael Tlusty, and Andrew Rhyne, reviews how wildlife trade datasets are derived and examines their limitations and characteristics by looking at the Convention on the International Trade in Endangered Species of Wild Fauna and Flora (CITES) and the United States Fish and Wildlife Service's Law Enforcement Management System

(LEMIS) databases. The insights provide wildlife trade researchers with contextual information that will enable them to navigate these issues when conducting their investigations.

Part III Culture, Politics and Social Constructions

Part III emphasizes the importance of investigating green crimes and ecojustice through the lenses of culture, politics, and social constructions. This part explores how cultural narratives and political frameworks shape our understanding and examines the role of social constructions in defining what constitutes a green crime and how these definitions influence norms and policies.

Chapter "Cockroaches Races and Crickets Fights", by Dina Siegel, explores the intersection of green and cultural criminology by examining cockroach races and cricket fights. This chapter highlights the paradoxical relationship between humans and insects, where insects are regarded as pets, friends, and athletes, as well as sources of food, profit, and entertainment. It underscores that gambling remains a significant element of both cockroach racing and cricket fighting. The legitimacy of these events is rooted in long-standing traditions but also involves ethical considerations influenced by the ongoing extinction crisis and animal welfare principles.

Chapter "The Devastation of Sea Otters in the American Northwest: A Look from Eco Global History and Criminology", by Matilde Souto Mantecón and Inés Arroyo Quiroz, analyzes the devastation of sea otters in the North Pacific, where their skins became valuable commodities in international trade during the 18th and 19th centuries. This chapter examines these processes through the lenses of conservation biology, green criminology, and history to foster critical consciousness. Understanding this historical phenomenon is crucial, as similar eco-global crimes continue today, affecting many species worldwide.

Chapter "What Mercy? A Study of Mercy Release in Hong Kong", by Rebecca Wong, analyzes the ritual of mercy release, where captured animals are released into the wild to generate merit and good karma. The study explores the demographics and motivations of participants, based on interviews and observations conducted in various mercy release hotspots across Hong Kong and examines how cultural norms shape this practice. The empirical data reveals that interviewees dismissed any harmful implications and believed their good intentions were misinterpreted, indicating they will continue their practices influenced by cultural beliefs.

Chapter "Photovoice for Ecojustice: Following the European Eel from the Sargasso Sea to the River Severn", by Alison Hutchinson, adopts a visual green-cultural criminological point of departure to examine how threats towards the European eel are conceptualized and responded to. Building on a green criminological focus on harm rather than explicit crime, the article follows the eel's migratory route and compiles photovoice interviews and observations to illustrate the intersections of (non) human and environmental harms. This approach provides a nuanced understanding of the multifaceted threats facing the European eel and highlights the importance of visual methodologies in capturing these complex dynamics.

Chapter "A Source of Disease, Health, or Both? Wildlife Consumption in China after the Emergence of COVID-19", by Jessica Bell Rizzolo, Annah Zhu, and Ruishan Chen, examines research in China on attitudes towards wildlife consumption, wildlife consumption bans, and zoonotic diseases in the wake of COVID-19. They explore how support for wildlife consumption bans intersects with perceptions of zoonotic risk, the source of wildlife products (farmed or wild-caught), and the use of wildlife supplements in Traditional Chinese Medicine. Using China as a case study, they outline how researchers and practitioners should consider analyzing attitudes towards zoonotic risk and health when instituting bans or other regulatory mechanisms on wildlife consumption.

Part IV Conflict, Organized Crime, and Crime Convergence

Part IV draws attention to conflict, organized crime, and crime convergence in the context of green crime and ecojustice. This section examines how environmental degradation and resource scarcity can exacerbate conflicts and foster organized criminal activities. It also explores the intersections between green crimes and other forms of illicit activities, highlighting the complex networks that facilitate these crimes.

In Chapter "Green Crimes, Illicit Economies and the Emergence of a Novel Mekong Crime-Conflict Complex" Mark Brown draws on empirical data from Myanmar and proposes a new conceptual approach to environmental crimes and illicit economies by introducing the idea of a Mekong crime complex. This crime complex operates on a regional scale that is more than the additive sum of national parts and exhibits crime dynamics unique to its scale, posing challenges to conventional approaches towards crime control. The chapter underscores the importance of regional cooperation and innovative strategies to effectively address these complex crime dynamics.

Chapter "Environmental Crimes in East Siberia and Activities of Chinese Mafia", by Andrey Anisimov, discusses environmental crimes in East Siberia in Russia and the activities of the Chinese Mafia. Based on official statistics from the Main Directorate of the Ministry of Internal Affairs of Russia for the Irkutsk region, official reports from the Irkutsk regional police, and surveys and interviews with local Russian and Chinese entrepreneurs, the chapter examines how Chinese criminal organizations are the main facilitating agents, taking advantage of the illegal benefits and explores the cooperation between Russian and Chinese criminal groups in facilitating green crimes.

Chapter "The Climate Change-Migrant Smuggling Nexus: Smuggling Services as the Only Way to Reach Europe" by Manuel Cabezas-Vicente, analyzes the consequences of climate change, compelling millions to abandon their homes. This leads to international economic and climate migrants often resorting to smuggling services for perilous journeys. The study focuses on the relationship between climate change and smuggling along Spanish migratory routes, highlighting its impact on Europe in terms of numbers and mortality. It examines climate change as a motivating factor in international mixed migration and smuggling networks, which are primary means of entering Europe.

Chapter "Environmental Crimes in Ukraine Under War Conditions by Anna Markovska, Oleksii Serdiuk, Angus Nurse and Iryna Soldatenko, critically analyzes the severe environmental impact of the

Russian invasion of Ukraine on its atmosphere, water, soil, and ecosystems. It examines Ukraine's legal framework for quantifying this damage, explores the potential of the Rome Statute and International Criminal Court (ICC) and advocates for a comprehensive approach to environmental restoration through international cooperation to ensure accountability for environmental devastation caused by armed conflicts.

Chapter "The Hidden Toll: Women and Girls in the Intersection of Climate Change and Domestic Servitude" by Nieves Sanz Mulas, addresses how the current climate crisis is significantly altering migration patterns, particularly affecting vulnerable populations such as rural and indigenous communities. She highlights the migration of millions of women due to severe climatic consequences, a clear effect of the global feminization of poverty. This 'genderization' of climate migration results in disparities rooted in the social position of women within patriarchal cultures and economic processes, leading to various forms of violence, abuse, trafficking, and enslavement.

References

Beck, U. (1992). *Risk society: Towards a new modernity*. Sage.

Bedford, L., McGillivray, L. & Walters, R.(2020). Ecologically Unequal Exchange, Transnational Mining, and Resistance: A Political Ecology Contribution to Green Criminology. *Critical Criminology* 28, 481–499.

Beirne, P. (1999). For a nonspeciesist criminology: Animal abuse as an object of study. *Criminology, 37*(1), 117–148. https://doi.org/10.1111/j.1745-9125.1999.tb00481.x

Beirne, P., & South, N. (2007). *Issues in green criminology: Confronting harms against environments, humanity and other animals*. Willan Publishing.

Benton, T. (1998). Rights and justice on a shared planet. *Theoretical Criminology, 2*(2), 149–175. https://doi.org/10.1177/1362480698002002002

Brisman, A., & South, N. (2020). *Routledge international handbook of green criminology*. Routledge.

Crook, M., Short, D., & South, N. (2018). Ecocide, genocide, capitalism and colonialism: Consequences for indigenous peoples and glocal ecosystems environments. *Theoretical Criminology*, 22(3), 298-317.

Crutzen, P. J., & Stoermer, E. F. (2000). The 'Anthropocene'. *Global Change Newsletter, 41*, 17–18.

Dirzo, R., Young, H. S., Galetti, M., Ceballos, G., Isaac, N. J. B., & Collen, B. (2014). Defaunation in the Anthropocene. *Science, 345*(6195), 401–406. https://doi.org/10.1126/science.1251817

Duffy, R. (2010). *Nature crime: How we're getting conservation wrong.* Yale University Press.

Elliott, L., & Schaedla, W. (2016). *Handbook of transnational environmental crime*. Edward Elgar.

Gore, M. (2017). *Conservation criminology*. Wiley.

Goyes, D. R. (2019). *Southern green criminology: A science to end ecological discrimination*. Emerald Publishing Limited. https://doi.org/10.1108/9781787692299

Hall, M. (2013). *Victims of environmental harm: Rights, recognition and redress under national and international law*. Routledge.

Halsey, M., & White, R. (1998). Crime, ecophilosophy and environmental harm. *Theoretical Criminology, 2*(3), 345–371. https://doi.org/10.1177/1362480698002003003

Johnson, C. N., Balmford, A., Brook, B. W., Buettel, J. C., Galetti, M., Guangchun, L., & Wilmshurst, J. M. (2017). Biodiversity losses and conservation responses in the Anthropocene. *Science, 356*(6335), 270–275. https://doi.org/10.1126/science.aam9317

Le Billon, P. (2012). *Wars of plunder: Conflicts, profits and the politics of resources.* Columbia University Press.

Maher, J., Pierpoint, H. & Beirne, P. (2017). *The Palgrave International Handbook of Animal Abuse Studies.* Palgrave.

Michalowski, R., & Kramer, R. (2006). *State-corporate crime: Wrongdoing at the intersection of business and government*. Rutgers University Press.

Moreto, W. D., & Pires, S. (2018). *Wildlife crime: An environmental criminology and crime science perspective*. Carolina Academic Press.

Nurse, A. (2015). *Policing wildlife*. Palgrave Macmillan UK. https://doi.org/10.1057/9781137400017

Passas, N., & Goodwin, N. (2005). *It's legal but it ain't right.* University of Michigan Press. https://doi.org/10.3998/mpub.11472

Petrossian, G. A. (2019). *The last fish swimming: The global crime of illegal fishing*. Praeger.

Ross, M. (2003). The natural resource curse: How wealth can make you poor. In I. Bannon & P. Collier (Eds.), *Natural resources and violent conflict.* World Bank.

Ruggiero, V., & South, N. (2013). Green criminology and crimes of the economy: Theory research and praxis. *Critical Criminology, 21*(3), 359–373. https://doi.org/10.1007/s10612-013-9191-6

Siegel, D. (2020). Diamond mining, organized crime, and corruption. In Y. Zabyelina & D. P. van Uhm (Eds.) *Illegal mining: Organized Crime, Corruption, and Ecocide in a Resource-Scarce World* (pp. 185–203). Springer International Publishing. https://doi.org/10.1007/978-3-030-46327-4_7

Smith, S.L. and Sahramäki, I. (2024). *Research Handbook on Environmental Crimes and Criminal Enforcement.* Elgar.

Sollund, R. (2008). *Global harms: Ecological crime and speciesism.* Nova Science Publishers.

Sollund, R. (2019). *The crimes of wildlife trafficking: Issues of justice, legality and morality*. Routledge.

Spapens, T., White, R., van Uhm, D. P., & Huisman, W. (2018). *Green crimes and dirty money*. Routledge.

Steffen, W., Richardson, K., Rockström, J., Cornell, S. E., Fetzer, I., Bennett, E. M., Biggs, R., Carpenter, S. R., de Vries, W., de Wit, C. A., Folke, C., Gerten, D., Heinke, J., Mace, G. M., Persson, L. M., Ramanathan, V., Reyers, B., & Sörlin, S. (2015). Planetary boundaries: Guiding human development on a changing planet. *Science, 347*(6223). https://doi.org/10.1126/science.1259855

Stretesky, P., Long, M., & Lynch, M. (2013). *The treadmill of crime political economy and green criminology*. Routledge.

UNEP & INTERPOL. (2016). *The rise of environmental crime: A growing threat to natural resources peace, development and security*. UNEP and INTERPOL.

van Uhm, D. P. (2023). *Organized environmental crime: Black markets in gold, wildlife, and timber*. Bloomsbury.

van Uhm, D. P. (2024). Green Criminology and Biodiversity Loss: Crimes and Harms against Flora and Fauna. In *Oxford Research Encyclopedia of Criminology*. Oxford University Press.

van Uhm, D. P., & Nijman, R. C. C. (2022). The convergence of environmental crime with other serious crimes: Subtypes within the environmental crime continuum. *European Journal of Criminology, 19*(4), 542–561. https://doi.org/10.1177/1477370820904585

van Uhm, D. P., & Siegel, D. (2021). Organised crime and animals. *Trends in Organized Crime, 24*(4), 419–424. https://doi.org/10.1007/s12117-021-09440-z

Welsch, H. (2008). Resource abundance and internal armed conflict: Types of natural resources and the incidence of 'new wars'. *Ecological Economics, 67*(3), 503–513. https://doi.org/10.1016/j.ecolecon.2008.01.004

White, R. (2011). *Transnational environmental crime: Toward an eco-global criminology*. Willan.

White, R., & Pink, G. (2016). *Environmental crime and collaborative state intervention*. Palgrave.

Whyte, D. (2020). *Ecocide: Kill the Corporation Before It Kills Us.* Manchester University Press.

Wilson, E. (1993). Biophilia and the conservation ethic. In S. Kellert & E. Wilson (Eds.), *The biophilia hypothesis*. Island Press.

Wong, R. (2019). *The Illegal Wildlife Trade in China.* Palgrave.

Wyatt, T. (2013). *Wildlife trafficking: A deconstruction of the crime, the victims, and the offenders.* Palgrave Macmillan UK. https://doi.org/10.1057/9781137269249

Zabyelina, Y., & van Uhm, D. P. (2020). The new Eldorado: Organized crime, informal mining, and the global scarcity of metals and minerals. In Y. Zabyelina & D. P. van Uhm (Eds.) *Illegal mining: Organized crime, corruption, and ecocide in a resource-scarce world* (pp. 3–30). Palgrave. https://doi.org/10.1007/978-3-030-46327-4_1

Zhu, A.L. (2022). *Rosewood: Endangered Species Conservation and the Rise of Global China.* Harvard University Press.

Green Crime and Ecojustice

Planetary Health and Eco-Justice: Contributions from Green, Southern, Cultural and Decolonial Criminologies

Nigel South

Planetary Health in the Age of the Anthropocene

Speaking in May 2023, Joyeeta Gupta, a professor of environment at the University of Amsterdam and the co-chair of the Earth Commission, an independent international group of scientists, suggested that if our planet was visiting the doctor for an annual check-up, the result would be a warning that "the Earth is really quite sick right now and it is sick in terms of many different areas or systems and this sickness is also affecting the people living on Earth" (Borenstein, 2023). Of course, there are problems with the 'personification' of the planet but trying to make sense of its complexity by representing it in certain ways has been enriching cultures and prompting enquiry throughout the history of mythology and art, literature and philosophy (DuRocher, 1996). It is unsurprising, then, that modern science should

N. South (✉)
Department of Sociology and Criminology, University of Essex,
Colchester, UK
e-mail: n.south@essex.ac.uk

D. van Uhm, D. Siegel (eds.), *Global Green Crime and Ecojustice*, Palgrave Studies in Green Criminology, https://doi.org/10.1007/978-3-031-91557-4_2

also resort to using the familiar metaphorical form of the 'body' when trying to convey 'findings' and enhance popular understanding (Giblett, 2008, pp. 1–18; Jackson, 1983). The main preoccupation about the body, of course, is what 'state of health' it is in and hence, given the dependence of all life on a sustaining and sustainable planet, it seems that it might be useful to employ a 'health' perspective when thinking about the planet, eco-justice, environmental harms, and the interdependence of species.

This chapter explores the theme of 'planetary health' via a structure that follows a set of personal research interests in a way that, eventually, comes full circle. Much of my early career was focused on both drug use and control as matters that cut across criminal justice *and* public health systems (Dorn & South, 1985; South, 1993), and the interest in public health questions grew further as a result of later teaching of health professionals and in thinking about environmental harms and the health consequences of pollution. This led to the development of ideas about a 'green' perspective for criminology and how to write and teach about related, wide-ranging issues of justice and exclusion, environmental rights, and cultural representations (South, 1998; Beirne & South, 2012; South & Brisman, 2013; Brisman & South, 2013a). On this basis, for me, a green criminology has always been interdisciplinary and its subject matter inevitably draws us to scholarship and insights from other social sciences and the natural sciences, as well as the arts and humanities, supporting work on, for example, threats to non-human species, climate change, and Indigenous cultures, cosmologies and relationships with nature. It is this latter area of work that will take me full circle here with a concluding discussion of the impact of a zoonotic infectious disease on the experiences of four Colombian Indigenous communities during the COVID-19 pandemic. But first, a note on the idea of the 'Anthropocene'.

The Anthropocene, the Earth, and Our Waste(d) Planet

The Anthropocene—or the Capitalocene—or …?—it seems that there are many other terms that might be used here as, according to Chwalczyk (2020), there are already around eighty alternative names for the

Anthropocene. Debates about the strengths and weaknesses of these various alternatives are generating a substantial literature, some of which is usefully provocative (see e.g. Carrabine, 2025) while other formulations, it seems to me, may simply divert us from the urgent task of widening public support for measures that address the crisis. As Hecht (2018, p. 111) helpfully puts it: "Wrestling with these counterpoints is invigorating, but alternative nomenclatures seem unlikely to gain much traction in natural-science disciplines or the public sphere." Hecht adds that while some "Critics of the Anthropocene concept rightly worry that the idea can be used to revive fantasies about humans' ability to control nature", for most its real utility lies in offering "a way of signalling human responsibility, not of asserting control" and inviting a "broad conversation about our earthly condition, to make friends rather than foes (Jonsson, 2015; Latour, 2015; Rockström, 2015)".

Regardless of some critiques, this is not to deny that such responsibility lies more with some (states, corporations, consumers) than with others. Such debates need a sense of perspective however because, as Amster (2015, p. 40) observes, "the earth and its life-giving capacities are resilient and will almost certainly (at least on a geological time scale) survive whatever we throw at it short of total nuclear pulverization. In fact, many other life forms would flourish without us here, with nature rapidly rewilding even the concrete jungles we have created." It may be more intellectually realistic and politically urgent to pursue a path that, as Amster puts it, is more "consistent with our actual place in the web of life" and do what we can to save ourselves, other species, and our shared planet (Brisman & South, 2018a). So, if, as the Earth Commission suggests, "the Earth is really quite sick right now", to save the planet we need to improve its health.

In the field of public health, proposals for perspectives that take a holistic view of the earth, ecosystems and the diversity of species have been gaining ground for over two decades.[1] Several positions have been set out. 'One Health' and 'Eco-Health' arguably take ecocentric and biocentric perspectives, while some suggest the 'Planetary Health' position takes a more anthropocentric viewpoint (Lerner & Berg, 2017), but

[1] I am grateful to Aitor Ibáñez Alonso for our discussions about these different perspectives.

articulates well with principles of environmental justice and "offers an opportunity to advance the 2030 Agenda for Sustainable Development, including identification of co-benefits across targets, encouraging effective cross-sector action and partnerships, and ensuring policy coherence" (Pongsiri et al., 2019, p. e402). COVID-19 has stimulated the rapid development of the One Health concept (Gibbs, 2014) because it has had more of a concern with animal health and zoonoses (emerging infectious diseases and transmission) whereas Planetary Health has "focused more on the environment, particularly climate change and human health, and on social determinants of human health" (de Castañeda et al., 2023, p. e109). However these are, as de Castañeda et al. (2023, p. e110) argue, "highly complementary fields of scientific inquiry with solid leverage for translation into policy and practice"; they offer "an opportunity to build a stronger research community and transdisciplinary evidence to collectively address pressing public and global health issues in a truly integrated way"; and "above all" they address the key point that "life on Earth is at risk, and our diversity will only make us stronger". Work within these converging categories has the valuable characteristics of being transboundary, open to interdisciplinarity, and accepting of the fundamental *ecological interdependence* of life on the planet. To echo Chakrabarty's (2009, p. 213) argument about our priorities at this moment in "the climate of history", "we need to bring together intellectual formations that are somewhat in tension with each other". In Chakrabarty's discussion, he finds the term 'planetary' as or more useful than 'global', and in what follows I also choose to use the term 'Planetary Health' as, for my purposes here, it most closely seems to signal alignment alongside the idea of Eco-Justice.[2]

It is unsurprising and obvious that the global COVID-19 pandemic has pushed forward further thinking in this area and, on this point, MacDonald (2021) notes proposals for a 'global pandemic treaty' linking the essentials of: prevention and preparedness; animal welfare; environmental protections; health care systems; justice; and equity. Following

[2] Elsewhere, the prospects for drawing green criminology together with the broad 'One Health' perspective are explored in relation to the threat of a rising number of emerging infectious diseases, see Alonso and South (2025).

the line of discussion so far, the next question is whether 'green criminology' has already laid foundations that can contribute to such visions and proposals? Obviously I am saying 'yes' and next turn to a brief outline of some of these established foundations as well as some more recent directions in the fields of Green Criminology and Eco-Justice.

Three 'Principles' are fundamental to building upon these foundations. First, recognising the *interdependence* of species and planet—the vital importance of biodiversity (Carrington, 2018). Second, accepting the profound significance of *intergenerational* justice, which represents the partnership between the living, the dead, and the unborn whereby each generation receives a natural and cultural legacy from previous generations that it holds in trust for succeeding ones (Ruggiero & South, 2010). This entails a duty "to pass on to succeeding generations a planet at least as healthy as the one it inherited so that each generation will be able to enjoy its fruits" (Kofele-Kale, 2006, p. 324). Third, pursuing and valuing *interdisciplinarity* in our work. As pointed out in an early attempt to explain what a 'green' perspective could contribute to criminology, "[g]reen issues open up a wide range of possibilities for interdisciplinary work, both within the social sciences and with disciplines in the natural sciences, offering the potential for collaboration between criminologists and economists, geographers, biologists, health specialists, philosophers, human rights workers, lawyers and others" (South, 1998, p. 226). Indeed, methodologically and theoretically, as White (1998, p. 214) suggests, the "investigation of environmental harm *requires* analysis which is wide-ranging and multi-disciplinary, and which is sensitive to the interconnectedness of social and ecological phenomena" (emphasis added).

'Green Criminology', Environmental Harms and Health: Not So 'New' Connections?

As Lynch (2020, p. 51) notes, "Long before GC [Green Criminology], similar issues were examined in public health, toxicology, and sociology. The first laws addressing environmental harms emerged from public health research that predated GC by centuries." There are many examples of such precursor, environmentally-informed, studies on harms, crimes, power, and exploitation *before* green criminology. Very importantly, many were overlooked in the dominant English language literature because they were undertaken in the Global South and reported in other languages (Rodríguez Goyes & South, 2017). The same cautionary reminder applies to important recent developments responding to such neglect. Hence, despite making a real 'new' impact, a 'Southern Criminology' (Carrington et al., 2016) also needs to be accompanied by recalling how much significant work had already been done to set out a southern critical perspective. I have suggested elsewhere (South, 2022, 2023), in writing of the legacy of the Venezuelan criminologist Rosa del Olmo (Rodriguez Goyes et al., 2023), that even as we celebrate exciting new directions that emerge in critical criminology, we should also ensure that we acknowledge earlier contributions—particularly if not part of the usual transatlantic literature reviews. In pioneering works about state crimes and political conflicts across Latin America, del Olmo laid foundations for a truly Southern critical criminology, written *from* the South and absolutely representative of populations and cultures that had faced subjugation and exploitation since colonisation. Two particular essays (del Olmo, 1987, 1998) are, however, particularly relevant here as they show how the USA-led 'drugs wars' directed against Colombia, Mexico and other nations, were forms of 'state crime' and 'corporate crime', which pursued a strategy of ecocide—which del Olmo termed 'eco-bio-genocide'—by using chemical defoliants and herbicides such as Glyphosate. Such chemical warfare was known to be a cause of deadly *health and eco-harms*—the two are correlates and interrelated.

Green Criminology 'New'...

Green criminology is now expansive and a well-recognised feature of criminological inquiry and curricula (Brisman & South, 2023), concerned with crimes and harms affecting the natural environment, the planet as a whole, and the associated impacts on human and non-human life (Sollund, 2024). It considers causes and consequences of environmental degradation and threats to the sustainability of life on the planet, as well as responses. It is concerned with environmental disruptions based on illegalities and omissions as well as those arising from legal activities that are 'taken for granted' in 'everyday life' and in 'business as usual'. Since the 1990s (Lynch, 1990; South, 1998), the field has seen significant growth and become an important signal of how criminology can show its relevance in a changing and threatened world (Lynch, 2020). Interestingly, one of the most insightful criminological arguments for an 'environmental turn' in the subject came not from a specialist in this area but in the presidential address of Margaret Zahn (1999) to the American Society of Criminology as she surveyed the 'future' of the field. Health and mortality factors are central here: "pollution of our rivers and oceans is causing global mass extinctionsc and a significant reduction in biodiversity. The impacts for life on this planet are far-reaching. ... Particulate pollution annually kills … four times [more people than] … homicide." Zahn argued that, "With more focus on environmental crime will come a new definition of victims to include species other than humans and a definition of offenders to include those who pollute for convenience … [and] for profit" and therefore "Just as Sutherland's white-collar crime expanded our crime paradigm, … environmental crime will change it in the future."

If the 'crime paradigm' has not brought green criminology to the core of the discipline it has at least—as Zahn predicted—expanded the discipline in various directions, now accommodating a variety of topics and bodies of research previously under- or un-explored. New directions and subfields have emerged that cut across wildlife crime and conservation issues (e.g. Beirne, 1999; Brisman & South, 2018b; Ellefsen et al., 2012; Gibbs et al., 2010; Hall, 2013; Sollund, 2008; van Uhm, 2016; Walters, 2011; White, 2008; Wyatt, 2013), and importantly, in terms of intellectual history and rationale, these expressions of a 'greening' of criminology were inevitable because they "reflected scientific interests and political challenges of the moment, carried forward the momentum of critical non-conformist criminology, and offered a point of contact and convergence" (South, 2014, p. 7).

'Contact and convergence' can also be important and stimulating *within* a field, bringing together complementary perspectives or directions that (hopefully) enrich the criminological imagination, and this has occurred with the bringing together of green and cultural criminologies.

Cultural criminology explores how the events and processes of crime and crime control are represented, given meaning, and critiqued, all via various forms of popular culture (for example, media and subcultures). As a framework, it is also explicitly open and inclusive (Ilan, 2019). This is a characteristic shared with green criminology and the idea of 'Green Cultural Criminology' (Brisman & South, 2014, 2025), builds on both bodies of work to pursue a threefold agenda.

First, considering the way(s) in which environmental crimes, harms, and disasters are constructed and represented by news media (e.g., online, print), social media, and in popular cultural forms (e.g., art, books, films); second, paying attention to patterns of consumption, constructed consumerism, commodification of nature, and related market processes; and third, examining the contestation of space, transgression, and resistance, in order to analyse the ways in which environmental harms are resisted and opposed in/on the streets and in day-to-day life.

It is worth noting that much of the early work within Green Criminology shared a generalised focus on examples of 'what we DO to nature', so it is helpful that some more recent directions have been about how we 'live with nature', through better 'understanding' and engagement with 'representations'. Examples would include developments that try to 'give voice' to non-human species that are victims of environmental crime (MacBeath & Whitfort, 2024) as well as analysis and interpretation of the cultures, cosmologies, and sensory experiences associated with nature (Brisman & South, 2014; Rodríguez Goyes et al., 2021a; 2021b; McClanahan & South, 2019). However, one of the most momentous turning points in 'understanding' and 'representation', impacting our knowledge and imagination across all disciplines, came with a new image of the planet as a fragile blue sphere or small 'marble' hanging in space (Moran, 2018; Tanzer et al., 2015).

The planet is Blue—a feature humans had not fully appreciated until we could see its image from space and begin an age of new exploration—but for old purposes: the search for resources, ultimately colonisation and, along the way, dumping redundant waste 'where no waste had been before'. Human-created matter in outer space has been increasing exponentially in the Anthropocene (Lampkin & South, 2024) but, like many other threatening outcomes of anthropocentric technological advances, the problem of *space pollution* (Brito et al., 2013) was not unforeseen. In 1979, for example, the science-fiction writer Arthur C. Clarke, credited with first conceiving of the idea of the geo-synchronous satellite that enables mobile phone signals, had written in his novel of the future, *The Fountains of Paradise* (Clarke, 1979): 'For two hundred years, satellites of all shapes and sizes, from loose nuts and bolts to entire space villages, had been accumulating in Earth orbit. … Three-quarters of this material was abandoned junk, much of it long forgotten. Now it had to be located, and somehow disposed of.'

The warnings of science fiction are often dismissed as precisely that, 'fiction', yet instead of simply turning predictive discussion and debate into the plots of apocalyptic-themed entertainment, we should also be prompted by imaginings of a future of a damaged and abject earth to do more to prevent such futures materialising (Brisman & South, 2013b, 2025). Whether as space colonies, or at sea through attempts to create independent and sovereign states as islands (South, 2019), or via AI-managed, robot mining settlements on the ocean floor, ultra-wealthy individuals and corporations are using their technological, legal and financial machinery to 'max-out' the planet (Lam et al., 2022). This gives us a future vision of the earth as one shaped by "a tiny minority of ultra-wealthy individuals and private-sector companies" whose speculative ventures are "ultimately about dictating policy and ownership" of the planet and beyond (Shaw, 2021).

Meanwhile, back on the Blue Planet, policy and research about sustainability and 'planetary boundaries' have frequently focused on the terrestrial, and only belatedly are more new initiatives addressing crises in the oceans and seas. The oceans and waters that produce the distinctive profile of our 'blue marble' are being subjected to intense and accelerated (Jouffray et al., 2020), forms of disregard, damage, harm and crime (García Ruiz et al., 2020), including, for example: breaking of national, regional and international environmental law (Bueger & Edmunds, 2020); illegal fishing and trawling that destroys sea ecosystems (Belhabib & Le Billon, 2020, p. 1); deep sea mining, offshore fracking, and chemical dumping (Jouffray et al., 2020); depletion of coastal ecosystems (WWF, 2020); and impacts of light pollution on marine fish and zooplankton (Berg et al., 2020). So much of the planet is threatened, so many species are at risk of extinction, so many human populations are suffering—it is easy to overlook, to forget, and to exclude.

Criminological Exclusion

The construction of scientific representation generally means the selection and inclusion of a limited number of phenomena and disregard of others (Santos, 2014; Rodríguez Goyes & South, 2017), so the dominant western/northern criminological literature has tended to hold up a mirror and, with a few exceptions, write of the society reflected. As Franko Aas (2012, pp. 6–8) has argued, "the collective imaginary in western criminology tends to 'see' in a particular way, assuming that criminology has flourished only in Anglophone countries. Consequently, efforts to detect criminological literature of relevance are always directed to established literatures and databases that reflect this assumption." Studies of the crimes and harms of, for example, oil, chemical, and pharmaceutical companies would therefore generally be focused on misdeeds in consumer societies of the global north, and also be described in terms of 'corporate crime'. In the past, this has often meant that descriptions of the damage done by, for example, energy companies, rarely paid great attention to the pollution of land and water or biodiversity loss due to resource extraction in locations of oil or gas exploration, while the biopiracy and theft of nature resulting from the activities of Big Pharma and agribusiness in, for example, cattle and soy farming, were also usually overlooked (Rodríguez Goyes et al., 2017; Lymbery & Oakeshott, 2014; Whyte & Tombs, 2023). In recent years, a compensatory reaction within criminology has been recognition of "the global south" as a context in which exploitative—Northern-controlled—global forces have exercised power and engineered massive ecological withdrawals as part of the post-World War II intensification of the 'treadmill of production' (Stretesky et al., 2014) and the rise of hyper-consumer societies (Brisman & South, 2014). The beneficiaries have been economies of the global north, accelerating the anthropogenic conditions that are causing climate change and creating species vulnerabilities, with the global south being the least culpable but the most vulnerable. To take the whole continent of Africa as an example, as Hindley (2024) observes, "African countries are responsible for an insignificant proportion of the greenhouse gas (CO2e) emissions that are driving global climate change" with a "share of global emissions" at just 3.9% in 2021 and "estimates of cumulative historic emissions ... even lower at less than 3%." Putting aside the emissions of a small number of oil-, gas- and coal-producing states (e.g. South Africa, Egypt, Algeria, Libya, Morocco) leaves "the per capita and total emissions of most other African countries" as extremely small, yet "the continent is suffering disproportionately from the effects of climate change". These conditions, as Hecht (2018, p. 1) argues, have also made the Anthropocene the 'apotheosis of waste' as a result of a massive increase in the "quantity, extent, and durability of discards", for example, "the planetary production of particulates, the millennial futures of microplastics and radioactive waste". The dumping ground for disposal of much such waste has been—no surprise—the nations of the global south.

Within criminology, significant responses to this 'state of the world' have been calls to 'Southernize' and 'decolonize' the field (Carrington et al., 2016; Cunneen & Tauri, 2017; Dimou, 2021). In the work of Rodríguez Goyes, for example, this movement has led to an inspirational combination of 'green' and 'southern' criminologies (Rodriguez Goyes, 2019), and also led to studies exploring 'ways of seeing', understanding and representing nature and wildlife within Colombian Indigenous communities (Rodríguez Goyes et al. 2021b), thereby making an invaluable contribution to the usefulness of a 'green cultural' criminology. This latter work has been particularly revealing of the connections that we must investigate between planetary health and eco-justice, from the point of view of convergence between Green, Southern, Cultural and Decolonial Criminologies. A series of projects (reported in Rodríguez Goyes, Abaibira, et al., 2021a, Rodríguez Goyes, South, et al., 2021b, Rodríguez Goyes et al., 2022) employed a 'peer research methodology' to explore the cultural representations of nature and wildlife within four Colombian Indigenous communities: the Barí, the Huitoto, the Nasa, and the Ticuna. This methodology involves 'peers' of the researched communities at all the stages, and aims to: (1) empower groups (the Indigenous researchers and the research communities); (2) enhance closer understanding of issues than would be gained by methods reliant on outsider researchers; and (3) gain deeper access to information (interviewees being more willing to engage in discussion with peer researchers than with 'outsider' academic researchers). Epistemologically and ontologically, the projects sought inspiration from Indigenous environmental cosmologies and representations of the natural environment, aiming to draw out guidance on ways of living with and relating to nature (Rodríguez Goyes et al. 2021b). This also required a description of the painful genocidal and ecocidal undermining of Colombian Indigenous environmental ontologies and ways of life that has been occurring since colonisation (Rodríguez Goyes et al. 2021b), involving atrocities committed by armed groups (army, guerrillas, paramilitaries, hitmen, mercenaries, organised crime and so on) but also forms of 'slow violence' (Nixon, 2011) and 'silencing' (Mathiesen, 2004). These are interventions with traumatic consequences that have eroded trust in many forms—trust in external others, in northern medicine and, sometimes, in others of their own Indigenous communities. This process was dramatically demonstrated in a period and under conditions that return me to the issue of health and the human/nature relationship.

In 2020, during the COVID-19 global pandemic we undertook a case study that examined the interaction between Indigenous ontologies, historical trauma, and health system inequalities in a moment of enormous stress (Rodríguez Goyes et al., 2022). COVID-19 made more visible the intersection and mutual reinforcement of the many injustices already affecting Colombian Indigenous communities. Analysis of newspapers highlighted reports that Indigenous communities lacked water (Redacción Bogotá, 2020), and access to appropriate health services (El Colombiano, 2020) to help combat the Coronavirus. Some news reports suggested the government treated Indigenous peoples as second-class citizens and this was echoed in our project findings in which respondents expressed the view that government (national and local) did not support Indigenous people in relation to the pandemic. According to Daniel, a Uitoto healer: "I feel that the state has abandoned us; it is only concerned with the big cities. We have to confront the pandemic with the help of God and of our doctors. We have Mother Earth and Nature, and our elders" (Rodríguez Goyes et al., 2022, p. 11). For the communities, the lessons of their history suggest forces of external power are not to be trusted and indeed are often very clearly to be regarded as 'the enemy', which easily translates into suspicion about, and resistance to, external medical authority and expertise. Hence community representatives said that they distrusted the government, doctors and vaccines: "they turned the disease into the business and they turn the vaccine into a business. We will not take the vaccine because we have our own medicine." Thus, there is (1) suspicion of motives, (2) resistance to imposition, and (3) strengthening of internal cultural beliefs, as reflected in these quotes from healers within different communities: "....we gathered the traditional knowledge of each clan, ... The pandemic pushed us to unite our thought" (Romualdo, Uitoto); "We were forgetting traditional medicine, but the pandemic awakened that knowledge that was asleep" (Wilson, Tikuna); "The pandemic ... made people see that the knowledge traditional doctors have works. ... thanks to this, people believe more in traditional and ancestral medicine ..." (Hervin, Nasa) (quoted in Rodríguez Goyes et al., 2022).

The relationship between Nature and Health is valued here in a spiritual and cultural way but, as explained by Dr. Pedro Pachaguaya, an anthropologist working on Indigenous health matters in Bolivia (quoted in Graham, 2022), this is not "'denialism' about COVID, nor is it the typical anti-vaccine denialism"; rather, "there is a lack of trust in the healthcare system, and a strong preference for traditional medicine". Dominant healthcare systems have tended to ignore or reject traditional medicine and "Instead of being understood, these systems have been made invisible, erased, in a violent way", and that, says Dr. Pachaguaya, is the "fundamental problem the healthcare system has here". We are left wondering whether, in the future, the global south and the global north could learn from each other about productive fusions of forms of medicine and better ways of living with nature? Or are we, as always in the past, trapped by global divisions and the politics of economic, ideological, and ontological conflict?

Conclusion

Globally, the state of human health improved dramatically between 1950 and 2010 (Gupta et al., 2019)—not equally or uniformly, and with the persistence of close links between poverty and poor health in countries around the world whether in the global north or south (Deaton, 2013). Indeed, some advances in health may now be in reverse. Human health and global life-support systems are now under serious threat and these same years—through the late twentieth and into the twenty-first centuries—have also seen unprecedented environmental degradation. Wars and conflicts have exacerbated and accelerated such impacts and we can expect further global pandemics, quite probably driven by climate change. More than 200 infectious diseases and dozens of non-transmissible conditions have already been exacerbated by climate change. Rising temperatures risk insect-borne diseases being pushed further out of the tropics; severe droughts force animal and human populations to live closer together fuelling zoonotic infection transmission; more floods spread disease and cause displacement and deaths (Mora et al., 2022).

We need a more holistic view of global public health. If developed alongside a green criminology perhaps this could enable both fields to develop better understanding and methodologies in relation to overlooked harms and threats to the health and well-being of humans, non-human beings and ecosystems. As well as the case of zoonotic Emerging Infectious Diseases, this hybrid framework could also examine trends in harm, health and criminal activity related to climate change or the toxic contamination of our air, seas and land, and draw on interdisciplinary sources of expertise in, for example, epidemiology, biology, veterinary science, virology, and ecology (Alonso & South, 2025).

A holistic approach would appreciate the interdependence of species on the planet, and be inclusive of not only all humans but also the health of non-human species and the natural environment. As Armstrong et al. (2020) assert, "The health, social and economic consequences of COVID-19 should act as a wake-up call for all governments to take stock … and ensure post COVID-19 responses reverse our war on nature … a war on nature is ultimately a war against ourselves" (Armstrong et al., 2020; Beirne, 2021; Humboldt-Dachroeden et al., 2020). If the planet is 'sick' it is not difficult to understand why—or who is to blame.

References

Alonso, A. I., & South, N. (2025). Incorporating a One Health approach into the study of environmental crimes and harms: Towards a 'One Health Green Criminology'. *British Journal of Criminology, 65*(2), 328–343.

Amster, R. (2015). *Peace ecology*. Paradigm Publishers.

Armstrong, F., Capon, A., & McFarlane, R. (2020). Coronavirus is a wake-up call: Our war with the environment is leading to pandemics. *The Conversation*, 30 March. Retrieved February 7, 2023, from https://theconversation.com/coronavirus-is-a-wake-up-call-our-war-with-the-environment-is-leading-to-pandemics-135023

Beirne, P. (1999). For a nonspeciesist criminology: Animal abuse as an object of study. *Criminology, 37*(1), 117–147.

Beirne, P. (2021). Wildlife trade and COVID-19: Towards a criminology of anthropogenic pathogen spillover. *The British Journal of Criminology, 61*(3), 607–626.

Beirne, P., & South, N. (2012). Greening criminology. In S. D. Fassbinder, A. Nocello, & R. Khan (Eds.), *Greening the academy: Ecopedagogy through the liberal arts*. Sense Publishers.

Belhabib, D., & Le Billon, P. (2020). Editorial: Illegal fishing as a trans-national crime. *Frontiers in Marine Science, 7*(19 March), 162. https://doi.org/10.3389/fmars.2020.00162

Berg, J., Geoffroy, M., Daase, M., Cottier, F., Priou, P., Cohen, J. H., Johnsen, G., McKee, D., Kostakis, I., Renaud, P. E., Vogedes, D., Anderson, P., Last, K. S., & Gauthier, S. (2020). Artificial light during the polar night disrupts Arctic fish and zooplankton behaviour down to 200m depth. *Communications Biology, 3*, 1–8. https://doi.org/10.1038/s42003-020-0807-6

Borenstein, S. (2023). Earth is 'really quite sick now' and in danger zone in nearly all ecological ways, study says. *Associated Press*, 31 May, https://apnews.com/article/earth-environment-climate-change-nature-sick-2dded06915af4645253f5c29abff4794

Brisman, A., & South, N. (2013a). For a green-cultural criminology. *Crime, Media, Culture, 9*(2), 115–135.

Brisman, A., & South, N. (2013b). Conclusion: The planned obsolescence of planet Earth? How green criminology can help us learn from experience and contribute to our future. In N. South & A. Brisman (Eds.), *The Routledge international handbook of green criminology* (pp. 409–417). Routledge.

Brisman, A., & South, N. (2014). *Green cultural criminology: Constructions of environmental harm, consumerism, and resistance to ecocide*. Routledge.

Brisman, A., & South, N. (2018a). Autosarcophagy in the Anthropocene and the obscenity of an epoch. In C. Shearing & C. Holley (Eds.), *Criminology and the Anthropocene* (pp. 25–49). Routledge.

Brisman, A., & South, N. (2018b). Perspectives on wildlife crime: The convergence of 'green' and 'conservation' criminologies. In W. Moreto (Ed.), *Wildlife crime: From theory to practice* (pp. 17–37). Temple University Press.

Brisman, A., & South, N. (2023). Green criminology. In A. Liebling, S. Marunna, & L. McAra (Eds.), *The Oxford handbook of criminology* (7th ed., pp. 521–541). Oxford University Press.

Brisman, A., & South, N. (2025). *Monstrous nature, abjection, and the end of the world: A green cultural criminological perspective*. Temple University Press.

Brito, T. P., Celestino, C. C., & Moraes, R. V. (2013) A brief scenario about the 'space pollution' around the earth. *Journal of Physics Conference Series*. [Online]. Retrieved February 5, 2024, from https://iopscience.iop.org/article/10.1088/1742-6596/465/1/012020/pdf

Bueger, C., & Edmunds, T. (2020). Blue crime: Conceptualising transnational organised crime at sea. *Marine Policy, 119*, 104067.

Carrabine, E. (2025). Grave matters: Ghost criminology, necropolitics and the Anthropocene. In E. Carrabine & A. Di Ronco (Eds.), *Criminological connections, directions, horizons: Essays in honour of Nigel South*. Routledge.

Carrington, D. (2018) Briefing: What is biodiversity and why does it matter to us? *The Guardian*, 12th March. https://www.theguardian.com/news/2018/mar/12/what-is-biodiversity-and-why-does-it-matter-to-us

Carrington, K., Hogg, R., & Sozzo, M. (2016). Southern criminology. *The British Journal of Criminology, 56*(1), 1–20.

de Castañeda, R. R., Villers, J., Guzmán, C. A. F., Eslanloo, T., de Paula, N., Machalaba, C., Zinsstag, J., Utzinger, J., Flahault, A., & Bolon, I. (2023). One Health and planetary health research: Leveraging differences to grow together. *Lancet Planetary Health, 7*(2), e109–e111. https://doi.org/10.1016/S2542-5196(23)00002-5

Chakrabarty, D. (2009). The climate of history: Four theses. *Critical Inquiry, 35*(2), 197–222.

Chwalczyk, F. (2020). Around the Anthropocene in eighty names: Considering the urbanocene proposition. *Sustainability, 12*, 4458.

Clarke, A. (1979). *The fountains of paradise*. Gollancz.

Cunneen, C., & Tauri, J. (2017). *Indigenous criminology*. Policy Press.

Deaton, A. (2013). *The great escape: Health, wealth, and the origins of inequality*. Princeton University Press.
del Olmo, R. (1987). Aerobiology and the war on drugs. *Crime and Social Justice, 30*, 28–44.
del Olmo, R. (1998). The ecological impact of illicit drug cultivation and crop eradication programs in Latin America. *Theoretical Criminology, 2*(2), 269–278.
Dimou, E. (2021). Decolonizing southern criminology: What can the "decolonial option" tell us about challenging the modern/colonial foundations of criminology? *Critical Criminology, 29*, 431–450.
Dorn, N., & South, N. (1985). *Helping drug users*. Gower.
DuRocher, R. J. (1996). The wounded earth in "Paradise Lost". *Studies in Philology, 93*(1), 93–115.
El Colombiano. (2020). Cinco grupos de población en los que pesa la covid-19 [Five groups for which COVID-19 is a heavy burden]. El Colombiano. https://www.elcolombiano.com/coronavirus-en-colombia-contra-el-miedo/cinco-grupos-de-poblacion-en-los-que-pesa-la-covid-IF13437197.
Ellefsen, R., Sollund, R., & Larsen, G. (Eds.). (2012). *Eco-global crimes: Contemporary problems and future challenges*. Ashgate.
Franko Aas, K. (2012). 'The Earth is one but the world is not': Criminological theory and its geopolitical divisions. *Theoretical Criminology, 16*(1), 5–20.
García Ruiz, A., South, N., & Brisman, A. (2020). Eco-crimes and ecocide at sea: Toward a blue criminology. *International Journal of Comparative Criminology and Offender Therapy, 66*(4), 407–429.
Gibbs, E. (2014). The evolution of One Health: A decade of progress and challenges for the future. *Veterinary Record, 174*(4), 85–91.
Gibbs, C., Gorer, M., McCarrell, E., & Rivers, L., III. (2010). Introducing conservation criminology: Towards interdisciplinary scholarship on environmental crimes and risk. *British Journal of Criminology, 50*(1), 124–144.
Giblett, R. (2008). *The body of nature and culture*. Palgrave Macmillan.
Graham, T. (2022). Misinformation and distrust: Behind Bolivia's low Covid vaccination rates. *The Guardian*. https://www.theguardian.com/global-development/2022/feb/06/misinformation-and-distrust-behind-bolivias-low-covid-vaccination-rates?CMP=Share_iOSApp_Other
Gupta, J., Hurley, F., Grobicki, A., Keating, T., Stoett, P., Baker, E., Guhl, A., Davies, J., & Ekins, P. (2019). Communicating the health of the planet and its links to human health. *The Lancet Planetary Health, 3*(5), e204–e206.

Hall, M. (2013). *Victims of environmental harm: Rights, recognition and redress under national and international law*. Routledge.

Hecht, G. (2018). Interscalar vehicles for an African Anthropocene: On waste, temporality, and violence. *Cultural Anthropology., 33*(1), 109–141.

Hindley, J. (2024). *Review essay: African handbook of climate change adaptation.* Walter Leal Filho (Editor-in-Chief), Springer 2021. *International Journal of Crime, Justice and Social Democracy, 13*, 1.

Humboldt-Dachroeden, S., Rubin, O., & Frid-Nielsen, S. S. (2020). The state of One Health research across disciplines and sectors—A bibliometric analysis. *One Health, 10*, 100146.

Ilan, J. (2019). Cultural criminology: The time is now. *Critical Criminology, 27*(1), 5–20. https://doi.org/10.1007/s10612-019-09430-2

Jackson, M. (1983). Thinking through the body: An essay on understanding metaphor. *Social Analysis: The International Journal of Social and Cultural Practice, 14*, 127–149.

Jonsson, F. (2015). Planetary politics. *Public Books*, 15 June. https://www.publicbooks.org/planetary-politics/

Jouffray, J.-N., Blasiak, R., Norström, A., Österblom, H., & Nyström, H. (2020). The blue acceleration: The trajectory of human expansion into the ocean. *One Earth, Perspective, 2*(1), 43–54. https://doi.org/10.1016/j.oneear.2019.12.01

Kofele-Kale, N. (2006). *The international law of responsibility for economic crimes*. Ashgate.

Lam, A., South, N., & Brisman, A. (2022). A convergence of crises: COVID-19, climate change and bunkerization. *Crime, Media, Culture, 19*(3), 327–344.

Lampkin, J., & South, N. (2024). Space junk and the protection of the space environment. In R. White (Ed.), *Elgar encyclopedia of environmental crime.* Edward Elgar.

Latour, B. (2015). Fifty shades of green. *Environmental Humanities, 7*(1), 19–25.

Lerner, H., & Berg, C. (2017). A comparison of three holistic approaches to health: One health, ecohealth, and planetary health. *Frontiers in Veterinary Science, 4*, 163.

Lymbery, P., & Oakeshott, I. (2014). *Farmageddon: The true cost of cheap meat.* Bloomsbury Publishing.

Lynch, M. (1990). The greening of criminology: A perspective on the 1990s. *Critical Criminologist, 2*(3–4), 11–12.

Lynch, M. (2020). Green criminology and environmental crime: Criminology that matters in the age of global ecological collapse. *Journal of White Collar and Corporate Crime, 1*(1), 50–61.

MacBeath, A., & Whitfort, A. (2024). *Species victim impact statements: Giving a voice to the unheard victims of environmental crime*. Global Initiative against Transnational Organised Crime.

MacDonald, M. (2021). Emerging from Covid19: A new, rights-based relationship with the non-human world. *Health and Human Rights Journal, 23*, 2.

Mathiesen, T. (2004). *Silently silenced*. Waterside Press.

McClanahan, B., & South, N. (2019). 'All knowledge begins with the senses': Toward a sensory criminology. *British Journal of Criminology, 60*(1), 3–23.

Mora, C., McKenzie, T., Gaw, I. M., et al. (2022). Over half of known human pathogenic diseases can be aggravated by climate change. *Nature Climate Change, 12*, 869–875.

Moran, J. (2018). Earthrise: The story behind our planet's most famous photo, *The Guardian*, 22nd December. https://www.theguardian.com/artanddesign/2018/dec/22/behold-blue-plant-photograph-earthrise

Nixon, R. (2011). *Slow violence and the environmentalism of the poor*. Harvard University Press.

Pongsiri, M., Bickersteth, S., Colón, C., DeFries, R., Dhaliwal, M., Georgeson, L., Haines, A., Linou, N., Murray, V., Naeem, S., Small, R., & Ungvari, J. (2019). Planetary health: From concept to decisive action. *The Lancet Planetary Health, 3*(10), E402–E404.

Redacción Bogotá. (2020). Al menos siete indígenas Embera, asentados en parque Tercer Milenio, Tienen COVID-19 [At least seven Embera indigenous persons, living at Tercer Milenio park, have COVID-19]. El Espectador. https://www.elespectador.com/bogota/al-menos-siete-indigenas-embera-asentados-en-parque-tercer-milenio-tienen-covid-19-article/

Rockström, J. (2015). The Anthropocene, control, and responsibility: A reply to Andy Stirling. *STEPS Centre Blog*, October 29th. https://steps-centre.org/blog/johan-rockstrom-on-the-anthropocene/

Rodriguez Goyes, D. (2019). *Southern green criminology. A science to end ecological discrimination*. Emerald Publishing.

Rodríguez Goyes, D., & South, N. (2017). Green criminology before 'Green Criminology': Amnesia and absences. *Critical Criminology, 25*(2), 165–181.

Rodríguez Goyes, D., Mol, H., Brisman, A., & South, N. (Eds.). (2017). *Environmental crime in Latin America: The theft of nature and the poisoning of the land*. Palgrave.

Rodríguez Goyes, D., Abaibira, M., Baicué, P., Cuchimba, A., Ñeñetofe, D., Sollund, R., South, N., & Wyatt, T. (2021a). Southern green cultural criminology and environmental crime prevention: Representations of nature

within four Colombian indigenous communities. *Critical Criminology.*, *29*, 469–485.
Rodríguez Goyes, D., South, N., Abaibira, M., Baicué, P., Cuchimba, A., & Ramos Ñeñetofe, D. (2021b). The erosion of a way of life and memory: Genocide and ecocide in four Colombian indigenous communities. *British Journal of Criminology, 61*(4), 965–998.
Rodríguez Goyes, D., South, N., Abaibira, M., Baicué, P., Cuchimba, A., & Ñeñetofe, D. (2022). 'An incorporeal disease': Covid 19, social trauma and health injustice. *The Sociological Review, 71*(1), 105–125.
Rodriguez Goyes, D., South, N., Scott, J., & Creagh, T. (2023). Rosa del Olmo Prize: Introductory essay. *International Journal for Crime, Justice and Social Democracy, 12*(4), 1–10. https://doi.org/10.5204/ijcjsd.3223
Ruggiero, V., & South, N. (2010). Green criminology and dirty-collar crime. *Critical Criminology, 18*(4), 251–262.
Santos, B. D. S. (2014). *Epistemologies of the South: Justice against epistemicide.* Paradigm Publishers.
Shaw, M (2021). Billionaire capitalists are designing humanity's future. Don't let them. *The Guardian*, 5 February. https://www.theguardian.com/commentisfree/2021/feb/05/jeff-bezos-elon-musk-spacex-blue-origin
Sollund, R. (Ed.). (2008). *Global harms: Ecological crime and speciesism.* Nova Science.
Sollund, R. (2024). Speciesism and the non-human. In *Oxford research encyclopedia of criminology* https://oxfordre.com/criminology/view/10.1093/acrefore/9780190264079.001.0001/acrefore-9780190264079-e-748
South, N. (1993). Criminal justice versus public health: Decriminalisation, legalisation and harm reduction. *Criminal Justice Matters*, 12, Summer.
South, N. (1998). A green field for criminology?: A proposal for a perspective. *Theoretical Criminology, 2*(2), 211–234.
South, N. (2014). Green criminology: Reflections, connections, horizons. *International Journal for Crime, Justice and Social Democracy, 3*(2), 6–21.
South, N. (2019) Arcologies, eco-shelters and environmental exemption: Constructing new divisions and inequalities in the Anthropocene. *International Journal of Crime, Justice and Social Democracy.* online (2019): (2020), 9(2), 60–73.
South, N. (2022). Arguing with Stan and revisiting Rosa: Some thoughts on new criminologies in a changing world. In M. Duggan (Ed.), *Papers from the British Criminology Conference* (pp. 6–19), 21. https://www.britsoccrim.org/wp-content/uploads/2022/12/BSC-Online-Journal-2022.pdf

South, N. (2023). Revisiting Rosa: Eco-bio-genocide, drug wars and southern green criminology. In D. Rodriguez Goyes (Ed.), *Green crime in the global South: Essays on southern green criminology*. Emerald.

South, N., & Brisman, A. (2013). Critical green criminology, environmental rights and crimes of exploitation. In S. Winlow & R. Atkinson (Eds.), *New directions in crime and deviance* (pp. 99–110). Routledge.

Stretesky, P. B., Long, M. A., & Lynch, M. J. (2014). *The treadmill of crime: Political economy and green criminology*. Routledge.

Tanzer, J., Phua, C., Jeffries, B., Lawrence, A., Gonzales, A., Gamblin, P., & Roxburgh, T. (2015). *Living blue planet report: Species, habitats and human well-being*. WWF.

van Uhm, D. (2016). *The illegal wildlife trade: Inside the world of poachers, smugglers and traders*. Springer.

Walters, R. (2011). *Eco crime and genetically modified food*. Routledge.

White, R. (1998). Environmental criminology and Sydney water. *Current Issue Criminal Justice, 10*, 214–219.

White, R. (2008). *Crimes against nature: Environmental criminology and ecological justice*. Willan.

Whyte, D., & Tombs, S. (2023). Food: Crime, harm and regulation. *Studi sulla questione criminale, 3*, 19–40. https://doi.org/10.7383/109483

WWF. (2020). Mangrove forests are one of the world's most threatened tropical ecosystems. https://wwf.panda.org/our_work/oceans/coasts/mangroves/mangrove_threats/

Wyatt, T. (2013). *Wildlife trafficking: A deconstruction of the crime, the victims, and the offenders* (2nd ed., 2023). Palgrave Macmillan.

Zahn, M. (1999). Thoughts on the future of criminology—American Society of Criminology 1998 presidential address. *Criminology, 37*(1), 1–16.

Greening Justice: Ecojustice and Remedying Environmental Harm

Angus Nurse

Introduction

Environmental harm manifests itself in various ways, from the direct harms of capitalist activity (e.g. pollution caused by industrial activity), to indirect or unintentional harms caused by accidents and upset incidents, and the exploitation of non-human nature. How to address and achieve redress for environmental harms is a potentially contentious issue. Traditional forms of criminal justice are predominantly situated within notions of retributive punishment, applying sanctions to offenders whilst achieving little for victims and often failing to repair the harms (Wenzel & Okimoto, 2016). By contrast, restorative justice approaches provide for reparation; in principle, putting the victim (the community or the environment) back in the position it would have been in had the harm not occurred (Sherman et al., 2007). Whilst this is not always possible, restorative approaches have been attempted in some cases of

A. Nurse (✉)
Anglia Ruskin University, Cambridge, UK
e-mail: angus.nurse@aru.ac.uk

D. van Uhm, D. Siegel (eds.), *Global Green Crime and Ecojustice*, Palgrave Studies in Green Criminology, https://doi.org/10.1007/978-3-031-91557-4_3

environmental harm by, for example, requiring polluting companies to make some form of environmental restoration to affected communities.

The reliance on criminal law mechanisms to deal with environmental harms is potentially flawed given the reality that much environmental harm is caused by ostensibly legal activity. For example, pollution caused by industrial activity is a major threat to human health and significant cause of environmental degradation yet is arguably outside of the remit of mainstream criminology and criminal justice except in extreme cases of deliberate polluting activity. Yet, arguably there are far more deaths from polluting activity and associated harms than is the case from criminal activity (Lynch & Stretesky, 2014).

How best to deal with environmental offending and environmental harms? Criminal law approaches are situated within a perspective of detention apprehension and sanction which primarily deals with offending behaviour after it has happened rather than adopting a preventative approach (Nurse, 2015). Civil and administrative approaches tend to be regulatory in nature, arguably allowing corporations to continue their illegal and harmful operations, with criminal prosecution being the last resort. In the case of major corporations, there is evidence that the costs of compliance (e.g. fines and other regulatory sanctions) can simply be absorbed within running costs and become the cost of doing business (Crain & Crain, 2014; Nurse, 2022). As an example, in 2008 the Sierra Club filed a federal lawsuit against the Shell oil company and its subsidiaries over an estimated 1000 incidents of pollution that occurred between 2003 and 2007 at the company's Deer Park refinery and chemical plant in Texas. Shell had been cited by regulators and had paid fines for some of the incidents but had allegedly failed to resolve issues at the plant leading to the Sierra Club and Environment Texas pursuing civil claims to enforce provisions of the US Clean Air Act. Joshua Kratka of the National Environment Law Centre alleged that 'Shell is factoring these fines into its costs of operating these facilities' (Seba, 2008). This is not an uncommon perspective; implicit in the regulatory justice approach is the notion of encouraging compliance rather than pursuing a retributive justice approach. Almost by definition, this allows continued activity until either negotiated compliance takes place or criminal prosecution is used as a last resort.

Dealing effectively with contemporary environmental problems arguably requires a new, hybrid approach to justice and regulatory responses. Ecojustice refers to both the notion of applying concepts of justice to environment and non-human nature (including animals), and to the linking of environmental concerns to social justice issues (Bowers, 2009; White, 2014). This includes considering how our justice and regulatory systems deal with the impact of environmental degradation on vulnerable groups and how best to regulate polluting industries and environmental harm caused by corporations and other actors. With a focus on corporate environmental harms and the behaviour of multinational corporations (hereafter MNCs), this chapter argues that effective ecojustice requires an approach based in remedying environmental harms and that takes into account the needs of the environment and non-human nature rather than continued reliance on the punitive approach of the criminal law. The chapter identifies that mainstream policing is mostly reactive and concerned with retributive punishment, whereas the approach applied by environmental regulators is generally concerned with encouraging compliance and applying civil penalties. Arguably neither is entirely satisfactory nor do they fully provide for effective ecojustice. Instead this chapter argues for an approach combining punitive elements with restorative and preventative elements, with the harm caused to nature at the centre.

Definition of Environmental Crime

For the purposes of this chapter's discussion, environmental crime is to be taken as a broad range of non-compliant activity situated within environmental and regulatory law discourse and incorporating a broad conception of illegal activities aimed at the environment and non-human nature (including non-human animals living in the wild). This definition goes beyond the somewhat reductive definition of environmental crime, which generally defines it as:

> An unauthorized act or omission that violates the law and is therefore subject to criminal prosecution and criminal sanctions. This offence harms or endangers people's physical safety or health as well as the environment

itself. It serves the interests of either organizations—typically corporations—or individuals. (Situ & Emmons, 2000, p. 3)

In principle, this definition is useful as it clarifies that environmental crime is a social construction, predicated on whatever is the current status of the law within a particular jurisdiction. Thus, definitions of environmental crime change within their social and geographic location as occurs, for example, with more lenient corporate laws in some Global South countries which are colloquially referred to as 'pollution havens' due to their lax environmental regulation and relatively lenient enforcement regimes (Neumayer, 2001). Global North conceptions may also be relatively friendly to MNCs for a host of political and socio-legal reasons which dictate that environmental laws should not place an undue regulatory burden on business (discussed later in this chapter) and that arguably make a calculation of the socio-economic benefits derived from continued business activity compared with the environmental benefits of a strong regulatory system (Nurse, 2022). This strict legalistic conception on environmental crime contends that any behaviour not specifically prohibited by the law is not a crime. In contrast, green criminology contends that it is the effect of the action and its harmful nature that should determine how we define crimes, rather than the specifics of the justice system categorisation. Thus, green criminology is more concerned with identifying and addressing the harm being caused than it is with a strict adherence to criminal law definitions of prohibited action, retributive justice and who might be defined as a 'criminal'. Another important aspect of the legalistic determination is that behaviour which is currently legal can become illegal (and vice versa) simply by the passing of the relevant legislation. However, what is classified as illegal is subject to a complex determination, drawing again on the weighing of competing economic, social, political and ideological considerations. Such considerations may weigh a range of subjective and objective criteria meaning that 'what gets defines as a "problem"—its severity, its nature, its identification—is contingent upon the capacity of sectional interests to secure its definition' (White & Heckenberg, 2014, p. 78). Thus, for example, historically there has arguably been a reluctance to consider the actions of MNCs to be criminal, in part because of the perceived benefits (including jobs, tax revenues and philanthropic activity) associated with MNC practices. However, in recent years, the greater level of public attention and awareness brought to harmful corporate activity (e.g. tax avoidance, incidents like the Gulf Oil Spill and the harms caused by major pollution incidents and oil spills in Africa) have begun to shine a critical lens on the legitimacy of corporate activity (Nurse, 2022, 2024). In addition, the activities of protest groups like Just Stop Oil and Extinction Rebellion have brought corporate activities into the mainstream of news coverage and, combined with social media activism, have the potential to impact public opinion (and thus political opinion) against corporate activity deemed harmful to public interests (Walter, 2024). The conception on how environmental crimes are defined and what becomes classed as criminal is a complex one but it does include the following:

- **Focal considerations**—a concern with identifying key actors in environmental harm, including issues concerning who or what is an environmental victim and assessing the nature of their victimisation (Hall, 2011, 2013).
- **Geographical considerations**—assessing how particular environmental harms may have geographical connotations, for example, the nature of air pollution in the industrialised Global North or the impacts of severe weather patterns and effects (e.g. tsunamis and droughts) in the Global South. Geographical concerns may situate some environmental harms either at a local level (e.g. the illegal discharge of sewage by a UK utilities company, or an oil spill) or at a national or international level (e.g. deforestation or wildlife trafficking, air pollution).
- **Locational considerations**—a consideration of 'proximity of human habitation to the sources of harm themselves' (White & Heckenberg, 2014, pp. 68–69). This distinguishes geographical areas from the extent to which place can be both a source of environmental harm and a factor in the nature of environmental harm that takes place and how it might be addressed. For example, the Love Canal incident, which was initially viewed as a local toxic waste health-related.[1]

[1] Love Canal is often cited as an example of how grassroots activism was able to mobilise and achieve effects in dealing with what might otherwise have been considered a local problem. For further details, see for example: Thomson (2016) and Gibbs (2010).

- **Temporal considerations**—the impact of changes over time and the perceived imminent nature of any threat. Thus, the calls for greater control over the illegal wildlife trade, and for greater action to limit activities that cause climate change, are stronger in a contemporary setting given the greater level of foreseeability of species extinction and imminent climate disaster.
- **Ideological considerations**—in part what gets criminalised reflects the dominant ideology in place within a given jurisdiction. Thus, one could see that in those societies with conservative administrations reliant on and friendly towards big business, the regulatory regime may be unwilling to excessively criminalise environmental harm (e.g. American and middle Eastern oil-producing states). However, where socialist governments are elected with large or workable majorities, and where public opinion is in favour of greater business control, this is likely to change.

A greater level of scientific knowledge can inform many of these considerations such that, for example, the precautionary principle might be deployed allied to temporal, locational, or other considerations (Hickey & Walker, 1995). Thus, what gets criminalised and how is variable, socially constructed and subject to jurisdiction specific considerations of who and what is considered to be an environmental victim. The enforcement approach is also dependent on societal considerations on how the enforcement regime should deal with environmental offending. Enforcement policy and approaches are discussed later in this chapter; before doing so, however, the nature of environmental offences should be clarified.

Types of Corporate Environmental Harm

The demands of modern consumerism and the operation of markets result in production practices that inevitably cause environmental harms. For example, 'industrial pollution of the air, water and land, toxic waste sites, deforestation, species extinction, excessive pesticide use and pollution, climate change, the excessive use of fossil fuels, acid rain, a growing reliance on coal and oil, the environmental effects of drilling for oil or mining coal…are the problems of the modern world', many of which are linked to corporate activity (Lynch & Stretesky, 2014, p. 11). Several of these harms are the result of legal corporate activity, although the legal and the illegal often go side by side with corporations, for example, going beyond permitted levels of pollution. Green criminology recognises that 'eco-global crimes are often transnationally organized, with damage being caused by legal transnational corporations, such as British Petroleum in the Gulf of Mexico, and the destruction of the soil and forest in the Ecuadorian Amazon caused by Chevron and Texaco' (Sollund, 2012, p. 3). International systems of production, distribution and consumption generate, reinforce and reward diverse environmental harms and those who perpetrate them. The reality is that 'the systematic causal chains that underpin much environmental harm are located at the level of the global political economy – within which the transnational corporation stands as the central social force' (White, 2012, p. 15). Corporations are responsible for various harms to the natural environment, including 'illegal emissions to air, water and land; hazardous waste dumping; and illegal manufacturing practices' (Tombs & Whyte, 2015, p. 47).

Corporate environmental crimes have both simple and complex explanations. In simple terms, corporate environmental crimes take place when corporations choose to break the law in order to maximise profits; taking direct and deliberate action to bypass or ignore environmental laws. This might include, for example, the deliberate and illegal dumping of toxic waste that would be a clear violation of the law and a criminal act. Lawful industries, such as oil and gas extraction and even the operation of permitted polluting factories, can cause harmful consequences for ecosystems but will fall outside the strict legalistic definition of being 'crimes' where these are unintended or accidental consequences of a lawful regulated activity. But examples of corruption, the bribing of officials and deliberate illegal harm caused to communities by, for example, ignoring environmental regulations, deliberately polluting water sources and ignoring regulations on repairing or mitigating harm can also amount to corporate crime. In this regard, corporate environmental crime is 'a product of motivation and opportunity conditioned by the quality of law enforcement' (Situ & Emmons, 2000, p. 67). In other words, where the inherent conflict between drivers to make money and the responsibility of a corporation to make profits for its shareholders and investors meets the costs of environmental compliance and the negative impact on potential profits, then crime is likely to occur. Corporate environmental crimes can also occur when corporations are negligent in respect of their obligations under environmental protection legislation. But there is also a more complex explanation that can apply when the very nature of corporate activity and the drive for profits places undue pressure on staff who have no choice but to break the law. For example, the 'Dieselgate' scandal illustrated how cheating emissions controls had become normalised in the motor industry with staff being under pressure to bypass regulations (Spapens, 2018). In this regard criminological control theory indicates that the lack of informal controls and the nature of encouragement to bypass environmental regulation, whether direct or indirect, create a situation where crime is almost inevitable (Hirschi, 2002).

Thus, contrary to the received wisdom that corporate offending is an aberration in the behaviour of an otherwise law-abiding entity, rational choice theory allows us to examine the extent to which corporations weigh up the cost of environmental compliance and regulatory obedience and then choose the alternative likely to provide the most profitable and satisfactory form of behaviour and outcome for them. Baumol identifies a distinction between 'productive corporate innovation' and 'unproductive activities such as organized crime' (1990, p. 893). In reality, however, corporate compliance with environmental regulations operates along a continuum consistent with Hobsbawm's view that private enterprise has a bias only towards profit (1969, p. 40).

Thus, in some cases corporate environmental crime arguably represents an appropriate, logical choice and can be actively pursued as a form of innovation designed to facilitate maximum profit whilst minimising risk to the corporation (Nurse, 2022, p. 69). McBarnet (2006) used the term 'creative compliance' to describe the use of ostensibly legal techniques that are used to undermine the purpose of regulation. In practical terms, this includes action such as operating within pollution havens to operate a more polluting factory/operation where the jurisdiction chosen might allow for lower operating costs, higher levels of pollution and thus higher profits than might be allowed in a Global North setting with appropriate environmental regulation and enforcement. Corporations might also develop Corporate Social Responsibility (CSR) mechanisms that shift responsibility away from the parent company and onto subsidiaries or even individual staff (Nurse, 2016, p. 54). Thus, when an environmental harm issue occurs, the company can point to its policies and argue that the subsidiary or the 'rogue' employee is at fault. McBarnet identifies that it is in the nature of creative compliance that practices that might be illegal, if not criminal, can be appropriately structured so as to be claimed to be lawful (2006, p. 1092).

Constructive compliance is another tool used to make a corporation appear compliant with environmental regulation, when this is not in fact the case. CSR reporting and sustainability initiatives can create the impression of compliance and environmentally responsible behaviour when in reality it becomes impossible for individual staff to comply with the policies and meet corporate productivity targets. Thus, staff may have little choice but to ignore regulatory controls. Indeed, the nature of constructive compliance is that it risks exploiting staff propensity for risk-taking and a desire to push against traditional rules and structures, particularly in environments where staff are encouraged to be innovative (Obschonka et al., 2013). When caught in environmental wrongdoing, MNCs may dispute the nature of the alleged wrongdoing, deny the nature of any victimisation or condemn their condemners by deploying neutralisation techniques similar to those applied by other offenders to deny or minimise their culpability for offending (Sykes & Matza, 1957).

These issues raise questions about the appropriate response to environmental offending and environmental harm. White (2008, p. 182) identified three broad approaches that can broadly be defined as: (1) the Socio-Legal Approach—where the emphasis on use of current criminal law which would involve the detection of harms and the prosecution of offenders; (2) the Regulatory Approach—where the emphasis is on social regulation (self-regulation, for example, allowing corporations to sign up to voluntary codes of practice and agree standards to abide by, and use of NGOs for regulation and monitoring for example through certification schemes); and (3) the Social Action Approach—a long-term approach with an emphasis on fundamental change, social transformation mostly achieved through democratic institutions and citizen participation. For example, societal acceptance of the need to move away from use of fossil fuels and wood products.

The reality is that effective ecojustice likely requires more than a single approach, and this raises some questions about the nature of the justice approach.

The Environmental and Ecological Justice Arguments

A green criminological approach identifies that 'members of other animal species and the rest of non-human nature urgently need to be protected from destructive human activities' (Benton, 1998, p. 149). Justice systems are generally anthropocentric in nature, concerned primarily with human harms and human interests with the focus on any remedy generally also concerned with a remedy for human victims. Arguably, this reflects the anthropocentric perspective that the environment lacks intrinsic value and is viewed primarily in the context of how environmental harms negatively impact human interests. Yet the green criminological perspective thinks otherwise, arguing not just for the intrinsic value of the natural environment but also for the interests of non-human animals and the impact of environmental harms on non-human animals to be recognised.

Two theoretical concepts are arguably at play here: environmental justice and ecological justice. Environmental justice is concerned in part with environmental rights as an extension of human rights and intergenerational responsibility such that achieving environmental justice requires justice systems that consider environmental harm in the context of achieving equity for both present and future generations (White, 2008, p. 15). Conceptually, environmental justice has a broad reach originating in the use of environmental justice as a term 'used to cover at least two overlapping parts of the grassroots environmental movement; the antitoxics movement and the movement against environmental racism' (Schlosberg, 2007, p. 46). This conception of environmental justice is synonymous with environmental justice being concerned with equity and the marginalisation of vulnerable groups and those historically denied access to environmental resources and the benefits derived from exploiting these. Thus, within environmental justice scholarship considerable attention has been paid to 'environmental racism' to denote the manner in which ethnic minorities and indigenous groups disproportionately suffer from a lack of access to environmental resources. Ethnic minority groups are also disproportionately affected by the location of toxic factories, pollution and waste sites in their neighbourhoods and the degradation of their environment. Environmental racism discourse contends that negative environmental consequences of business impacts communities of colour more than Caucasians. Kale (2020) also identifies that while people of colour are likely to suffer the effects of climate change more than their white counterparts, they are sometimes erased from debates about environmental issues. Anecdotally, it appears that few Black and Minority Ethnic (BAME) professionals are visibly in position as senior staff in the UK's major environmental organisations. This arguably lessens the voice of BAME communities in discussions on the environment, whereas environmental justice discourse has identified that communities of colour generally suffer more from environmental harm and are denied access to environmental justice. For example, much like the debate in the US, the UK has also recently engaged with the idea that the coronavirus disproportionately affects its BAME citizens. Official UK data released in June 2020 found that the highest age standardised diagnosis rates of COVID-19 per 100,000 population were in people of Black ethnic groups (486 in females and 649 in males) and the lowest were in people of White ethnic groups (220 in females and 224 in males). The data also went on to state that citizens of 'other Asian, Caribbean and Other Black ethnicity had between 10 and 50% higher risk of death when compared to White British'. The data also showed that 'death rates from COVID-19 were higher for Black and Asian ethnic groups when compared to White ethnic groups. This is the opposite of what is seen in previous years, when the all-cause mortality rates are lower in Asian and Black ethnic groups' (Public Health England, 2020, p. 4). In addition, the data shows that more than 90% of doctors who died during the early stages of the COVID-19 pandemic were from BAME backgrounds. An environmental justice and ecojustice perspective would explore why this was the case and would consider the extent to which racial injustice is a factor in environmental and health equalities in much the same way that the antitoxics movement in the US identified its disproportionate impact on communities of colour as a core part of developing an environmental justice approach.

The UK Government initially appeared to resist calls for an inquiry into the impact of COVID-19 on BAME communities despite evidence from the data suggesting that BAME communities suffered worst. Yet the reality is that disproportionate rates of infection, hospitalisation and deaths among BAME communities continued to raise questions to such an extent that Public Health England conducted the analysis outlined above. BAME citizens are considered more likely to work in those occupations that are unable to work from home and are considered key/essential workers. BAME citizens are also more likely to either live in high-pollution environments (the inner cities) and are susceptible to a range of pre-existing health conditions that impact on COVID-19 survival. Public health issues such as these highlight a wider conception on environmental harms and the negative impacts of environment that should also fall within our ecojustice conception.

While it is beyond the scope of this chapter to provide a comprehensive analysis of environmental justice approaches, scholarship in this area extends into other areas such as procedural environmental justice, distributive environmental justice and decolonial and recognitional justice. Procedural discourse is concerned with fairness in decision-making (Heydon, 2018) and appropriate access to legal mechanisms and remedies ideally designed to address the harm caused to affected individuals and communities (Nurse, 2022). Distributive justice is concerned with the distribution and experience of harms. Thus, 'for a distributive justice claim, it is not necessary to point to a specific actor causing injustice; disproportionate harms, or intolerably intense harms, trigger distributive justice concerns even if they are not caused by direct discrimination' (Kaswan, 2021, p. 22). Unlike the specific nature of a procedural claim which may require consideration of a specific event or legislative breach a distributive claim could, for example, consider the cumulative impact of pollution activity or a failure to deal with such an activity (discussed further later in this chapter). Thus, a distributive claim might consider the lived experience of a group that has experienced environmental harm. A decolonial approach to environmental justice recognises that coloniality persists 'as a form of power that creates structural oppression over marginalised sectors of society, such as Indigenous peoples, whose alternative worldviews become devalued, side-lined and stigmatized in development and environmental management practice' (Rodriguez, 2021, p. 85). Thus, we see colonial power exercised by MNCs in their relationships with Global South governments, through dominant Eurocentric colonial/modern values and worldviews that are institutionalised and disseminated through education, the media and state-sanctioned practices.

Our second conception, ecological justice, is ecocentric in nature, situated in a discourse that considers the environment and non-human nature to be important. Ecological justice argues that the environment should be protected because nature has intrinsic value, irrespective of any human benefits or interests. Ecological justice also incorporates the idea of species justice which arguably contends that justice systems should explicitly consider harms against non-human animals and provide effective redress where such harms occur (Nurse, 2013). Three main types of the ecological justice approach are relevant to a discussion of ecojustice:

- an ecological justice approach predicated on addressing ecological harms affecting humans;
- an ecological justice approach that examines ecological harms against the environment; and
- a species justice approach that is concerned with exploring ecological harms against non-human animals. (Lynch et al., 2019, pp. 1–2)

Ecological justice thus seeks to provide for an expanded notion of justice ideally incorporating the environment and non-human animals as crime/harm victims and incorporating a form of ecological jurisprudence (or earth jurisprudence) that brings nature to the centre of legal thinking and ensures that laws are interpreted to provide for better environmental protection. However, ecojustice is not just a function of writing appropriate laws; there is also a need to give consideration to how different types and function of laws work to provide not just protection but to remedy harms. This being the case an ecojustice perspective might also argue for some consideration of what rights need to be protected and how they might be enforced. This chapter contends that this broader perspective also requires some consideration of environmental and human rights within our ecojustice toolkit.

Environmental Rights in International Law

Increasingly, human rights litigation is being used to address issues of environmental harm, particularly those situations relating to state inaction and public authority failure to observe and uphold human rights. It is beyond the scope of this chapter to provide a comprehensive overview of human rights instruments and their relevance to environmental protection. However, environmental rights and particularly a right to live in a healthy environment are arguably spread across a range of international human rights instruments including: the Universal Declaration on Human Rights—Articles 22, 25, and 27; the International Covenant on Economic Social and Cultural Rights—Articles 1, 6, 7, 11, 12, 13, and 15 ecologically sustainable development; the UN Convention on the Rights of the Child—Articles 24 and 27; various case law and national legislation, and the UN Declaration on the Environment (1972) and the World Charter for Nature (1982).

Human rights law may not explicitly state that there is a right to live in a healthy environment, although some state constitutions explicitly provide for this right. Elsewhere, rights such as the right to life, which is enshrined within the European Convention on Human Rights (ECHR) and national legislation (e.g. the UK's Human Rights Act 1998), have been used to address state failures to deal with environmental problems. More recently, human rights laws have been directly applied to contemporary environmental issues such as climate change.

Case Example: *Verein KlimaSeniorinnen Schweiz and Others* v *Switzerland* [2024]

In *Verein KlimaSeniorinnen Schweiz and Others v Switzerland* [2024], the European Court of Human Rights found that Switzerland failed to comply with its climate change convention obligations. The case was brought by a Swiss association on behalf of a group of older women who claimed that state inaction on climate change had negative impacts on their health and living conditions. The claim was based on an argument that the Swiss authorities were not taking action required by the climate change convention to mitigate the effects of climate change.

The court dismissed the claims made by four individual applicants as inadmissible because they did not meet the victim-status criteria of Article 34 of the European Convention. But the association that brought the case was considered to have the right (locus standi) to bring a complaint arising from the threats caused by climate change on behalf of those individuals who could claim to be subject to specific threats or adverse effects of climate change on their life, health and well-being and quality of life.

The court concluded that the Swiss Confederation had failed to meet its positive obligations under the climate change convention, including a failure to quantify national greenhouse gas emissions (GHG). The Court found that Article 8 of the European Convention on Human Rights provides for a right to effective protection by the state from the serious adverse effects of climate change on lives, health, well-being and quality of life. The Court ordered Switzerland to pay costs of 80,000 Euros. While the Court did not specify specific measures that the Swiss authorities now need to take, the judgement will likely lead to a review of climate change policies to avoid further legal action.

Such cases allow individuals to sue states for climate change action failures that directly impact on their human rights rather than seeking to pursue individual companies or industries for their polluting activity. Potentially, this is a more effective form of climate justice because it directly considers the harms suffered by individuals and recognises climate change not just as an environmental policy issue but as a human rights one that states are required to take action on. A United Nations report (2021) specifies an increase in climate change litigation to 1550 cases in 38 countries (39 including the European Union courts). Increasingly, the climate emergency is framed as a human rights issue which the courts are required to address.

Prosecuting the Criminal Corporation: The Problems

As indicated earlier in this chapter, the reality of criminal law approaches to environmental harm are primarily that they punish the offender but do not necessarily remedy harm. While this may be acceptable in the case of mainstream crimes such as street crime and burglary, green criminological approaches consider that the focus should be on the harm caused to the environment and non-human nature irrespective of the extent to which a particular harm is considered to be distinctly criminal or is dealt with by another justice system approach. Repairing harm is an integral part of green criminological approaches situated within eco-justice principles. So too is the role of justice systems in changing behaviour.

Practical problems sometimes exist in respect of the capacity and willingness of states to treat corporations as 'criminal'; the liability of corporations for some crimes (such as homicide) may be limited within justice systems. In addition, the nature of corporations as 'legal persons' means that identifying the person or entity to prosecute can sometimes be problematic. A corporation may be involved in an environmental crime or an infringement of laws, but identifying whether the decision to commit the harmful act was taken at a sufficient level to be indicative of the corporation as a whole may be problematic. This raises questions concerning whether company directors, the company itself or the staff member who commits the harmful act should be prosecuted. For example, where an oil spill pollutes the marine environment and the spill can be traced back to a faulty repair by a rig worker, should this worker be prosecuted? Arguably, this individual is directly responsible for the harmful act, but they may be a victim if the corporation has put them under extreme pressure to commit the work. The power dynamics and complexity of corporate structures within the modern transnational corporation also mean that complex bureaucratic structures exist which can involve layers of management, directors, senior operating executives, and other staff through to lower management and staff on the ground making it difficult to locate the 'controlling mind' behind an environmental harm incident. Drawing clear lines of culpability between the staff member committing the offence, their immediate line manager and others at higher levels within the company through which it becomes possible to legally identify the manager or executive responsible for the decisions leading to the environmental harm.

Given that corporations are often significant sources of employment and tax revenues, governments generally target the criminal law at individual employees rather than the corporation itself. In addition, within environmental law, regulators such as the Environmental Protection Agency (EPA) and the Environment Agency are often deployed as the default option to deal with wrongdoing and non-compliance. Environmental statutes may contain criminal and civil sanctions although 'civil penalties are most typically involved as the legal remedy against environmental wrongdoers. Criminal penalties are usually invoked only when environmental statutes have been wilfully or knowingly broken' (Situ & Emmons, 2000, p. 38). Environmental regulation is generally designed to allow the corporation to continue its operations with the object of regulatory justice being to change behaviour and encourage compliance. Accordingly, environmental regulation is corrective rather than punitive, differing from the criminal law. Regulators often have considerable discretion over how best to pursue regulatory breaches and evidence suggests that civil and administrative law tools are generally preferred over criminal law tools with criminal prosecution often being a last resort. The Gulf Oil Spill (also known as the Deepwater Horizon Spill) serves as a useful example where, although the response incorporated some elements of criminal penalty, the core remedy involved a compensation scheme for those affected by the pollution, civil claims and a mandated clean-up following an oil spill that covered 28,958 square miles. The case 'was the subject of an estimated $20 billion settlement which included $5.5 billion in civil Clean Water Act penalties and billions more to cover environmental damage and other claims by the five Gulf states and local governments' (Nurse, 2022, p. 59).

Regulators have tools that allow them to require remedial action, such that corporations may need to modify their processes to reduce environmental harm. But the reality of regulatory justice is that it can allow continued offending and be ineffective in respect of changing behaviour. Secondly, the regulatory process itself, predicating on allowing corporations to achieve compliance, can be problematic. For example, a corporation exceeding its permissible pollution limits might be served with a warning notice that identifies the nature of their offence and gives them an opportunity to fix the problem and achieve compliance. If the corporation fails to fix the problem it may be served with a further notice and allowed more time to achieve compliance—only in the most serious cases or after repeated failings will criminal prosecution be pursued. The reality is that major corporations can withstand repeated prosecution and factor in legal costs and fines as simply the cost of doing business and the level of fines may be such that it is more cost-effective to pay these fines rather than carry out costly repairs or modifications to meet the requirements of environmental compliance.

In some cases, civil claims, such as those raised by an NGO or as a class action, have proved to be more effective in dealing with corporate environmental harms than criminal prosecution or regulatory action. One advantage of such claims is that they are situated within the 'polluter pays' principle which seeks to make corporations liable for the various harms they cause. This can include environmental damage, human rights breaches, and the wider impact of corporate activity. The following case example illustrates this.

Case Example: *Okpabi and Others* v *Royal Dutch Shell Plc and Another* [2021] UKSC 3

In this case, the UK Supreme Court considered whether and in what circumstances the UK-domiciled parent company of a multinational group of companies may owe a common law duty of care to individuals who allegedly suffer serious harm as a result of alleged systemic health, safety and environmental failings of one of its overseas subsidiaries as the operator of a joint venture operation. In *Okpabi and others v Royal Dutch Shell* [2021] UKSC 3, the Supreme Court overturned the Court of Appeal in a dispute concerning whether the English courts had jurisdiction over mass tort claims brought against Royal Dutch Shell Plc by Nigerian claimants. The claim against Shell was based on the notion that Shell owed the claimants a duty of care, because it exercised significant control over material aspects of the operations of Shell's Nigerian subsidiary. In this case, the Supreme Court considered the question of whether Royal Dutch Shell (RDS), the parent company incorporated in London, owed a duty of care in negligence to claimants based in Nigeria regarding the activities of its subsidiary company and/or it assumed responsibility for the subsidiary's operations.

The Supreme Court concluded there was a real issue to be tried and that it was at least arguable that Royal Dutch Shell owed a duty of care to claimants in Nigeria who were affected by the actions of its subsidiary, opening the door for further legal action. The recent decisions of *Alame & Ors* v *Shell plc & Anor* [2023] EWHC 2961 ("*Alame 1*") and [2024] EWHC 510 ("*Alame 2*"), the Bille and Ogale group litigation, which subsequently bring together four related sets of proceedings, reinforce the importance of class actions and group litigation as a means of redressing corporate harms committed overseas.

These cases can be seen as significant contributions to developing international law dealing with corporate harms, which allows multinationals to be held accountable in their home jurisdiction for damage caused by, or related to the operations of, their foreign subsidiaries. This is significant in part because the use of subsidiaries in countries with (perceived) weak judicial systems is a mechanism through which corporations seek to protect themselves from costly environmental litigation and divert responsibility towards an 'arm's length' company.[2]

Perhaps encouragingly, in both the UK and US, multinationals have been sued based on what is referred to as 'foreign direct liability' in which a claimant may bring proceedings before a European or a US court

[2] See, for example, the Bhopal chemical disaster which took place on 3 December 1984 and in which a toxic cloud of methyl isocyanate (MIC) gas enveloped the pesticide plant in Bhopal, India. The initial tabulation of injuries indicated that 1430 people died in the immediate aftermath of the disaster and by 1991 the official Indian government figure was more than 3800 dead and approximately 11,000 with disabilities. The parent company, Union Carbide, largely denied culpability and placed the blame with its subsidiary Union Carbide India Limited. In 2010, seven former employees of Union Carbide India Ltd. were found guilty of causing death by negligence while the US-based parent company was able to largely avoid the jurisdiction of the Indian courts. The nature of the corporate response, and the way in which corporations may divert responsibility to an overseas subsidiary, has been the subject of considerable academic discussion. See, for example: Bharghava (1986), Pearce and Tombs (1989), Nurse (2024).

(Prussia, 2006; Jawger, 2010).[3] A claimant may seek to bring proceedings before a European or US court rather than locally for a variety of reasons. These include: dissatisfaction with local proceedings; lack of independence or expertise of the courts; or extensive possibilities for delay. One additional reason for organisations such as Friends of the Earth Netherlands or Global North law firms such as Leigh Day (acting on behalf of Global South clients) to bring a claim against both the subsidiary and the parent company in the jurisdiction where the parent company has its headquarters will be to draw attention to the issue of CSR in the home market of the multinational. While not a perfect solution, this potentially takes us beyond restrictive approaches based on criminal law and regulatory enforcement and towards making the polluter pay.

As long ago as 1992, Lord Woolf, at the time one of the UK's most senior judges, commented, questioned whether the judiciary was 'environmentally myopic' (Woolf, 1992) and noting a range of problems with the prosecution of environmental cases and judicial consideration of these. These problems included: lack of available scientific support; lack of specialist environmental and legislative knowledge; perceived loopholes in legislation; and difficulties in bringing cases to court due to lack of prosecutorial expertise. In our contemporary setting, whether it is possible to achieve transnational justice is a debatable point (Nurse, 2015, p. 146; Woolf, 1992). The human rights cases referred to earlier in this chapter highlight how ecojustice can sometimes be achieved through a creative approach to which legislation should be used to pursue redress for environmental harm. In addition, there is a movement towards widening the use of specialist environmental courts which already exist in some jurisdiction and are able to hear a range of environmental cases including: land use; planning law cases; cases concerning pollution events

[3] The principle of foreign direct liability is enshrined within the US Alien Torts Act 1789, which provided 'a legal basis for non-US-citizens ("aliens") to bring transnational tort claims before the US federal courts in relation to international human rights violations perpetrated anywhere in the world' (Enneking, 2014, p. 44). The principle allows action to be taken against MNCs for human rights and environmental violations perpetrated overseas, particularly in Global South states where victims and affected communities are potentially denied opportunities to obtain (enforceable) remedies. The reasons for this can include poorly functioning legal systems and limited access to legal remedies, corruption (within both enforcement and judicial mechanisms) and state involvement and investment in the MNC presence and operation (see, for example, Nurse, 2024).

and regulatory environmental offences.[4] Environmental courts can compel compliance with environmental law through both civil and criminal enforcement and provide for the overlapping of remedies (civil and criminal) and interests (public and private). Given the specialised nature of environmental treaties and law, there is an argument for specialised courts that are able to develop certain fundamental environmental principles such as the precautionary approach, 'polluter pays', prevention at source, and procedural transparency and provide models for consideration of environmental problems that can then be adopted by other (mainstream) courts and also develop policy and interpret law where there are conflicts between government, NGOs and the environment. The benefits of recent developments highlighted here are that courts have clearly signalled that action can be taken against parent companies for the actions of their subsidiaries overseas. This also opens the door to law firms taking cases on a 'no win, no fee' basis and for both law firms and NGOs to take civil actions on behalf of communities who might otherwise be unable to take action to pursue a remedy. Where specialist environmental courts exist, the criticisms identified by Lord Woolf are potentially negated by the existence of environmental expertise. However, where they do not, the existing expertise of environmental law firms and NGOs who bring cases to court on behalf of affected communities ensures that these issues are addressed. Clearly, not all cases will come to court or be suitable for civil claims and class actions but this is an important development in the expanding ecojustice field and one that deploys the green criminological perspective of seeking to address harm.

The Future of Ecojustice

As this chapter's discussion illustrates, resolving environment harms requires the development of a holistic approach that considers both crime and harm and engages with a wide notion of the purpose and function of justice systems. This requires active consideration of what should be done

[4] For example, Australia's New South Wales, Land and Environment court. See https://lec.nsw.gov.au/

when corporations harm the environment, or states ignore their environmental obligations. It also requires extending our consideration of crime and deviance to incorporate a zemiological approach and one that is geared towards considering how best to engage with and resolve harms in addition to issues of (retributive) punishment. Green criminology has already identified that 'the form of criminal justice criminologists ordinarily examine to discuss the control of crime is a narrow form of justice' (Lynch & Stretesky, 2014, p. 47). This narrow conception is disproportionately focused on individual victims of crime and use of the criminal law whereas green criminology's focus on wider harm and a wider range of offenders requires a more effective form of justice where policy approaches are geared towards effective crime prevention and changing human behaviour rather than what is arguably a more restricted form of crime management (Benton, 1998).

The reality of environmental harm is that it incorporates a range of offending and non-compliant behaviour, much of it dominated by the demands and operations of neoliberal markets. Corporate environmental harm is an area of complexity; activities can be accidental or deliberate while the harm caused can range from the relatively minor, such as a small increase in emissions that can be easily rectified, through to serious incidents, such as major oil spills or the dumping of toxic waste that can have serious impacts on communities, lead to loss of life and have long-lasting and even irreversible impacts on the environment. This requires a justice response that is flexible and recognises the varied nature of environmental offending (Nurse, 2013, 2015) and also a conceptual response to justice that provides more than just punishment; one that actively considers repairing environmental harm. This means that penalties and sanctions for environmental harm need to do more than just punish an offender. They should also consider the environment as a victim and employ 'polluter pays' principles so that, as much as possible, offenders should try to put affected communities and the environment itself back in the position it would have been in had the harm not occurred. Where this is not possible, some element of mitigating the 'value' or cost of environmental harm should form part of the remedy. This could, for example, include planting an equivalent number of lost trees in the event of a timber-trafficking/deforestation incident, the clean-up of a river and surrounding area in the event of a pollution incident that impacts on a watercourse or appropriate payment to local communities to mitigate the effects of smog or other pollutants.

Accordingly, this chapter's conception on ecojustice adopts a wide view of environmental victims and victimisation. It considers that both environmental justice and ecological justice considerations should be incorporated into justice and regulatory systems that incorporate preventative, reparative, and retributive elements according to the nature of the offence and the offending behaviour. From a green criminological perspective, ecojustice is situated within conceptions of ecological justice and species justice and incorporates the fundamental concepts of both seeking to prevent environmental harm before it occurs and remedying the impact of environmental harms and adopting a wide conception on victimisation caused by environmental harm. This wider conception considers all those affected by environmental harms, including future generations. At the heart of our future ecojustice system, there should be an approach based in remedying environmental harms and that considers the needs of the environment and non-human nature rather than continued reliance on the punitive approach of the criminal law.

References

Baumol, J. (1990). Entrepreneurship: Productive, unproductive and destructive. *Journal of Political Economy, 98*(5), 893–921.

Benton, T. (1998). Rights and justice on a shared planet: More rights or new relations? *Theoretical Criminology, 2*(2), 149–175.

Bhargava, A. (1986). The Bhopal incident and union carbide: Ramifications of an industrial accident. *Bulletin of Concerned Asian Scholars, 18*(4), 2–19.

Bowers, C. A. (2009). Rethinking social justice issues within an eco-justice conceptual and moral framework. *Journal of Educational Controversy, 4*(1), Article 3.

Crain, W. M., & Crain, N. V. (2014). *The cost of federal regulation to the US economy, manufacturing and small business* (p. 1). National Association of Manufacturers.

Enneking, L. (2014). The future of foreign direct liability? Exploring the international relevance of the Dutch Shell Nigeria case. *Utrecht Law Journal, 10*(1), 44–54.

Gibbs, L. (2010). *Love Canal: And the birth of the environmental health movement*. Island Press.
Hall, M. (2011). Environmental victims: Challenges for criminology and victimology in the 21st century. *Journal of Criminal Justice and Security, 4*, 371–391.
Hall, M. (2013). *Victims of environmental harm: Rights, recognition and redress under national and international law*. Routledge.
Heydon, J. (2018). Sensitising green criminology to procedural environmental justice: A case study of first nation consultation in the Canadian Oil Sands. *International Journal for Crime Justice and Social Democracy, 7*(4), 67–82.
Hickey, J. E., & Walker, V. R. (1995). Refining the precautionary principle in international environmental law. *Virginia Environmental Law Journal, 14*(3), 423–454.
Hirschi, T. (2002). *Causes of delinquency*. Transaction. (Original work published 1969).
Hobsbawm, E. J. (1969). *Industry and empire from 1750 to the present day*. Penguin.
Jawger, K. (2010). Environmental claims under the Alien Tort Statute. *Berkeley Journal of International Law, 28*, 519.
Kale, S. (2020). 'We need to be heard': The BAME climate activists who won't be ignored. *The Guardian*. https://www.theguardian.com/environment/2020/mar/09/we-need-to-be-heard-the-bame-climate-activists-who-wont-be-ignored
Kaswan, A. (2021). Distributive environmental justice. In B. Coolsaet (Ed.), *Environmental justice*. Routledge.
Lynch, M., & Stretesky, P. (2014). *Exploring green criminology*. Ashgate.
Lynch, M. J., Long, M. A., & Stretesky, P. B. (2019). *Green criminology and green theories of justice*. Springer International Publishing.
McBarnet, D. (2006). After Enron will 'whiter than White collar crime' still wash? *British Journal of Criminology, 46*, 1091–1109.
Neumayer, E. (2001). Pollution havens: An analysis of policy options for dealing with an elusive phenomenon. *The Journal of Environment & Development, 10*(2), 147–177.
Nurse, A. (2013). *Animal Harm*. Ashgate.
Nurse, A. (2015). *Policing wildlife*. Palgrave Macmillan.
Nurse, A. (2016). *An introduction to green criminology and environmental justice*. Sage.
Nurse, A. (2022). *Cleaning up greenwash*. Lexington.

Nurse, A. (2024). Corporate environmental crime: A contested notion? *New Law Journal, 174*(8061), 9–11.

Obschonka, M., Andersson, H., Silbereisen, R. K., & Sverke, M. (2013). Rulebreaking crime and entrepreneurship: A replication and extension study with 37 year longitudinal data. *Journal of Vocational Behaviour, 83*, 386–396.

Pearce, F., & Tombs, S. (1989). Bhopal: Union carbide and the hubris of the capitalist technocracy. *Social Justice, 16*(2), 116–145.

Prussia, K. S. (2006). NAFTA & The Alien Tort Claims Act: Making a case for actionable offenses based on environmental harms and injuries to the public health. *American Journal of Law & Medicine, 32*(2–3), 381–404. https://doi.org/10.1177/009885880603200211

Public Health England. (2020). *Beyond the data: Understanding the impact of COVID-19 on BAME groups*. Public Health England.

Rodriguez, I. (2021). Decolonial environmental justice. In B. Coolsaet (Ed.), *Environmental justice*. Routledge.

Schlosberg, D. (2007). *Defining environmental justice*. Oxford University Press.

Seba, E. (2008). *Sierra Club sues Shell over refinery pollution*. Reuters/Environment News Network.

Sherman, L. W., Strang, H., Barnes, G., Bennett, S., Angel, C. M., Newbury-Birch, D., Woods, D. J., & Gill, C. E. (2007). *Restorative justice: The evidence*. The Smith Institute.

Situ, Y., & Emmons, D. (2000). *Environmental crime*. Sage.

Sollund, R. A. (2012). Introduction. In R. Ellefsen, R. Sollund, & G. Larsen (Eds.), *Eco-global crimes: Contemporary problems and future challenges* (pp. 3–14). Ashgate.

Spapens, T. (2018). The 'Dieselgate' scandal: A criminological perspective. In *Green crimes and dirty money* (pp. 91–112). Routledge.

Sykes, G. M., & Matza, D. (1957). Techniques of neutralization: A theory of delinquency. *American Sociological Review, 22*, 664–673.

Thomson, J. (2016). Toxic residents: Health and citizenship at Love Canal. *Journal of Social History, 50*(1), 204–223. https://doi.org/10.1093/jsh/shv105

Tombs, S., & Whyte, D. (2015). *The corporate criminal*. Routledge.

Walter, N. (2024). Jail for holding a placard? Protest over the climate crisis is being brutally suppressed. *The Guardian*. https://www.theguardian.com/commentisfree/2024/apr/13/climate-crisis-protest-activism-repression

Wenzel, M., & Okimoto, T. G. (2016). Retributive justice. In C. Sabbagh & M. Schmitt (Eds.), *Handbook of social justice theory and research*. Springer.

White, R. (2008). *Crimes against nature: Environmental criminology and ecological justice*. Willan.

White, R. (2012). The foundations of eco-global criminology. In R. Ellefsen, R. Sollund, & G. Larsen (Eds.), *Eco-global crimes: Contemporary problems and future challenges* (pp. 15–31). Ashgate.

White, R. (2014). Eco-justice and problem-solving approaches to environmental crime and victimisation. In T. Spapens, R. White, & M. Kluin (Eds.), *Environmental crime and its victims*. Routledge.

White, R., & Heckenberg, D. (2014). *Green criminology: An introduction to the study of environmental harm*. Routledge.

Woolf, H. (1992). Are the judiciary environmentally myopic? *Journal of Environmental Law, 4*(1), 1–14.

Living, Dying, and Sharing a Planet

Ragnhild Sollund

Introduction

Hvaldimir[1] is dead. Hvaldimir was a Beluga whale who appeared on the north of the Norwegian coast in 2019. He was found with several holes in his body—injuries that indicated he could have been murdered. Still, the police stated after the preliminary autopsy had been made that he had not been shot. It is difficult to evaluate on what grounds the police concluded, and how they explained the injuries. There are still strong suspicions among non-governmental organizations (NGOs) that his death was not due to natural causes, but that humans played a role in it. While suspicions were strong that he had been shot, the police reported that the Veterinary Institute which carried out the autopsy had found a 35 cm-long stick stuck in his mouth. The police made a statement based

[1] Hval means whale in Norwegian, Hvaldimir is a wordplay on whale and Vladimir (Putin).

R. Sollund (✉)
University of Oslo, Department of Criminology and Sociology of Law, Oslo, Norway
e-mail: ragnhild.sollund@jus.uio.no

D. van Uhm, D. Siegel (eds.), *Global Green Crime and Ecojustice*, Palgrave Studies in Green Criminology, https://doi.org/10.1007/978-3-031-91557-4_4

on this finding which has been interpreted in some media that he had died from starvation, but Beluga whales can survive for a long time without eating, since they live on the fat on their body, and Hvaldimir was found to be in good shape. This does beg the question of how Hvaldimir could have gotten a stick stuck in his mouth. He was reported to having been playing fetch with humans and there were also some reports of people putting objects in his mouth. Did he fetch a stick for a human, or did a human purposefully stick it in? Should this latter explanation be the cause for his demise, Hvaldimir's fate is yet another testimony that 'wild' animals[2] who move too close to humans in Norway risk their life.

When Hvaldimir first appeared, he wore a tight harness that looked like it had a small camera mount; the harness was marked with the words "Equipment St Petersburg" in Russian. This led to suspicion that Hvaldimir had been trained by the Russian military as a spy whale. Hvaldimir sought people, he was playful and charming, and his behaviour, in combination with the adoring appearance characteristic of Beluga whales with their 'Disney-style' high forehead and permanent 'smile', stirred love and fascination in many people who would go on excursions to see him, as he travelled up and down the Norwegian coast, principally in the north. When he died, however, he was in the south-west of the country, near Stavanger, and the NGOs OneWhale and NOAH, in cooperation with the Norwegian authorities, had drawn up plans to relocate Hvaldimir further away from Norway, up to the north, where he could be with other Beluga whales and out of harm's way.

Now many grieve for Hvaldimir. This was expressed by spokesperson Regina Haug, from OneWhale, in the following words: "The whale I knew and loved was a strong, capable, and energetic animal yet so gentle, with a sweet and cheeky personality. It was gut-wrenching to see this

[2] With the term 'wild' here I refer to free-born animals. The term is alienating, however, since it constructs an abyss between humans and free-born animals and has connotations to unpredictability and 'savageness', implying such animals are 'ferocious' and 'untamed'. The term should therefore be avoided. The term 'wildlife' equally poses a problem since it fails to acknowledge each individual, and their personality, abilities, and interests, since it is a mass term. In addition, it also usually includes plants, adding even more to diminishing the interests of sentient beings. The word 'animal' is also problematic, since it categorizes all animals other than the human animals as the same, disregarding the huge differences between species, of which the human species is just one. It also constructs an unnatural border between the human species and other animal species. The concept 'non-human animal' suffers from the same weakness. For lack of good alternatives and for simplicity, I retain the word 'animal' in this text.

beautiful animal now lifeless. It felt like my heart had fallen into my stomach. The pain of losing him to a senseless killing is an anguish which, right now, feels impossible to fix" (Talmazan & Aggarwal, 2024).

Hvaldimir's death followed that of Freya the walrus. She was killed by the Norwegian authorities in Oslo in 2022 after having moved around the Norwegian coastline succeeding her visits to other countries like Britain, Denmark, and the Netherlands, where she was welcomed and loved (Sollund, 2022, 2023). Attempts were made to protect her from humans, and vice versa. There was a discussion of how she could be moved to prevent the close encounters that were regarded as dangerous for the humans who continued to disrespect her. This spurred the decision by the Norwegian authorities to set out in a boat to kill her in the Oslo harbour in the early morning before the realization of the plans to protect her. Her death also created international uproar and grief among many people (Horowitz, 2022).

Freya and Hvaldimir were both boundary crossers, transgressing the border for where animals "belong", and where they do not "belong". The human species is just one of many, but it still defines the terms of belonging or non-belonging for other species. This chapter focuses on how can we succeed in sharing space with other animals. In doing so, I will provide further examples of three categories of shared space/habitats of humans and other animal species and how such crossing of space can be perceived: (1) when animals enter space humans regard as theirs; (2) when humans enter space animals regard as theirs; and (3) when humans and animals share space. Thereafter I will discuss how humans transgress the boundaries of other animals in their continuous destruction and encroachment of the steadily shrinking habitats of non-human species. While Freya and Hvaldimir have become 'mascots', they may still represent and symbolize humans' ongoing destruction and victimization of other animals amidst the ongoing crisis in nature.

When Free-Born Animals Come Too Close to the Humans' Habitat

Humans create different moral spheres in which animals are located that determine their value to humans, and thereby to which degree they are accorded intrinsic value and consequently respect and protection

(Sollund, 2024). In the core are the human animals and their relatives and friends whose intrinsic value and individual human rights are expected to be respected, as should human rights generally. In the first circle surrounding the core are the companion animals, whom to some degrees are respected and thus even included among the humans in the core, although they are usually perceived as inferior. In the second circle are the animals we *use* for different purposes; for food, clothing, entertainment, and research—animals to whom we recognize we owe a minimum of concern for having taken them into our lives to be used for our benefit. They are therefore supposed to be protected in animal welfare legislation in most countries. In the third circle are the free-born 'wild' animals, those we hunt, fish, catch or watch and photograph—while they are often regarded as catch and prey, we may also admire them as tokens of nature and the wild.

Hvaldimir did, when he was captivated by humans and trained by them. He crossed the boundaries, and entered the first circle, where he could be admired (almost) like a person. While legal rights apply only to human personhood, animals whom we place within this circle may be ascribed certain moral rights (Regan, 2016). This boundary crossing was unlikely to have been initiated by Hvaldimir; he was probably captivated as very little for the training of the Russian intelligence service. When he was first discovered at Norwegian shores, he had a harness strapped to his body. However, the social interaction to which he had become accustomed, and consequently his evident joy in socializing with humans he met on his way, was very much appreciated by humans, giving Hvaldimir and the people he met feelings of mutual joy and led to the recognition of him as a non-human person. Although Freya remained 'untamed' and 'wild' to a far greater degree than Hvaldimir, through the naming of her and the attention she received e.g. in the news media, she also came closer to a recognition of her as a non-human person, more so than if she had remained in her natural habitat in the North, out of view and thereby also the social boundaries of the human world.

Hvaldimir and Freya crossed both physical and geographical boundaries which led them to gradually also cross the social and moral boundaries that often separate humans from other free-born animals. In contrast, animals of other species are often alienated and cast as enemies, which can be seen in the treatment of species such as the wolf, which is frequently vilified, being described as 'wild' and dangerous. Other species are denominated as 'pests', such as, for example, parrots, who ravage the crops of farmers in Argentina. However, parrots often also figure frequently as family members, fulfilling a social value (Sollund, 2017), a treatment which indicates that it is the use value we ascribe to other animals that determine how we categorize and treat them, rather than their physical and cognitive capacities or their social abilities, and definitely not their own interests.

While Freya and Hvaldimir were publicly named and consequently regarded as *someone*, other free-born animals are regular victims of hunting, even when they do enter human habitats. In the west part of Oslo, adjacent to the famous Vigeland park, is a cemetery, which is a relatively large and peaceful space. I used to take daily walks there while I worked and lived in the neighbourhood and gradually grew to know a man working at the cemetery. He told me to my considerable surprise that there were two hares living at the cemetery, and he pointed them out to me, showing under which bushes and trees and behind which tombstones they used to rest. It became a huge pleasure to walk in the cemetery then. Hares which are so seldom seen in the forests surrounding Oslo because they are hunted so brutally were present at the cemetery in the city!

The joy was short-lived; after approximately six months, a man—a hunter—took his dogs and his rifle to the cemetery, where he killed the hares. Of course, hunting is not allowed on a cemetery or in the city; but this man apparently could not tolerate that the hares—a specie categorized as huntable—should not be hunted. To him, presumably, the hares could not transgress the border of their use value as a huntable species, even though they had transgressed the borders of their natural habitats and entered ours; thus, he killed them. While I and many others grieved the deaths of Hvaldimir and Freya, and I myself also grieved over the deaths of the urban hares, why do people react more to Hvaldimir's and Freya's deaths than to the deaths of the thousands of animals who are killed in 'recreational hunts' in Norway and many other countries on a regular basis? This is a question to which I return to after presenting what occurs when humans are the transgressors.

When Humans Come Too Close to the Habitats of Free-Born Animals

With regard to the question of boundary crossing, there is a central question: who crosses the boundaries of whom? Is the 'problem' that animals cross the borders to the habitats of humans, or is it the other way around? Two examples can illustrate this. In 2023, I bought a cabin in Jeløya, an island not far from Oslo, with dedicated nature reserves. The first night in the cabin I was lying awake in bed at around three o'clock in the morning—it is summer so it is daylight outside beaming in through the window blind—when I hear a strange roar. It is clearly an animal, but which? I get up and look down at the footpath that goes alongside the cabin, assuming it must be a fox, before I finally look up to see a majestic roedeer buck with big antlers positioned about three meters away from me, up the hill. He faces me and roars again—it is as if he had a message for me, but what was he saying? He seemed very determined, so I assumed he said that this was his territory and that I ought to respect that as I entered his home.

In the second example, I am on a beach in Oslo, it is a warm day so there are plenty of (too many!) people, but also a lot of seagulls (and ducks, sparrows, and crows). Most of the seagulls in Oslo have become vulnerable species because the Oslo fiord is nearly dead as a result of human impact, such as pollution, which has destroyed the fragile ecosystems, and fishing; there is little left over to eat for the seabirds. This makes the seagulls seek other opportunities, such as snapping the food out of the hands of humans in the streets, or seeking to people at the beaches, knowing they have food. They are, however, unwelcome in the view of most of the people at the beach. Seagulls are generally regarded as a nuisance; like the wolves they are vilified and disrespected, even hated, as exemplified by the song by a Norwegian singer Odd Børretsen; *Jeg hater måker!*, meaning "I hate seagulls".[3]

I personally do not hate seagulls; I find them charming and willingly help them in their daily struggles to find enough food by feeding them. This is not an act welcomed by the other people at the beach, so when I

[3] https://www.youtube.com/watch?v=2YSVJjIPDoA

arrive and the seagulls spot me and come to gather around me knowing I have something to offer, I am about as disparaged as the seagulls themselves and scolded by the people around me. They shout that they do not want the seagulls on the beach, but whose beach is it? Does the beach belong to the people who come to there on a warm day, or is it the home of seagulls who desperately try to survive despite the anthropogenic destruction of their natural habitats? And if we accept that the beach is the home of the seagulls, more than the home of the humans, what right do the humans have to decide on how the seagulls should behave in their own home?

By feeding the seagulls, I played a role in the conflict that arose. If the seagulls are not fed by the humans (or if they do not have the opportunity to snatch their food), they may keep a greater distance. This would perhaps mean that they would be less subject to stigmatization and animosity by humans. On the other hand, we have destroyed their food sources, leaving them to be more vulnerable and most seagull species in Norway are in steep decline (Birdlife Norway, 2024). This could mean that we *owe* them to help, to compensate for the damage we have done, as custodians (White, 2013). However, the help we offer must be balanced with the reactions and consequences of such help; for example, making animals with whom we share habitat dependent on our help may not be to their advantage in the long run if there are alternatives that support their autonomy.

Clearly, the seagulls and the humans share space; but only temporarily. During the winter, those seagulls who have not migrated will remain at the beach while the people snuggle up at home. While the roedeer must share space with us during summer, when winter comes he will have the ground surrounding my cabin to himself. But what of the animals who share space with humans on a more permanent basis?

When Humans and Free-Born Animals Share Space: Species Cohabitation

There are several species that have developed alongside the human species and who are dependent on us for their survival, but whom we do not keep in captivity and exploit for food or clothing or for medicinal purposes—such as house sparrows, rats and pigeons. They have been very plentiful and are regarded with varying degrees of attention, respect and disrespect; while the sparrows are generally accepted around the world, pigeons and rats are often regarded as 'vermin' or 'pests'. The sharing of habitats has worked well for the humans and the sparrows, leaving the sparrow to be one of the most successful species on earth, spread across every continent.[4] Now, however, the species is suffering a steep decline: There were per 2021, an estimated 247 m million fewer house sparrows in Europe than there were in 1980. The house sparrows have vanished from many cities for unknown reasons; however, the reasons are likely to include shortages of food, diseases such as avian malaria, and air pollution (Barkham, 2021). While the house sparrows may have benefitted from their co-existence with humans, humans are also causing their decline.

While the house sparrows receive food in numerous birdfeeders around the world, rats and pigeons are victims of prejudice, alienation, and drastic reducing measures. But why? If you take a close look at a pigeon, you will see their radiant colours and beauty; if you care to read about them, you will learn that they are highly intelligent, with incredible navigating skills and are faithful to their partners (Ackerman, 2017). Free-born rodents are also highly intelligent and also very cute, which is the reason why many people keep them as 'pets'. Should this not call for respect?

The pigeons are part of our physical, geographical space, so why then are they generally treated with such disrespect and not included in our moral sphere? The common pigeon, or rock dove, *Columba livia*, was one of the first domesticated species of birds and has been cast in numerous roles. They have been used to carry messages in wartimes (Ackerman, 2017), as lab animals, as food, racing birds, show birds, and 'pets'[5], aside from being regarded as 'pests' and 'vermin' (Monson et al., 2016). Pigeons are estimated to have been the first domesticated bird, with this occurring around 3000 B.C.E. (Humphries, 2008).

[4] https://en.wikipedia.org/wiki/House_sparrow

[5] The word 'pet' implies that the animal is included in the human's social sphere. It is illustrative of how the animal may be used instrumentally for coddling, care and company through which the hormone oxytocin—the love hormone—is produced in mammals, strengthening bonds, and also strengthening the 'pets' position in the moral sphere system.

The pigeons in the streets not only transgress the space of humans, but they also transgress, perhaps even violate, their roles as birds, which demands a certain distance to humans for them to be regarded as properly 'wild' (Monson et al., 2016). This transgression of both space and role creates a degree of disrespect in humans. As previously suggested, we humans place other animals in different categories according to the use value or social value they hold for us; or, as phrased by Arluke and Sanders (1996), according to the social construction of 'the meaning of animals' which we pass on between generations in which social designations are placing them on a "sociozoologic ladder" (see also Douglas, 2003; Nussbaum, 2022). On the sociozoological ladder, species are ranked depending on how well they fit into the human society and act in relation to us. Control systems, such as breeding and 'domestication' processes as well as the culling of free-born animals who 'misbehave' maintain rankings whereby species at the bottom of the ladder can be justifiably exploited, oppressed and killed, while those at the top of are privileged. Humans are, of course, at the top, followed by the 'best animals', particularly 'pets' who can be 'domesticated' and adapted to share their lives with humans. Further down (or further away[6]) come animals who are under our control. The lowest rank of animals are those who are 'out of place', 'vermin', such as rats and pigeons, but also the 'wild' animals, like wolves, which may be regarded as threatening and disturbing and therefore killed.

While 'usable' animals, like 'pets' and animals used for meat, can be viewed as 'moral', immoral animals are useless and consequently viewed as 'vermin' and 'pests'. However, it is also the pigeons' roles as scavengers, eating our garbage in the same way as rats do, which is what implies that both rats and pigeons themselves are regarded as 'dirty', largely overlooking their important role in cleaning up after us. And when they come in big numbers, for many people they are just 'too much'.

There are other free-born animals who share our space, which are regarded in a more appreciative way, illustrated by the following example: My daughter Vilde is out with her cat Bella in central Oslo on a late

[6] The sociozoological ladder can also be perceived as a model of moral circles, as described above, which also defines the 'moral' value of other animals and what we owe to them. See above and Sollund (2024).

autumn evening. Bella is on a leash; she is not allowed to roam freely, but is taken out every day to sniff, walk, listen: to experience and explore the life outdoors in safety, and without being a risk to birds and other small animals. Bella is in a bush, and Vilde is standing with the leash in her hand outside the bush on the sidewalk, when a fox suddenly comes walking at a good pace. After the fox, about two, three metres behind her, comes a man on a bike. He says enthusiastically to Vilde that he has been following the fox for three kilometres, he is so fascinated by the fox and wants to know where s/he is heading. In the same area live a group of badgers, also 'wild' animals, who do not interact with their human neighbours. Both the fox and the badgers are in the outer circle of moral concern; they all have the ability to surprise and charm us when we meet them unexpectedly in what we consider a human habitat, but which, in fact, is also theirs.

There are many reasons why foxes and badgers live in the city; they may have had their natural habitats there before the humans came and settled; something which became clear during the COVID-19 pandemic when the people were at home during lockdown. During this period wildlife in the cities became abundant: There were jackals in the streets of Tel Aviv, racoons in Central Park in New York; goats and sheep in Istanbul; deer in Nara in Japan; a sealion was observed in the Mar de Plata harbour in Argentina; a fallow deer was seen grazing in London; grey langurs were roaming the streets of Ahmedabad; a wild boar grazed in a residential garden in Ajaccio in Corsica; seabirds swam in the canals of Venice; and buffalos walked the streets of New Delhi.[7] So, were they always there, just in hiding, or did they seize the opportunity to take back what was rightfully theirs when the opportunity came?

That there are foxes in London is well known and they seem to have found a form of cohabitation with the humans in the big city (Brisman & South, 2025); they are taking advantage of the humans' littering and their gardens. Their choice to migrate into the cities may also be the choice for the lack of opportunities where they used to live, because of the paving of roads and monocultural farming practices that deteriorate

[7] https://www.theguardian.com/world/gallery/2020/apr/22/animals-roaming-streets-coronavirus-lockdown-photos

the ecosystems and reduce their prey and other food sources (Irwin, 2017). In 2017, there were estimated to be 150,000 foxes in urban areas of England, a rise from 33,000 in the 1990s. In Bournemouth, there were 23 foxes per km^2, while in London there were an estimated 18 per km^2, and in Brighton, the population is 16 per km^2. There is, on average, 1 fox per every 300 human citizens in urban areas. At the same time, sightings of foxes in rural areas have declined, so are the foxes simply abandoning their rural habitats and becoming urban foxes? According to the researchers behind the study, there are anthropogenic causes for their migration to the cities connected to a lack of food in the country. Earthworms are an important source of food for foxes, but they are increasingly being killed by pesticides. The rabbit population has also decreased (Irwin, 2017). When the animals share habitats with us, how are we to regard them, and how can we ensure that cohabiting with other species allows space both for us and for them? What should be our guidelines for how to treat them?

Citizenships, Rights, and Moral Concerns: Issues of Belonging

In the book *Zoopolis*, Donaldson and Kymlicka (2011) discuss different degrees of citizenship based on the group to which animals belong: 'domesticated[8] animals', whom they feel should have citizenship; 'wild animals', regarded as sovereigns in their habitats, where they will try to avoid humans' presence or the group they denominate; and 'liminal animals', which cover several of the examples described above, and to which I return.

The first group of these animals—the 'domesticated'—are animals I placed within the first moral circle; they have obligations (or rather are forced) to render their bodies for human use, or, if lucky, they are 'pets' who are treated as family members, but who nonetheless must not step

[8] The word 'domesticated' implies that the animals are made part of humans' homes; however, for many of these animals the reality is that they are incarcerated in barren cages in industrial, crowded complexes, very far from humans' domestic spheres. They have little opportunity to interact with humans, and both feeding and slaughtering is automatized (Sollund, 2008).

out of line. For instance, dogs cannot bite or be too loud; neither must parrots scream, bite or ruin curtains or books, etc. The distance from full acceptance in the inner circle to the expulsion and death at a veterinarian's hands or being dumped in a ditch may be a short one.

The 'domesticated' animals should, since they have been deprived of a natural life, be accorded membership rights (Donaldson & Kymlicka, 2011). These 'citizenship rights' are rights that go beyond those rights usually accorded to animals, which are non-existent since generally they are not accorded the right to life. What animals can hope for is to be killed in ways that humans define as 'humane', and to be kept under conditions that are according to regulations. However, such regulations do not acknowledge their need for movement, social life, etc. (e.g. Francione, 2008; Nussbaum, 2023; Sollund, 2008). Citizen rights should nonetheless include residency, inclusion among the sovereign people—the humans—freedom to agency. After all, they have not chosen this life; their situation is, above all, like slavery. However, counter to human slavery, many animal species have, through thousands of years, been bred to better adapt to this slavery than humans (Nussbaum, 2022). With this, they have also been made dependent on humans, which makes this human-orchestrated development not only a breach of their individual rights, but also a breach of species' rights to flourish and evolve based on natural principles, rather than 'animal husbandry science'. I will leave these groups of animals for now since my main concern in this essay are the groups of animals that are positioned in the second and third groups: the free-born 'wild' and the 'liminal'.

'Wild', or free-born, animals are sovereigns, according to Donaldson and Kymlica, and therefore ideas of citizenship and sovereignty relating to human communities should be employed: 'Wild' animals are suffering injustices relating to their deprivation of self-governance and sovereignty analogous to the suffering of many human communities of indigenous peoples when control of their own territory has historically (and currently) been denied them. The colonization of indigenous land is similar to the colonization of the natural habitats of many animals, through which their property rights are consistently disrespected, e.g. when in Norway local municipalities construct cabin villages in the mountain areas where the last population of wild reindeer have their migrating rout.

In some instances when humans occupy the land of other species, the animals may be 'relocated'. Should this be translated into an Anthro-rights discourse it would be called forced migration, which is inconsistent with human rights. While reserves and national parks designated to conserve and protect species of free-born animals from human impact may have very positive effects, they are still under human 'management', and in Norway this means that even in national parks animals are hunted, injured[9] and killed, so not even there do they have protection from human violence.

In the view of Donaldson and Kymlica, the most important element of sovereignty is autonomy, and this is something free-born animals fully possess: "Wild animals have legitimate interest in maintaining their social organisation on their territory, and sovereignty is an appropriate tool for protecting that interest against vulnerability to injustice" (Donaldson & Kymlica, p. 174). Donaldson and Kymlica proceed to state that any paternalistic intervention we (humans) undertake to protect wild animals will have unintended—and perhaps perverse—effects. This is a strong warning against nature 'management' which is often a euphemistically designed word that disguises that what really goes on is killing, or state-organized theriocide ('animal murder', Beirne, 2018). This is, for example, the case in Norway when the state licenses the mass theriocides of the critically endangered wolves (Sollund, 2017; Sollund & Goyes, 2021). On a regular basis, hunters around the world are allowed to kill thousands of animals in the name of conservation (Warwick, 2024), e.g. through trophy hunting (Ghasemi, 2021).

The Rights of the 'Liminal Animals'

While free-born animals who live in their natural habitats should be regarded as sovereigns of their own natural habitats, what should we say about the liminal animals, and to what degree do their rights apply when humans and animals share space? Many indigenous groups share their

[9] About 50% of animals who are shot during hunts survive and continue living with injuries, or die after being injured for many days (Martinsen, 2024).

habitats with animals and respect the animals, ascribing them intrinsic value and profound spiritual significance (e.g. Weatherdon, 2022). In view of the examples given above, the situation is different in modern societies, where humans are more alienated form nature, regarding animals as subordinated to themselves, and not acknowledging that they must come to terms with how they can share space in a world where the natural habitats of other animals are constantly shrinking due to human behaviour.

The 'liminal animals', while profiting from human presence, e.g. in finding shelter and food, are also harmed through the humans' lack of recognition of other species' needs when they share habitat, for example when spikes are set up in window posts on buildings to prevent seagulls and pigeons from nesting and resting there, and through blocking roads and routes that prevent them access. As shown above, these animals are often labelled and stigmatized, partly for 'not belonging' in the human sphere, for being 'out of place' (Douglas, 2003), since they live in what humans define as their own physical space, not that of 'wild' free-born animals. While a squirrel at the cemetery may be accepted, a pigeon or a rat on the sidewalk are not. Their presence is regarded as illegitimate.

In the case of the presence of Freya, this is even more the case, since she was not a liminal animal; a walrus per se is definitely wild, and this wildness was even more out of space and belonging than that of Hvaldimir, since in a sense Hvaldimir showed tokens of being 'one of us' since he was socialized to be among humans. While one can argue that free-born animals living in their natural habitats can simply be left alone to be out of humans' harm's way; the liminal animals with whom we share our habitats require more concern. While in my view they should be recognized as co-citizens, in the view of Donaldson and Kymlica this should not generate the same, full citizenship rights as given to 'domesticated' animals, with whom we more often mutually interact (but not always, as mentioned). Donaldson and Kymlicka (Donaldson & Kymlicka, 2011, p. 214) thus argue we should accord them part-citizenship: *denizenship*. With this they imply that we must accept that they are co-habitants, but not full citizens, but what if we turn this the other way around? Are we trespassing into their habitats; are the denizens us? For some of the liminal animals this is the case, such as with the seagulls at the beach, while others again have moved into our sphere precisely because of our presence. Whether 'opportunists' (e.g. peregrine falcons), niche specialists (corncrakes), introduced exotics (reptiles and parrots) or feral animals (e.g. dogs and cats), they all share our space and are entitled to our concern and to rights to live unharmed in their (and our) naturalized habitats.

To imply that the animals with whom we share habitats, particularly in the cities, only 'halfway belong' is to take an anthropocentric approach, not acknowledging their entitlement to living just there, with us. For example, they may have been there to begin with, *before* humans settled there and transformed their natural habitat into an urban space. Yet other 'liminal' animals are animals who were once in the core of the moral circles, as 'pets', but who have either been expulsed from their homely surroundings and continued to breed as 'wild'. These are the so-called feral animals which include a wide range of species in different countries, from camels, cats, pigs and rabbits to parrots who have settled in a range of European countries that are now regarded an alienated as 'pests' or invasive species.[10] Donaldson and Kymlica mean that 'liminal' animals are better off as denizens than as citizens, since 'making them citizens' would imply a domestication procedure that would not be to their advantage.

However, again I ask, who is liminal in relation to what and to whom? Only from our anthropocentric perspective are the feral animals liminal, and, then again, we can be accused of aggrandizing our role in their lives. One important right of the 'liminal' animals would be secure residency. This means that people cannot, as in the case of seagulls, throw rocks at them to make them withdraw from humans' presence, as I witnessed recently at the beach. It implies also that humans cannot set up traps or spread poison to take their lives; this is incompatible with safety and residency rights. Rather, we should assist them by offering them facilities that can make their lives easier, such as nesting places and, if necessary, feeding stations. While there is a risk that this could increase free-born animals' dependence on humans, it is our fault that their natural habitats cease to exist or are so deteriorated that they can no longer sustain them. Therefore, it is our duty to care for them in ways that allow them to

[10] See for example Government considering Ring-necked Parakeet cull—BirdGuides.

continue to be free while being offered better chances for survival and breeding.

When Humans Transgress Geographical, Social and Moral Boundaries

As shown, the free-born animals living in their natural habitats can be regarded as sovereigns. This means that their right to occupy their territory should be respected and acknowledged. This again should entail that humans should abstain from disturbing them on their territory. However, the opposite is the situation in many countries around the world, where animals are hunted and killed. In Norway, such theriocides even include the killing of foxes and wolverines when they have pups, and the killing of elks when they have calves. Such theriocides destroy families and very often the lives of the young who depend on their parents' nurturing and guidance to grow up and be capable sovereigns, knowing where to find food, rest, etc. When the calves are killed in the presence of the mothers, which is often the case, the mothers are left to grieve.

As mentioned, the killing of free-born animals is often referred to as part of 'management' practices that legitimizes it from an anthropocentric perspective (Sollund, 2017; Sollund & Goyes, 2021). In many countries, e.g. Costa Rica, natural reserves are proper reserves where the inhabitants are allowed to live free from human-inflicted harm. In Norway, however, even in so-called national parks and nature reserves, the hunters are in action on their 'sport' hunting missions, which they allegedly do to be "close to nature" and to live for the moment of excitement when they pull the trigger and shoot a bullet (or hail) against an animal (Presser & Taylor, 2011). Tens of thousands of deer, elk, fox, hare, grouse, crow, jay, squirrel, beaver, and many more die in hunts or in traps in Norway every year. There is even licensed hunting of critically endangered, or vulnerable, species like wolves, wolverines, lynx ('quota hunts') and occasionally brown bear (Sollund, 2017). While the goal (or the instrumental goal, since the kill itself may not necessarily be the main motive of the hunt) is to kill the animal, often they suffer for a long time after the shot. Studies of how long deer take to die after being shot show it can take an average of 12 minutes, with 19% living longer than 10 minutes (Mejdell, 2004). Clearly, 10 minutes is a very long time to suffer. Only 21% of elks died within seconds after a shot. 25% of elks who were shot kept on moving after being shot; of these, 29% were never found (NOAH, 2024). Similarly, in marine nature reserves in Norway, bottom trawls are allowed, which is a fishing method that eradicates all life at the bottom, destroying the ecosystems, while killing both targeted and untargeted species (Hopkins, 2003; Hommedal, 2023).

What status, then, should one attribute to the humans who enter the homes of the sovereigns to kill as their main motivation? Since the humans do not share these habitats with the animals, clearly the humans cannot be regarded as sovereigns therein. They do not occupy the territory permanently; they do not live there; therefore, they cannot have full citizen rights. Neither do they (as 'uninvited guests', like the 'liminal' animals) share living space with the sovereigns. Therefore, neither can they be defined as denizens, with limited rights. The humans in a modern society do not depend on sharing space with the animals.

The human intruders can best be described in terms of 'war' metaphors: They come on yearly invasions, heavily armed, and occupy the land of the inhabitants that should have the sovereign rights, not to be attacked in such a way. The humans in their yearly hunts create a one-sided war that the sovereigns are doomed to lose since they do not possess similar weaponry and power. They become refugees, with the humans leaving behind injured or dead animals, motherless offspring and grieving families who have suffered the rupture of their social bonds.

Summarizing Conclusion

Let us return to the two animals I introduced in the beginning of this essay and how they achieved their importance in relation to humans, although it did not save them. Both Hvaldimir and Freya were named, they were *someone* (e.g. Francione, 2008; Nussbaum, 2022), in contrast to how 'wildlife' individuals are often regarded, in reductionist ways as merely 'representatives of a species', as 'specimens' in the language of

CITES,[11] the convention that regulates trade in endangered species of wild fauna and flora.

Gradually, Hvaldimir and Freya moved away from this status, from the 'wild' sphere and into the social sphere of humans, becoming loved almost like 'pets' (also property). This was particularly true of Hvaldimir, who was so accustomed to humans, presumably after having been raised and trained by them. Freya, on the other hand, travelled around the Norwegian shores, rested on boats, and was killed because she allegedly constituted a danger to humans who did not respect her need for distance. She was still regarded as 'wild', and therefore dangerous and unpredictable, and someone the humans needed to be protected from.

As shown, free-born animals are often categorized and defined as 'huntable species'. This implies that their purpose is to be hunted and killed by humans. They are definitely not regarded as someone, but some *thing*, as 'huntable' property, whether of landowners or of the hunters who buy the rights to kill them for entertainment, for their trophy value, or for their meat value. Animals who are 'domesticated' or kept as 'livestock' are always the property of humans (Sollund, 2021, 2024).

Freya entered the human sphere uninvited; but who then encroaches on the habitats of whom, and how should such human–non-human crossing of space be dealt with? Granting residency rights and sovereignty would also mean that the animals with whom we share space have the right to live and flourish where *they* choose to live, whether they seek our natural habitats or were there to begin with. One could regard such rights as natural rights, rather than it being up to us to extend such rights because this is an anthropocentric approach setting ourselves as superior to other species. We need to accommodate our lifestyles to allow other species to flourish, whether on their own territory or in territory we share with them. The question, then, is how cohabitation may be facilitated in ways that do not stir conflict and destroy peaceful coexistence. If free-born animals pay us a visit, we should help and assist them if needed, rather than taking their lives, which is the usual practice in Norway. We should welcome and appreciate their visit and help them to leave if they are lost in our urban space.

---

[11] https://cites.org/eng

The roedeer buck, who so loudly commented on my arrival to the cabin, was clearly a sovereign; he was in his home. The seagulls likewise were at home at the beach, they were in their natural habitats and thereby sovereigns and in their full right to be there, more than the humans who were there and who wanted to expel them.

In contrast to Hvaldimir and Freya, and those we include as our companions, the free-born sovereigns are usually not ascribed personhood, although they, like us and like Hvaldimir and Freya, all have different personalities, autonomy, agency and interests. So, before discussing how *we humans* should perceive the rights of other animals, perhaps we ought to discuss our own lack of right to proceed as we do in our relations to them and regard these relations not only in political terms of rights, obligations and belonging, but also in terms of ethical-moral terms of care and species justice (Nussbaum, 2023; White, 2013).

References

Ackerman, J. (2017). *The genius of birds*. Penguin.

Arluke, A., & Sanders, C. (1996). *Regarding animals*. Temple University Press.

Barkham, P. (2021). House sparrow population in Europe drops by 247m. Retrieved September 20, 2024, from https://www.theguardian.com/environment/2021/nov/16/house-sparrow-population-in-europe-drops-by-247m

Beirne, P. (2018). Murdering animals: Writings on theriocide, homicide and nonspeciesist criminology. Springer.

Birdlife Norway. (2024). Måker i byene er kommet for å bli—BirdLife Norge. Retrieved October 2, 2024, from https://www.birdlife.no/naturforvaltning/nyheter/?id=2981

Brisman, A., & South, N. (2025). *Monstrous nature and representations of environmental harms: A green cultural criminological perspective*. Temple University Press.

Donaldson, S., & Kymlicka, W. (2011). *Zoopolis: A political theory of animal rights*. Oxford University Press.

Douglas, M. (2003). *Purity and danger: An analysis of concepts of pollution and taboo*. Routledge.

Francione, G. L. (2008). *Animals as persons: Essays on the abolition of animal exploitation*. Columbia University Press.

Ghasemi, B. (2021). Trophy hunting and conservation: Do the major ethical theories converge in opposition to trophy hunting? *People and Nature, 3*(1), 77–87.

Hommedal. (2023). Det fanger, forflytter og skader livet på havbunnen i enkelte leveområder. Havforskerne har oppsummert alt vi vet om bunntråling, men finner også kunnskapshull. *Havforskningsinstituttet.* Retrieved October 1, 2024, from https://www.hi.no/hi/nyheter/2023/januar/traling-pavirker-livet-pa-havbunnen

Hopkins, C. C. E. (2003). *The dangers of bottom trawling in the Baltic Sea.* Coalition Clean Baltic.

Horowitz, J. (2022). Freya, a famous Walrus, is killed and Norwegians are divided—*The New York Times.* Retrieved October 2, 2024, from https://nytimes.com

Humphries, C. (2008). *Superdove.* Smithsonian Books.

Irwin, A. (2017). There are five times more urban foxes in England than we thought. *New Scientist.* https://www.newscientist.com/article/2116583-there-are-five-times-more-urban-foxes-in-england-than-we-thought/

Martinsen, S. (2024). Ny viltlov: "Vitlressurser" eller medskapninger? Noahsark. Nr. 2 2024, pp. 4–15.

Mejdell, C. (2004). RAPPORT VESO Oslo Veterinary Science Opportunities *Konsekvenser av menneskelig aktivitet på dyrevelferd hos viltlevende dyr.* Retrieved October 1, 2024, from https://jaktogfiske.njff.no/files/2023/01/04/2004%20Mejdell%20Dyrevern-VILT-%20MENNESKELIG%20AKTIVITET_VESO.pdf

Monson, A. T., Andrea Button, M., & Beverly Stiles, M. (2016). The sociological significance of pigeons. *Metamorphosis.*

NOAH. (2024). Retrieved October 1, 2024, from https://www.villedyr.no/fakta/ville-dyr/jakt-pa-hjortedyr/

Nussbaum, M. C. (2022). *Justice for animals: Our collective responsibility.* Simon and Schuster.

Nussbaum, M. C. (2023). *Justice for animals: Our collective responsibility.* Simon and Schuster.

Presser, L., & Taylor, W. V. (2011). An autoethnography of hunting. *Crime, Law and Social Change, 55*, 483–494.

Regan, T. (2016). The case for animal rights. In *The animal ethics reader* (pp. 45–51). Routledge.

Sollund, R. (2008). Causes for speciesism: Distance, difference and denial. In R. Sollund (Ed.), *Global harms* (pp. 109–130). Nova Science Publishers.

Sollund, R. (2017). Perceptions and law enforcement of illegal and legal wolf killing in Norway: Organized crime or folk crime? *Palgrave Communications, 3*(1), 1–9.

Sollund, R. A. (2021). *Bare et dyr?* Solum Bokvennen.

Sollund, R. (2022). Wildlife trade and law enforcement: A proposal for a remodeling of CITES incorporating species justice, ecojustice, and environmental justice. *International Journal of Offender Therapy and Comparative Criminology, 66*(9), 1017–1035.

Sollund, R. (2023). The dark side of nature conventions: A call to end anthropogenic wildlife destruction. *Criminology & Criminal Justice, 17488958 231181309.*

Sollund, R. (2024). Speciesism and the non-human. In *Oxford research encyclopedia of criminology and criminal justice.*

Sollund, R., & Goyes, D. R. (2021). State-organized crime and the killing of wolves in Norway. *Trends in Organized Crime, 24*(4), 467–484.

Talmazan, Y., & Aggarwal, M. (2024). Who shot Hvaldimir? https://www.nbcnews.com/news/world/hvaldimir-russian-spy-whale-shot-dead-beluga-norway-rcna169496

Warwick, H. (2024). *Cull of the Wild: Killing in the Name of Conservation.* Bloomsbury Publishing.

Weatherdon, M. S. (2022). Religion, animals, and indigenous traditions. *Religions, 13*(7), 654.

White, R. (2013). *Environmental harm: An eco-justice perspective*. Policy Press.

Conceptualizing Ecocide: Interdisciplinary Perspectives

Daan van Uhm, Cedric Ryngaert, Susanne Knittel, and Marijke van Kuijk

Introduction

The changing ecological conditions in our globalized world are confronting human societies with new challenges. The collapse of the pre-industrial institutions and the development of new technologies have drastically changed the world's ecosystems; the twentieth and twenty-first centuries are characterized as a period in which the Earth and its atmosphere

The original version of this chapter was previously published without open access. A correction to this chapter is available at https://doi.org/10.1007/978-3-031-91557-4_21

D. van Uhm (✉)
Willem Pompe Institute for Criminal Law and Criminology, Utrecht University, Utrecht, The Netherlands
e-mail: d.p.vanuhm@uu.nl

C. Ryngaert
Public International Law, Utrecht University, Utrecht, The Netherlands

S. Knittel
Comparative Literature at Utrecht University, Utrecht, The Netherlands

M. van Kuijk
Tropical Forest Ecology at Utrecht University, Utrecht, The Netherlands

D. van Uhm, D. Siegel (eds.), *Global Green Crime and Ecojustice*, Palgrave Studies in Green Criminology, https://doi.org/10.1007/978-3-031-91557-4_5

experience major consequences due to the impact of human activities (Crutzen, 2002; Hamilton, 2013; Steffen et al., 2007). In the context of large-scale pollution, global deforestation, the disappearance of species, and climate change, discussions about criminalizing ecocide are on the international agenda, not only to protect the environment, but also to preserve the commons of nature, essential for the life conditions on Earth (Hardin, 1968; South, 2010).

International lawyers started to ponder the criminalization of ecocide quite soon after biologist Arthur Galston coined the term in 1970. In fact, as early as 1973, legal scholar Richard Falk proposed a draft International Convention on the Crime of Ecocide (Falk, 1973). This proposal was mainly informed by US herbicidal warfare in the Vietnam War by means of the defoliant Agent Orange, and boosted international legal efforts to criminalize military ecocide (Hough, 2022). These efforts resulted, most notably, in the insertion of an article about environmental war crimes into the Rome Statute of the International Criminal Court (ICC, art. 8(2)(b)(iv)). In these early decades, there were also some efforts to codify environmental crimes as international crimes outside war (i.e. in peacetime), notably at the UN's International Law Commission in the 1980s and 1990s (ILC, 1981, 1994), but these efforts were abandoned for reasons that are not fully clear.

In recent years, there have been new efforts to make ecocide a fifth international core crime under the Rome Statute of the International Criminal Court, alongside genocide, crimes against humanity, war crimes, and the crime of aggression. A renewed push for the international criminalization of grave environmental crimes came about in the 2010s, especially after the late Polly Higgins called for the criminalization of ecocide, which she defined as 'extensive damage to, destruction of or loss of ecosystem(s)' (Higgins, 2012, p.10). The campaign 'Stop Ecocide International', which had been established by Higgins, then went on to convene an Independent Expert Panel (IEP) for the Legal Definition of Ecocide. In 2021, this IEP proposed amending the ICC's Statute with a view to incorporating the fifth international crime of ecocide, being defined as 'unlawful or wanton acts committed with knowledge that there is a substantial likelihood of severe and either widespread or long-term damage to the environment being caused by those acts'

(IEP, 2021). Other definitions of ecocide have also been proposed recently, which slightly differ from the IEP's definition (ELI, 2023; UCLA, 2021).

The IEP's report has galvanized international legal discussion on ecocide, especially among international criminal lawyers (Sluiter & van Straaten, 2020). In the meantime, ecocide has also established itself in the public vocabulary to denote the unfolding environmental and climate crisis (Mehta & Jackson, 2021). Beyond its legal significance, the term also carries an important symbolic meaning. Ecocide taps into the powerful moral and cultural significance of the term genocide, and as such resonates both in the popular imagination and in culture. But the link between ecocide and genocide goes beyond semantics: there are also strong historical and structural connections between crimes against humanity and crimes against nature, which the resonance between the two terms can help make visible (Crook & Short, 2014).

Conceptualizing ecocide raises fundamental questions regarding the way we think about guilt, liability and the duty of care, about how we acknowledge human and non-human victims, and about the power of legal concepts. In addition to these theoretical and legal challenges, ecocide also poses a fundamental representational challenge: the processes and actions that constitute ecocide are often highly complex and moreover frequently occur at a scale that is difficult to fathom. And the victims of ecocidal violence, both human and non-human, often do not have a voice within established legal and political frameworks. Science, scholarship, activism, culture, and the media all have an important role to play in raising awareness about ecocide.

In order to understand the concept of ecocide in all its ecological, criminological, legal, and cultural dimensions, an interdisciplinary approach is called for. Such an approach is crucial to understanding ecocide because it integrates diverse perspectives, essential for comprehensively addressing the multifaceted nature of serious and large-scale environmental harm. Ecocide involves complex interactions between environmental, economic, legal and criminogenic factors, as well as social and cultural ones. An interdisciplinary approach allows for a more holistic understanding of these interactions and their impacts.

This chapter is based on our project Conceptualizing Ecocide, which we, as members of an interdisciplinary group of researchers at Utrecht

University in the Netherlands, have recently embarked upon.[1] This chapter focusses on the question how perspectives from ecology, criminology, international criminal law and humanities can contribute to a holistic understanding of ecocide. This chapter represents the various disciplinary perspectives we bring to bear on the topic of ecocide, with the aim of sketching some of the facets of this complex topic that any interdisciplinary investigation must consider.

The Ecological Dimension

Ecocide refers to the widespread destruction of the natural environment and requires a multifaceted understanding of the devastating impact of human activity on ecosystems. From an ecological point of view, ecocide refers to the destruction or damage of ecosystems, and encompasses a range of human activities, including deforestation, pollution, the overexploitation of natural resources, and climate change-induced phenomena. Since nature is connected to our social and economic systems, ecocide not only affects the environment, but can also have far-reaching consequences that extend beyond environmental degradation, affecting human societies and economies in various ways.

Ecosystems are dynamic and complex systems that play a crucial role in supporting life on Earth. In the context of ecocide we define an ecosystem as a community of living organisms (plants, animals, and microorganisms) interacting with each other and also with their physical environment. This interaction creates a complex web of relationships and processes, in which energy and nutrients flow through the system, sustaining life and maintaining ecological balance. Ecosystems can vary greatly in size and complexity, ranging from a small pond or forest to entire biomes like deserts, grasslands, or oceans. Each ecosystem has unique characteristics determined by factors such as climate, geology, topography, and the organisms that inhabit it. These diverse ecosystems provide essential services, such as oxygen production, water purification, climate regulation, and habitat provision, which are vital for all life on Earth.

---

[1] Project Ecocide has been funded by the strategic theme Pathways to Sustainability at Utrecht University.

Because of the complex interactions between the biotic and abiotic components in ecosystems, destruction or damage done to them can have many cascading effects with consequences not only for the environment but also for socio-economic systems, on various spatial and temporal scales.

Cascading Effects on the Environment

Environmental destruction can take various forms, including deforestation, pollution, the overexploitation of natural resources and climate change. The cascading effects of environmental destruction refer to the wide-ranging and interconnected consequences that arise from the degradation or alteration of natural ecosystems.

Natural ecosystems provide a wide range of services that are essential for human well-being, such as clean air and water, soil fertility, and climate regulation. Environmental destruction disrupts critical ecological processes and can erode ecosystem services, leading to declines in agricultural productivity, water scarcity, and increased vulnerability to natural disasters. The loss of these ecosystem services can have far-reaching consequences for human well-being, including impacts on food security and public health (Costanza et al., 1997).

The destruction of natural habitats through deforestation, urbanization, and industrial activities diminishes the availability of suitable habitats for wildlife and plants, leading to population declines and potential extinction events (Foley et al., 2005). The loss of species and genetic diversity caused by the destruction of habitats, but also by the overexploitation of species, pollution, and the introduction of invasive species, threatens the functioning, stability, and resilience of ecosystems. Loss of biodiversity can have cascading effects on ecosystem functioning by disrupting food webs, nutrient cycling, and other ecological processes (Dirzo et al., 2014). This disruption can destabilize ecosystems and make them more vulnerable to further degradation (Steffen et al., 2015).

Environmental destruction contributes to climate change through processes such as deforestation, the burning of fossil fuels, and land degradation. Climate change, in turn, can exacerbate environmental destruction by increasing the frequency and intensity of extreme weather events, altering precipitation patterns, and amplifying other environmental stressors (IPCC, 2014).

Ecocide thus represents a significant threat to the health and integrity of ecosystems worldwide, with profound consequences for biodiversity, ecosystem services and the climate.

Interconnectedness between Nature and Social and Economic Systems

Ecosystems and the goods and services they provide are often interconnected with social, economic, and cultural systems. Healthy ecosystems can provide a buffer against environmental shocks and provide natural defences against disasters, while degraded ecosystems can exacerbate vulnerabilities and increase exposure to risks. There are feedback loops between nature, social systems, and economies, where changes in one system can influence the dynamics of the others. For example, environmental degradation can lead to social unrest, economic losses, and public health crises, while economic development and population growth can exert pressure on natural resources and ecosystems. Ecocide often disproportionately impacts marginalized communities and exacerbates social inequalities, underscoring the interconnected nature of ecological and other systems.

Many communities rely on ecosystems for their livelihoods, including agriculture, fishing, forestry, and tourism. Ecocide disrupts these activities, leading to loss of income, unemployment, and poverty. Deforestation, for instance, destroys habitats and reduces biodiversity, depriving communities of resources and opportunities (Kissinger et al., 2012). Loss of biodiversity, soil fertility, and changes in climate patterns can lead to crop failures, food shortages, and malnutrition, especially in vulnerable communities (Godfray et al., 2010). Ecosystem degradation thus undermines food security by reducing agricultural productivity, depleting fisheries, and disrupting food supply chains.

Ecocide may also lead to air and water pollution, soil contamination, and the spread of diseases, posing serious threats to public health. Polluted air and water can cause respiratory illnesses, waterborne diseases, and other health problems. Communities near ecocide sites or exposed to environmental toxins are particularly vulnerable (Costello et al., 2009).

Ecocide affects water resources through pollution, deforestation, and changes in hydrological cycles; contaminated water sources jeopardize drinking water supplies, sanitation, and irrigation, posing risks to human health and agricultural productivity. Competition over scarce water resources can also lead to conflicts (Cosgrove & Rijsberman, 2014).

Many indigenous and traditional communities rely on intact ecosystems for their cultural identity, livelihoods, and traditional knowledge systems. Environmental destruction can lead to the loss of this cultural heritage, as well as the valuable ecological knowledge that these communities possess (Berkes et al., 2000). Environmental destruction can force communities to migrate in search of food, water, and shelter, leading to social upheaval and conflict. Displacement can also exacerbate poverty, food insecurity, and other social and economic challenges, particularly for vulnerable populations (Adger et al., 2005).

Ecocide imposes significant economic costs, including expenses for environmental clean-up, healthcare, the loss of ecosystem services, and damage to infrastructure. These costs burden governments, businesses, and individuals, leading to long-term economic consequences (Costanza et al., 1997).

Geographical and Temporal Boundaries of Environmental Destruction

The ecological impact of ecocide can extend far beyond the immediate area of destruction. It can transcend geographical and temporal boundaries, affecting ecosystems at local, regional, and global scales, with far-reaching consequences for biodiversity, ecosystem services, and human well-being. The immediate area where ecocide occurs experiences the most direct and severe ecological impact. Local ecosystems may suffer from habitat destruction, loss of biodiversity, soil degradation, and water pollution, among other issues. Ecological impacts can extend to surrounding regions, affecting ecosystems that are interconnected through ecological processes such as migration, nutrient cycling, and water flow. Contaminants from ecocide sites, for example, can spread through air and water, affecting ecosystems downstream or downwind. Some ecological impacts of ecocide can have global ramifications. For instance, deforestation in one region can contribute to climate change on a global scale by reducing carbon sequestration and altering weather patterns. Loss of biodiversity in one part of the world can disrupt global ecosystems and lead to cascading effects throughout the biosphere.

The ecological impact of ecocide can persist for generations, affecting the ability of ecosystems to recover and adapt. Loss of key species, the degradation of soil and water resources, and the disruption of ecosystem services can have lasting effects on the health and resilience of ecosystems, making them more vulnerable to future environmental stressors.

The Criminological Dimension

The overarching ecological effects of ecocide described above reflect the critical state of our planet, driven by several worrisome illegal trends in deforestation, wildlife exploitation, mining of metals and minerals, pollution, and fisheries. Environmental crime has become one of the largest criminal activities in the world, with estimates between US$51 billion and US$157 billion (UNEP & Interpol, 2016). The annual growth rate of environmental crime is at least 5–7% per year and the consequences of such environmental crimes include increased levels of pollution, reduction of biodiversity, the disturbance of ecological balances and risk to human health, among others (White, 2011).

Perpetrators Behind Ecocide

Perpetrators of ecocidal activities are not homogeneous groups. On the one hand, organized crime groups have become highly involved in environmental crime, alongside their 'traditional' criminal activities, such as drug or human trafficking. These are highly organized and disciplined groups that may use violence or corruption to control illegal goods and services for profit. For example, Colombian drug cartels are increasingly involved in large-scale mining operations in Latin America, Italian mafia groups control aspects of the waste industry, Congolese militias in Central Africa finance their weapons through trade in tropical timber, and

Chinese organized crime is involved in wildlife trafficking in Southeast Asia (van Uhm, 2023; Boekhout van Solinge, 2014).

On the other hand, ecocide is not solely the domain of criminal underworlds—the relationship between the 'underworld' and the 'upperworld' is an important characteristic of ecocide. Corporate crime groups use their legitimate structures to camouflage or facilitate ecocidal activities. These legal and illegal partnerships can be based on alliances between companies, criminal groups, and governments (Ruggiero & South, 2013; Whyte, 2020). For example, companies may use their legitimate infrastructures to camouflage environmental crimes or launder natural resources, facilitated by state officials, by concealing their illegal origins and harmful extractive processes of ecocide (Harvey, 2003). Such collaborations blur the boundaries between legality and illegality, making it difficult to distinguish between the two, and highlight how ecocide can be embedded within formal economic and political systems, rather than occurring on the margins of society.

These criminal networks and state-corporate crime groups may also work with local communities to outsource illegal or harmful activities against the environment, including mining, deforestation, or poaching (South, 2016). This serves as an example of how perpetrators behind ecocide can be both perpetrators and victims of ecocide (Brisman & South, 2014). In such contexts, local communities may be drawn into ecocidal practices not only through coercion or economic necessity, but also through promises of development or employment. Since ecocide is often a chain of events or stages, there is the potential for many different actors to be involved at each link in the chain. Such alliances could be considered to be *joint crime ventures* that facilitate ecocide, based on interactions between criminal groups, companies, local communities, and corrupt government officials (van Uhm, 2024).

Victims of Ecocide

Many environmental harmful activities committed by such perpetrators are not criminalized by law, or have only recently been considered to be criminal or at least seriously harmful (Beirne & South, 2007). For a long time, political and economic elites have played an extensive role in minimizing the regulation of environmental markets and excluding ecocidal activities from criminalization in order to continue environmental destruction (Sellin, 1938; Schwendinger & Schwendinger, 1970; Sutherland, 1949). The social construction of ecocide has been influenced by power dynamics and anthropocentric interests; consequently, nature-rich areas in the Global South are being legally, or illegally, exploited by actors from the Global North, resulting in numerous human and non-human victims (Goyes, 2019; White & Kramer, 2015). Therefore, rather than solely in anthropocentric binary terms of legal or illegal, victims of ecocide should be approached from ecocentric and biocentric perspectives (Higgins et al., 2013).

From an *ecocentric perspective*, it is important to recognize the dialectical relationship between humans and nature based on ecological principles (Benton, 1998; Halsey & White, 1998). Humans are socially and ethically responsible for minimizing harm, preserving the natural balance and the integrity of non-human entities (Hall, 2013). The harm of ecocide to ecosystems and ecosystem functions can be linked to large-scale mining or deforestation that threatens both local human and non-human communities as victims, but also the world in general due to the annual emissions of greenhouse gases into the atmosphere. For instance, the heavy machinery and toxic substances, such as mercury and cyanide, used to extract gold or timber affect the natural environments and their ecosystem functions, by polluting the ground and water and destroying the rainforests, which is essential to the animals and plants as well as the lives and cultures of Indigenous peoples (Lynch et al., 2018).

The relationship between humans and nature also raises questions related to the victimization of non-human species (Beirne & South, 2007; Nurse, 2013). From a *biocentric perspective* all species have an intrinsic value, which should be respected (Wilson, 1993). This biocentric approach can be complemented by non-speciesism, which emphasizes the animal being discriminated against by humans based on "inferior" characteristics and expressed in the social legitimacy of abuse or exploitative acts against other species (Beirne, 2009; Sollund, 2020). Activities like poaching or trafficking in endangered species can lead to significant individual harm and suffering and, eventually, the extinction of an endangered species as a form of ecocide. For example, in 2010, the last Javan rhinoceros in Vietnam was the victim of illegal hunting; the animal was found with a bullet in the leg and its horn had been removed (van Uhm, 2016).

The Criminogenic Nature of Ecocide

Ecocide does not stand alone; it often converges with other crimes. Ecocide not only has devasting effects on the environment; it can also be a potential driver of crimes, conflicts, and wars, affirming the interconnections between environmental and human rights (Le Billon, 2012; Zabyelina & van Uhm, 2020). Destroying or depleting natural resources can lead to tensions between communities competing for increasingly scarce resources, create refugee flows, and place strain on neighbouring regions. In more extreme cases, it can even fuel insurgencies or civil wars, especially where state institutions are weak or complicit in the exploitation. This illustrates how ecocide can be an intentional tool or "major accelerator" of conflict and war (Crook & Short, 2014; Lopez, 2007; Smith, 2013). This strategic targeting of ecologically vital areas often exacerbates existing inequalities and tensions, contributing to cycles of violence and displacement.

Ecocide can damage the environment in such a way that the fundamental rights of people are affected, for example by the contamination of soil, groundwater, and surface water via the chemicals emitted from extraction processes (Bruch, 2001). This can be illustrated by Indigenous populations who lost their way of life based on nature as a result of ecocide, as the destruction of forests, rivers, and other vital ecosystems not only disrupts their traditional practices and spiritual connections to the land, but also leads to cultural erosion and the breakdown of community structures (Weinstein, 2005). Ecocide can also lead to forced displacement of local communities by expulsion or other coercive acts from the area in which they are lawfully present (Borges, 2018).

The *criminogenic nature* of ecocide and the changing landscape of victimhood require a different look at the existing legal frameworks. Human societies are confronted with new forms of eco-violence and eco-victimization: scarcity of natural resources, increasing desertification, and air and soil pollution appear to be good predictors of conflict, crime, and ecocide. Ecocidal activities are therefore not only problems that affect our environment, but—in their origin and through their consequences—have become fundamental problems of people, their living conditions, the relationship to the world, and social, cultural and political situations.

The Legal Dimension

Having explained the ecological and criminological dimensions of ecocide, we now proceed to unpack a number of issues raised by ongoing legal efforts to address acts of ecocide via (international) criminal law. These efforts, as described in the introduction to this chapter, notably include the IEP's proposal to amend the Rome Statute of the International Criminal Court with a view to including the crime of ecocide in the Court's remit.

We identify four key legal issues that need to be addressed in the context of criminalizing ecocide: legal (un-)certainty regarding acts constituting ecocide; the tension between eco- and anthropocentrism; the difficulty of defining the mental element of the crime of ecocide; and the risk of epistemic and institutional bias when prosecuting ecocide. We discuss these issues in turn and suggest a way forward.

Legal Certainty

A key legal concern pertains to predictability of the law. As definitions of ecocide often contain vague terms, such as 'damage to the environment', 'wanton acts', 'substantial likelihood', or 'taking appropriate available measures', individuals and economic operators may have difficulties knowing beforehand whether particular activities are criminal or not. Especially in the criminal law, given its possibly harsh penalties, legal certainty is considered as of fundamental importance (*nullum crimen/poena sine lege*). Such concerns could, to a certain extent, be met by enumerating the precise acts of ecocide that are punishable, with a view to preventing that ecocide has a chilling effect on legitimate economic activity—to give just one example.

At the same time, legal language is necessarily open-ended. This enables laws to apply to a variety of situations. Indeed, as the legal theorist H.L.A. Hart said, the very concept of law is characterized by a 'duality of a core of certainty and a penumbra of doubt when we are engaged in bringing particular situations under general rules' (Hart, 1961, pp. 199–120). This penumbra of doubt can be restricted but cannot entirely be eliminated. Accordingly, legal drafters of an ecocide definition will have to find a middle way between generality and particularity (level of detail). They should bear in mind that an overly detailed definition of ecocide may fail to capture not-yet-foreseen instances of environmental destruction, and that a pithy definition of ecocide may have a stronger expressive effect.

Mental Element

Another key legal concern pertains to the 'mental element' (*mens rea*) of the crime of ecocide. In the criminal law, persons' liability can only be engaged if they have a 'guilty mind'. In international criminal law, this can, in principle, only happen if the criminal act has been 'committed with intent and knowledge' (ICC, Article 30). When persons commit crimes of ecocide, however, often they do not intend to cause environmental harm, nor may they know, with near certainty, that a particular act causes harm. Instead, they tend to be willfully blind to the adverse consequences of their decisions and economic activities.

Drafters of the ecocide definition are aware of this. They have proposed to use a lower *mens rea* standard of recklessness (IEP, 2021), or have even advocated strict liability, i.e., liability without fault, simply based on harm (Higgins, 2012). Such standards are not well known in criminal law, however, largely out of concern for overcriminalization. For instance, they are not used by the ICC (Minkova, 2023).

Domestic legal systems which already use standards of recklessness or negligence in general criminal law could nevertheless provide fertile ground for experimentation with ecocide prosecutions. Alternatively, of course, one could decide to stick to the classic mental element of knowledge, while conceding that this will only allow for the penalization of persons who actually *know* about the adverse consequences before embarking on a particular activity, e.g., where a corporation, corporate officer or government official knows, based on environmental impact assessments, that a mining project *will* cause severe, widespread and/or long-term damage to the environment.

Ecocentrism versus Anthropocentrism

In the international debate over ecocide, the main bone of contention arguably pertains to the normative focus of the prohibition of ecocide: should it centre on the rights/interests of the environment/nature, or on the rights/interests of humanity? This tension between ecocentrism and anthropocentrism also features prominently in legal debates over a suitable definition of ecocide (Negri, 2022).

The IEP's definition is a case in point: while this definition is at first sight ecocentric—criminalizing serious environmental harm *per se* ('acts causing severe and either widespread or long-term damage to the environment')—it allows for the justification of such harm if it is in the interest of humanity. Indeed, 'wanton' acts of ecocide are only criminal if committed 'with reckless disregard for damage which would be clearly excessive in relation to the social and economic benefits anticipated'. Regardless of how such benefits would actually be calculated and how a proportionality test would be applied in practice, the takeaway point is that socio-economic benefits of particular economic activities can outweigh the harm to the environment, i.e., an anthropocentric approach.

Perhaps ecocide drafters lack ambition, espousing an anthropocentric approach to ensure public and governmental buy-in (Winter, 2023). But the question is whether it could be otherwise. If humanity is to survive and flourish, inevitably it will have to engage in activities that are, one way or another, environmentally harmful, e.g., industrial production, extraction of natural resources, transport, housing, etc. (Robinson, 2022). The challenge is to *limit* our ecological footprint, while being cognizant that human activity will always have adverse environmental consequences. Perhaps the way forward is to only criminalize acts of serious environmental destruction that are *manifestly* unreasonable or unjustifiable in light of any claimed socio-economic benefit. This will also ensure that judges do not have to make politically sensitive trade-offs between ecology and humanity.

Epistemic and Institutional Bias

Following the IEP's proposal to include the criminalization of ecocide in the ICC Statute, legal debates have concentrated on the role of international criminal law, and the ICC in particular, as purveyors of environmental justice. International criminal trials can have important expressive and symbolic effects, as they express the international community's moral condemnation of acts constituting ecocide. Such trials can raise international awareness, deter future violations, and bring about behavioural change (Palarczyk, 2023).

At the same time, one should not overlook the blind spots of the international criminal justice project. For one thing, international criminal justice has developed as an accountability mechanism for violations committed in the context of armed conflict, violations of which human beings are victims. The preamble of the ICC Statute, for instance, reminds the reader of the fact that during the twentieth century 'millions of children, women and men have been victims of unimaginable atrocities that deeply shock the conscience of humanity' (ICC, preambular para. 2). This begs the question whether, given its anthropocentrism and institutional basis, as well as lack of track record and expertise, the ICC is the proper mechanism to prosecute environmental crimes in peace time (Kooijman, 2023; de Hemptinne, 2022a).

What is more, prosecuting ecocide under international criminal law may have the perverse effect of *legitimating* many forms of environmental destruction: international criminal law focuses on *spectacular* crimes (e.g., indiscriminate bombing of civilian areas), whereas environmental crimes tend to amount to slow, gradual, and structural violence (Cusato & Jones, 2023). Including ecocide in the ICC Statute may then raise expectations that may ultimately not be met (Mührel, 2022). On a comparative note, it is pointed out here that, as of 2024, the ICC Office of the Prosecutor had not acted on, or at least had not initiated investigations in relation to, its 2016 policy paper on prioritization of existing core crimes (e.g., war crimes) 'that are committed by means of (…) the destruction of the environment, the illegal exploitation of natural resources or the illegal dispossession of land' (The Office of the Prosecutor, 2016, para. 41). By the same token, the international criminalization of ecocide may remain a dead letter. This may convey the implicit message that environmental destruction is acceptable. In such a scenario, criminalizing ecocide will backfire badly.

Even if international investigations and criminal prosecutions regarding ecocide do materialize, there is the additional concern that international courts such as the ICC will limit their gaze to particular forms of environmental destruction. As the ICC is largely funded by the developed world and its staff have largely been educated there, there is a risk that the ICC will focus on those forms that are considered as ecocide in the eyes of the Global North—drowning out Global South and Indigenous perspectives (Branch & Minkova, 2023; Lindgren, 2018). For instance, it is not excluded that the ICC targets large development projects carried out by Global South actors, or even forestry practices of Indigenous peoples considered as unsustainable—while turning a blind eye to the environmentally destructive role played by Western corporations and consumers (van der Wilt, 2023).

Such risks need not materialize *per se*, however. Inculcating awareness of potential politicization may cause international criminal lawyers to be more self-reflective and to familiarize themselves with other perspectives on ecocide. And to the extent that the ICC suffers from a legitimacy crisis, having primarily targeted persons from the Global South (Africa in particular), one could confer the power to prosecute ecocide on *other institutions*. For instance, a new international convention on ecocide (separate from the ICC Statute) could be negotiated that mandates states to prosecute the crime of ecocide in their *domestic courts* (de Hemptinne, 2022b). Such a convention could also establish an international monitoring body, or even a new special international tribunal dedicated to prosecuting ecocide. Such bodies may be less encumbered by institutional baggage when prosecuting ecocide.

The Cultural Dimension

While legal scholars and criminologists are debating the definition of ecocide and the best way and venue to prosecute its perpetrators, the term has captivated the popular imagination and occupies a prominent place in the cultural sphere. A clear measure of this is the proliferation, especially over the past two decades, of films, novels, art projects, performances, music productions, and other cultural works that deal with environmental harm (Duvall, 2017; Johns-Putra, 2016). Cultural representations make a distinctive contribution to our understanding of ecocide through their exploration of its social, political, and cultural context and impact, and their reflection on questions of agency and responsibility. Furthermore, such representations can become vehicles for critical reflection on our values and our ways of life, on patterns of material consumption, and the relationship between individual and society, asking what sort of future we want ourselves and others to live in. Most importantly, they attempt to translate the complex and large-scale processes and acts involved in ecocide into stories and images that resonate with people and appeal to their emotions as well as their intellect.

The Speculative and Testimonial Modes

There are various modes in which culture responds to and engages with ecocide. Perhaps the most popular is the speculative mode, which often takes the form of science fiction, or more specifically climate fiction (cli-fi), and which focuses on the impact and aftermath of ecocide. Novels such as Paolo Baciagalupi's *The Wind-up Girl* (2009) or Margaret Atwood's *MaddAdam* (2013) imagine life in a post-apocalyptic near future or alternative timeline, foretelling the consequences of current actions or inactions, and thereby critiquing ecocidal practices in the present (Trexler, 2015; Vermeulen, 2020). One of the key functions of contemporary sci-fi and cli-fi is, in fact, to reframe the present as a future that has already arrived but that, in the words of sci-fi author William Gibson, is "just not evenly distributed yet" (Gibson, 2016; Heise, 2019, p. 282).

There is also a growing body of cultural works that focus on *actual* past or present instances of ecocide and environmental destruction. Many of these are in a testimonial mode and have a commemorative function. The question of testimony is an urgent one in the ecocide discussion, because it is tied to our ability to actively and ethically respond to it. But such testimony comes with fundamental representational challenges: Who or what counts as a victim of, or witness to, ecological violence and how can they be represented so that their voices may be heard? How can one witness the deaths of entire species or the widespread "slow violence" (Nixon, 2011) of pollution and the destruction of ecosystems? Due to their imaginative and affective capacities, literary and artistic works can help overcome the limitations of the traditional testimonial framework and can construct alternative forms of testimony. First-person accounts such as memoirs and other forms of life writing, e.g. by Indigenous authors and activists such as Davi Kopenawa (*The Falling Sky*, 2013) or Sheila Watt-Cloutier (*The Right to Be Cold*, 2016), showcase the important role of storytelling in articulating experiences that usually remain unheard in official legal or political discourses. Importantly, these memoirs are grounded in Indigenous epistemologies and land relations and thus reveal how ecological violence is bound up with the epistemic violence of (settler) colonialism and Western systems of knowledge-production.

Poetry, specifically eco-poetry, has recently emerged as one of the most vibrant forms of cultural expression around ecocide and environmental violence. Eco-poetry, for example the recent anthology *Ghost Fishing* (2018), is closely aligned with, and can be said to constitute, a form of environmental justice activism and resistance as it presents imaginative forms of testimony to ecocidal violence. Through poetic devices such as *prosopopoeia* (i.e. lending a voice to objects, animals, plants, the dead, etc.) and *apostrophe* (i.e. addressing an absent other), these poems give voice to the human and non-human victims of ecocide and invite readers to bear witness to silenced or repressed (hi)stories. For example, the Micronesian kingfisher, a bird driven to extinction in the wild and now restricted to a captive breeding program, is a recurring figure in the work of Chamorro poet Craig Santos Perez from Guåhan (Guam), who interweaves the bird's and the Pacific Islanders' histories of displacement, exploitation, and exposure to toxic US military testing.

The Documentary Mode

Besides the testimonial, there is also the documentary mode. Historical novels, documentary cinema or theatre play an important role in producing and transmitting knowledge about ecocide as they are often based on extensive research into historical as well as ongoing cases of environmental destruction and ecocide. These works often tap into an existing repertoire that is familiar from cultural responses to the major traumatic events and acts of mass violence of the twentieth century. These familiar frameworks of representation help to portray ecological harm as *violence* and provide a space for critical reflection on questions of guilt and responsibility.

One particularly salient example of this is tribunal theatre, a type of documentary theatre that revolves around the interrogation of the perpetrator, and that became important in the wake of World War II and the Holocaust when it served to publicly address the difficult question of responsibility for crimes against humanity (Arjomand, 2018). Post-WWII trials had to grapple with two problems: how to judge unprecedented crimes, and how to think about the relationship between judgement and law. The theatre became an important site for publicly addressing and working through these difficult questions. Post-war tribunal theatre drew on the transcripts of actual trials (the Nuremberg trials, the Frankfurt Auschwitz trials, the Eichmann trial) and, to a certain extent, *re-enacted* versions of these. Contemporary theatre about ecocide takes up the tradition of tribunal theatre and develops it further to address the unprecedented nature and complexity of ecological crimes. With an important difference: it *pre-enacts* trials that the authors feel *should* take place but which cannot be held because of the limitations of the legal system.

There are different kinds of ecocide performances. Some of these, such as Rebekka de Wit and Anoek Nuyens' *De zaak Shell* (2020) or

Andres Veiel's *Ökozid* (2020), are set in an alternative present or a fictionalized near future in which corporations and governments are put on trial for environmental crimes. They create a space where an ecocide case can literally be played out, and different arguments and viewpoints can be modelled. This is a type of tribunal theatre that is fully scripted and acted out, and thus closely resembles a traditional play, with a predetermined beginning, middle, and end. But there are also other forms, which blur the boundaries between performance and political activism, fiction and reality, and which take a more open-ended and participatory approach in the tradition of the people's tribunal. This format arguably began with the 1966 Russell–Sartre Tribunal, which investigated American war crimes in Vietnam and ultimately found the US and its allies guilty. Over the past decade, similar tribunals have emerged that are specifically dedicated to ecocide. *The Monsanto Tribunal* in The Hague, an international civil society initiative to hold Monsanto accountable for human rights violations, crimes against humanity, and ecocide; Milo Rau's *The Congo Tribunal* (2017), a theatre and film project about civil war and resource extraction in Eastern Congo, and Radha D'Souza and Jonas Staal's performative *Court for Intergenerational Climate Crimes* (CICC) (Amsterdam, 2021–2022) which held hearings against the Dutch state, ING, Airbus, and Unilever, are among the most prominent examples. Many of these experiment with alternative non-anthropocentric legal frameworks and empower the public to act as a jury. Importantly, through stage setting and the use of artistic and multimedia elements, performances like the CICC make visible the interconnections between different histories and structures of violence, and their long-term and cross-generational legacies.

Tribunal theatre plays with the double meaning of the word representation. At least in English, "representation" carries a double meaning: on the one hand, it means legal or political representation (as in speaking on behalf of someone or something) and on the other, it refers to cultural representation or depiction (as in producing an image, painting, etc. of something). This is a distinction postcolonial theorist Gayatri Spivak (1988) makes in her famous essay "Can the Subaltern Speak?", in which she discusses the impossibility for oppressed people to speak within existing structures of power and systems of knowledge. The double meaning of representation is important in the context of the ecocide discussion because it helps us think about the question of who gets to speak and in what context, and what role alternative spaces and forms of representation can play.

Culture not only represents the world, but also actively shapes how we make sense of it. Nowhere is this clearer than in these performative tribunals. While the law focuses on legal responsibility, the theatre focuses on moral responsibility. The theatre is a symbolic space where political ideals can be experimentally actualized in the 'here and now' rather than in some distant future. Tribunal theatre depends on a public, and it also creates a public in that it brings people together to witness and participate in the action of justice being done. In doing so, it presents an opportunity to rethink what kinds of institutions we need to be able to address the challenges we face today.

Conclusion and Discussion

This chapter has shown how an interdisciplinary approach can lead to a multifaceted understanding of the concept of ecocide, combining different interacting dimensions to form a more holistic picture. First, the ecological dimension illustrates how ecosystems are dynamic and complex systems with communities of living organisms interacting with each other and with their physical environment. Ecocide can therefore have cascading effects on the environment; the interconnectedness between nature and social and economic systems reflects the potentially unlimited geographical and temporal reach of ecocide. Second, the criminological dimension reveals the different perpetrators involved, including criminal networks, companies, local communities, and government officials, and how they collaborate and facilitate global ecocidal activities. From ecocentric and biocentric perspectives, the victims of ecocide are analyzed from a broad harm perspective, which includes human and non-human victims, such as ecosystems, animals, plants and rivers. Furthermore, ecocidal activities can have serious criminogenic implications in the context of crime, war, and conflict over natural wealth. Third, the legal dimension illuminates questions of legal certainty by looking at the acts of ecocide and the necessity to keep a legal definition open-ended to apply to a variety of situations from an ecocentric approach rather than an anthropocentric approach. The mental element in the context of recklessness and strict liability as well as the epistemic and institutional bias surrounding discussions about ecocide reflect the legal and institutional challenges of criminalizing ecocide. Fourth, taking the cultural dimension into account can add reflection on the important symbolic significance and moral dimension of ecocide, and on its role in the popular imagination. Films, novels, art projects, performances, music productions, and other cultural works help shape how we understand and think about ecocide. Ecocide brings with it fundamental representational challenges, and authors, artists, and filmmakers find various ways to deal with these challenges. Speculative approaches imagine post-apocalyptic future worlds and thereby show what the effects of ecocide can be. Testimonial and documentary approaches draw on familiar cultural forms and genres to tell the stories of human or non-human victims of, or witnesses to, ecocide, transmit knowledge about past or ongoing instances of ecocide, and reflect on questions of responsibility.

Taken together, these four dimensions merge into a rich and comprehensive picture of ecocide. Since ecocide can destroy large areas of the natural environment and diminish the peaceful enjoyment experienced by human and non-human inhabitants, it may have serious local and global consequences for all life on Earth. Therefore, and crucially, our interdisciplinary approach critiques the dualistic distinction between nature and humans that remains prevalent in the conceptualization of ecocide. The environmental crisis is not simply a question for scientists to solve; it is a fundamental societal, political, and cultural issue with far-reaching origins and consequences. Moreover, different disciplinary approaches warn of the tension between ecological and socio-economic interests embedded in the question of ecocide. By rejecting anthropocentric perspectives and instead emphasizing the fundamental entanglement of humans and nature, we argue that ecocide is an ecocentric crime that cannot be excused on the basis of any perceived socio-economic benefits. Ecocidal activities have become integrated, and to a certain extent normalized, within our social, political, and economic global dynamics—with cascading effects on both human and non-human victims. An interdisciplinary perspective is needed to gain deeper insights into how this can be undone.

References

Adger, W. N., Hughes, T. P., Folke, C., Carpenter, S. R., & Rockström, J. (2005). Social-ecological resilience to coastal disasters. *Science, 309*(5737), 1036–1039.

Arjomand, M. (2018). *Staged: Show trials, political theater, and the aesthetics of judgment.* Columbia University Press.

Beirne, P. (2009). *Confronting animal abuse law, criminology, and human–animal relationships.* Rowman & Littlefield Publishers.

Beirne, P., & South, N. (2007). *Issues in green criminology: Confronting harms against environments, humanity and other animals.* Willan Publishing.

Benton, T. (1998). Rights and justice on a shared planet. *Theoretical Criminology, 2*(2), 149–175.

Berkes, F., Colding, J., & Folke, C. (2000). Rediscovery of traditional ecological knowledge as adaptive management. *Ecological Applications, 10*(5), 1251.

Boekhout van Solinge, T. (2014). The illegal exploitation of natural resources. In L. Paoli (Ed.), *The Oxford handbook of organized crime.* Oxford Handbooks.

Borges, I. M. (2018). *Environmental change, forced displacement and international law: From legal protection gaps to protection solutions.* Routledge.

Branch, A., & Minkova, L. (2023). Ecocide, the anthropocene, and the international criminal court. *Ethics & International Affairs, 37*(1), 51–79.

Brisman, A., & South, N. (2014). *Green Cultural Criminology: Constructions of Environmental Harm, Consumerism, and Resistance to Ecocide.* Routledge.

Bruch, C. (2001). All's not fair in (civil) war: Criminal liability for environmental damage in internal armed conflict. *Vermont Law Review, 695*, 399–429.

Cosgrove, W. J., & Rijsberman, F. R. (2014). *World water vision: Making water everybody's business. eBook.* Routledge eBooks.

Costanza, R., D'Arge, R., De Groot, R., Farber, S., Grasso, M., Hannon, B., Limburg, K., Naeem, S., O'Neill, R. V., Paruelo, J., Raskin, R. G., Sutton, P., & Van Den Belt, M. (1997). The value of the world's ecosystem services and natural capital. *Nature, 387*(6630), 253–260.

Costello, A., Abbas, M., Allen, A., Ball, S., Bell, S., Bellamy, R., Friel, S., Groce, N., Johnson, A., Kett, M., Lee, M., Levy, C., Maslin, M., McCoy, D., McGuire, B., Montgomery, H., Napier, D., Pagel, C., Patel, J., & Patterson,

C. (2009). Managing the health effects of climate change. *The Lancet, 373*(9676), 1693–1733.

Crook, M., & Short, D. (2014). Marx, Lemkin and the Genocide–Ecocide Nexus. *The International Journal of Human Rights, 18*(3), 298–319.

Crutzen, P. J. (2002). Geology of mankind. *Nature, 415*(6867), 23.

Cusato, E., & Jones, E. (2023). The 'imbroglio' of ecocide: A political economic analysis. *Leiden Journal of International Law, 37*(1), 42–61.

de Hemptinne, J. (2022a). Concluding observations on the influence of international environmental law over international criminal law. *Journal of International Criminal Justice, 20*(5), 1287–1298.

de Hemptinne, J. (2022b). *Ecocide: An ambiguous crime?* EJIL:Talk!

Dirzo, R., Young, H. S., Galetti, M., Ceballos, G., Isaac, N. J. B., & Collen, B. (2014). Defaunation in the anthropocene. *Science, 345*(6195), 401–406.

Duvall, J. A. (2017). *The environmental documentary: Cinema activism in the twenty-first century*. Bloomsbury Academic.

European Law Institute. (2023). *ELI report on ecocide: Model rules for an EU directive and a council decision*. ELI.

Falk, R. A. (1973). Environmental warfare and ecocide—Facts, appraisal, and proposals. *Bulletin of Peace Proposals, 4*(1), 80–96.

Foley, J. A., DeFries, R., Asner, G. P., Barford, C., Bonan, G., Carpenter, S. R., Chapin, F. S., Coe, M. T., Daily, G. C., Gibbs, H. K., Helkowski, J. H., Holloway, T., Howard, E. A., Kucharik, C. J., Monfreda, C., Patz, J. A., Prentice, I. C., Ramankutty, N., & Snyder, P. K. (2005). Global consequences of land use. *Science, 309*(5734), 570–574.

Gibson, W. (2016). *Quoted in Heise, U. Imagining extinction: The cultural meanings of endangered species*. University of Chicago Press.

Godfray, H. C. J., Beddington, J. R., Crute, I. R., Haddad, L., Lawrence, D., Muir, J. F., Pretty, J., Robinson, S., Thomas, S. M., & Toulmin, C. (2010). Food security: The challenge of feeding 9 billion people. *Science, 327*(5967), 812–818.

Goyes, D. R. (2019). *Southern green criminology: A science to end ecological discrimination*. Emerald Publishing Limited.

Hall, M. (2013). *Victims of environmental harm: Rights, recognition and redress under national and international law*. Routledge.

Halsey, M., & White, R. (1998). Crime, ecophilosophy and environmental harm. *Theoretical Criminology, 2*(3), 345–371.

Hamilton, C. (2013). *Human destiny in the anthropocene—A speech to the conference "Thinking the Anthropocene"*. Sciences Po, Paris.

Hardin, G. (1968). The tragedy of the commons. *Science, 162*(3859), 1243–1248.

Hart, H. L. A. (1961). *The concept of law*. Oxford University Press.

Harvey, D. (2003). *Accumulation by dispossession*. Oxford University Press eBooks.

Heise, U. K. (2019). Science fiction and the time scales of the Anthropocene. *ELH, 86*(2), 275–304.

Higgins, P. (2012). *Eradicating ecocide*. Shepheard Walwyn Publishers.

Higgins, P., Short, D., & South, N. (2013). Protecting the planet: A proposal for a law of ecocide. *Crime, Law and Social Change, 59*(3), 251–266.

Hough, P. (2022). Military ecocide. In S. Sayapin, R. Atadjanov, U. Kadam, G. Kemp, N. Zambrana-Tévar, & N. Quénivet (Eds.), *International conflict and security law*. T.M.C. Asser Press.

Hübschle, A.M. (2016) A game of horns: Transnational flows of rhino horn [Doctoral Dissertation, Universität zu Köln].

Independent Expert Panel for the Legal Definition of Ecocide (IEP) (2021). *Commentary and Core Text*. Stop Ecocide Foundation.

Intergovernmental Panel on Climate Change (IPCC). (2014). AR5 synthesis report: Climate change 2014.

International Law Commission. (1981). Yearbook of the International Law Commission Report of the Commission to the General Assembly on the work of its thirty-second session (Volume II Part Two). United Nations.

International Law Commission. (1994). Yearbook of the International Law Commission 1991: Report of the Commission to the General Assembly on the work of its thirty-second session (Volume II Part Two). United Nations.

Johns-Putra, A. (2016). Climate change in literature and literary studies: From cli-fi, climate change theater and ecopoetry to ecocriticism and climate change criticism. *WIREs Climate Change, 7*, 266–282.

Kissinger, G., Herold, M., & de Sy, V. (2012). Drivers of deforestation and forest degradation: A synthesis report for REDD+ policymakers. CIFOR.

Kooijman, M. (2023). From Anthropos to Oikos in International Criminal Law: A critical-theoretical exploration of ecocide as an 'Ecocentric' amendment to the Rome Statute. In D. Dam-de Jong & F. Amtenbrink (Eds.). Netherlands Yearbook of International Law 2021.

Le Billon, P. (2012). *Wars of Plunder: Conflicts, Profits and the Politics of Resources*. Columbia University Press.

Lindgren, T. (2018). Ecocide, genocide and the disregard of alternative life-systems. *The International Journal of Human Rights, 22*(4), 525–549.

Lopez, A. (2007). Criminal liability for environmental damage occurring in times of non-international armed conflict: Rights and remedies. *Fordham Environmental Law Review, 18*, 231–271.

Lynch, M. J., Stretesky, P. B., & Long, M. A. (2018). Green criminology and native peoples: The treadmill of production and the killing of indigenous environmental activists. *Theoretical Criminology, 22*(3), 318–341.

Mehta, J., & Jackson, J. (2021, August 25). To stop climate disaster, make ecocide an international crime. It's the only way. *The Guardian.*

Minkova, L. G. (2023). The fifth international crime: Reflections on the definition of "Ecocide". *Journal of Genocide Research, 25*(1), 62–83.

Mührel, M. (2022). Ökozid als fünftes Kernverbrechen im Rom-Statut—Meilenstein oder Gefahr für das Völkerstrafrecht? Archiv des Völkerrechts (AVR) Jahrgang 60 Heft 3.

Negri, S. (2022). On meteors and comets: Is the crime of ecocide back to stay? Environmental crimes at the interface of philosophy, the law of state responsibility and international criminal law. *International Criminal Law Review, 23*(1), 1–30.

Nixon, R. (2011). *Slow violence and the environmentalism of the poor.* Harvard University Press.

Nurse, A. (2013). *Animal harm: Perspectives on why people harm and kill animals.* Ashgate.

Palarczyk, D. (2023). Ecocide before the international criminal court: Simplicity is better than an elaborate embellishment. *Crime Law Forum, 34*, 147–207.

Robinson, D. (2022). Ecocide—Puzzles and possibilities. *Journal of International Criminal Justice, 20*(2), 313–347.

Ruggiero, V., & South, N. (2013). Green criminology and crimes of the economy: Theory, research and Praxis. *Critical Criminology, 21*(3), 359–373.

Schwendingers, H., & Schwendingers, J. (1970). Defenders of order or guardians of human rights? *Issues in Criminology, 5*(2), 123–157.

Sellin, T. (1938). Culture conflict and crime. *American Journal of Sociology, 44*(1), 97–103.

Sluiter, G., & Van Straaten, B. (2020). Ecocide als internationaal misdrijf? Perspectieven op vervolging en berechting in Nederland. *Boom Strafblad, 1*(3), 127–135.

Smith, T. (2013). Creating a framework for the prosecution of environmental crimes in international criminal law. In W. A. Schabas, Y. McDermott, & N. Haves (Eds.), *The Ashgate Research companion to international criminal law* (pp. 90–108). Routledge.

Sollund, R. (2020). Wildlife management, species injustice and ecocide in the Anthropocene. *Critical Criminology, 28*(3), 351–369.

South, N. (2010). The ecocidal tendencies of late modernity. In R. White (Ed.), *Global environmental harm: Criminological perspectives* (1st ed.). Willan.

South, Nigel. (2016). "The contested planet: Global green criminology and environmental crime in transnational context." In *Environmental Crime in Transnational Context* (pp. 9–26). Routledge.

Spivak, G. C. (1988). Can the subaltern speak? In C. Nelson & L. Grossberg (Eds.), *Marxism and the interpretation of culture.* University of Illinois Press.

Steffen, W., Crutzen, J., & McNeill, J. R. (2007). The Anthropocene: Are humans now overwhelming the great forces of Nature? *Ambio, 36*(8), 614–621.

Steffen, W., Richardson, K., Rockström, J., Cornell, S. E., Fetzer, I., Bennett, E. M., Biggs, R., Carpenter, S. R., De Vries, W., De Wit, C. A., Folke, C., Gerten, D., Heinke, J., Mace, G. M., Persson, L. M., Ramanathan, V., Reyers, B., & Sörlin, S. (2015). Planetary boundaries: Guiding human development on a changing planet. *Science, 347*(6223).

Sutherland, E. H. (1949). *White collar crime.* The Dryden Press.

The Office of the Prosecutor (OTP) of the ICC. (2016). Policy paper on case selection and prioritisation. ICC.

The Promise Institute for Human Rights, UCLA School of Law. (2021). Proposed definition of ecocide promise institute for human rights (UCLA) Group of Experts.

Trexler, A. (2015). *Anthropocene fictions: The novel in a time of climate change.* University of Virginia Press.

UNEP & Interpol. (2016). *The Rise of Environmental Crime.* UNEP and RHIPTO.

van der Wilt, H. (2023). Climate change as the ultimate form of ecocide: Are producers and consumers "Partners in Crime"? Amsterdam Law School Legal Studies Research Paper No. 2023-05.

van Uhm, D. (2023). *Organized Environmental Crime: Black Markets in Gold, Wildlife, and Timber.* Bloomsbury Academic.

van Uhm, D. P. (2016). *The Illegal Wildlife Trade: Inside the World of Poachers, Smugglers and Traders (Studies of Organized Crime).* Springer.

van Uhm, D. P. (2024). Transnational Environmental Crime. In S. L. Smith and I. Sahramäki (Eds.), *Research Handbook on Environmental Crimes and Criminal Enforcement.* Edward Elgar.

Vermeulen, P. (2020). *Literature and the anthropocene.* Routledge.

Weinstein, T. (2005). Prosecuting attacks that destroy the environment: environmental crimes or humanitarian atrocities. *Georgetown International Environmental Law Review, 17*, 697–722.

White, R. (2011). *Transnational environmental crime*. Willan.

White, R., & Kramer, R. C. (2015). Critical criminology and the struggle against climate change ecocide. *Critical Criminology, 23*(4), 383–399.

Whyte, D. (2020). *Ecocide: Kill the Corporation Before It Kills Us*. Manchester University Press.

Wilson, E. (1993). Biophilia and the conservation ethic. In S. Kellert & E. Wilson (Eds.), *The biophilia hypothesis*. Island Press.

Winter, E. (2023). Stop ecocide international's blueprint for ecocide is compromised by anthropocentrism: A new architect must be found. *Israel Law Review, 57*(1), 175–209.

Zabyelina, Y., & van Uhm, D. P. (2020). *Illegal mining: Organized crime, Corruption, and Ecocide in a Resource-Scarce World*. Palgrave.

Flora and Fauna Markets

The European Illegal Puppy Trade & Organised Crime

Jennifer Maher

Introduction

Almost half of European households share their homes with a companion animals, of which there are estimated to be over 7000 different species (Davenport and Collins, 2016). According to Shahbandeh (2024), in 2018 there were an estimated 470 million companion dogs globally, making them the most popular companion animal. In Europe, their popularity is second only to that of cats, with an estimated 104 million companion dogs (FEDIAF,[1] 2023). The largest number of European dogs are in Russia (approximately 17.6 million), the UK (13 million), and Germany (10.6 million), with high numbers also estimated in Spain, France, Italy, and Poland (FEDIAF, 2023). According to GlobalPETS,

[1] FEDIAF is the pet food industry's governing body in Europe.

J. Maher (✉)
Centre for Criminology, University of South Wales, Cardiff CF371DL, UK
e-mail: jenny.maher@southwales.ac.uk

D. van Uhm, D. Siegel (eds.), *Global Green Crime and Ecojustice*, Palgrave Studies in Green Criminology, https://doi.org/10.1007/978-3-031-91557-4_6

the companion animal population grew by 11% between 2021 and 2022 (Rubio & Gupta, 2023), with 11 million new dogs sharing our homes (FEDIAF, 2023). These figures estimate[2] the population of companion dogs only; importantly, they do not include free-ranging dogs,[3] which make up the majority (approximately 75–85%) of the world dog population (World Organisation for Animal Health, 2023).

The desire for companion animals, in particular young animals (e.g. puppies), has given rise to an extensive pet trade across Europe, with local supply proving unable to keep pace with consumer demand (Wyatt et al., 2017). The trade in puppies is lucrative, with estimates from 12 EU countries in 2015 indicating an annual commercial value of €1.3 billion (IBF International Consulting et al., 2015). In the UK alone, sales were estimated to be between 0.8 and 1.3 million puppies, with online advertisements for puppy sales worth approximately £130 million per annum (Wyatt et al., 2017). In recent years, the increased demand for fashionable or designer dogs has made puppies a lucrative commodity in the legal and illegal trade alike.

Illegal trade is associated with social and economic harms, including economic loss and trauma for consumers, loss of taxes, and threat to legitimate business from unfair competition. In line with the green criminology critical harms-based focus (Beirne & South, 2007), a broad definition of harm is adopted herein to include the individual dog victims who suffer extensive harm in the trade. It is important, however, to recognise that legal trade does not ensure robust dog welfare and health, especially when dogs are supplied from large-scale commercial breeding operations (USPCA, 2021). Many popular breeds suffer from extreme conformation breeding practices, there is inadequate regulation of dog care and welfare standards, and there are concerns over breeders' ability to meet dog's behavioural and socialisation needs and over the disposal of breeding bitches at the end of their productive lives. In seeking to maximise profits and avoid detection, illegal and irresponsible breeders further

[2] Accurate dog population data is notoriously difficult to find, the reliability of estimates vary by country as they rely on different sources of data, and often the sale of commercial products (e.g. food or insurance), and registration/licensing data (e.g. microchipping) are used. Most countries, such as the UK, do not collect accurate official data on companion dog populations.

[3] Defined as non-confined stray, feral, wild, street or village dogs.

compromise dog welfare, resulting in significant harms to the breeding dogs and puppies involved (Wyatt et al., 2017). In treating dogs as mere commodities, the harms to the dogs involved in the trade are pervasive and severe and deserving of consideration alongside the particular issues relating to other serious and organised crimes.

The scale and severity of harms in the trade escalate when the illegal trade becomes organised (Maher & Wyatt, 2021). Understanding the role of organised crime in the trade is challenging, especially as both the legal and illegal commercial puppy trade are impossible to accurately measure. Distinguishing between legitimate, unscrupulous and illegal breeders and traders is challenging for both authorities and consumers. Despite the frequent and lucrative movement of puppies from poorer Eastern European counties and rural areas (e.g. Ireland) to wealthier, often urban Western European marketplaces, the European illicit pet trade is an overlooked organised crime. A study conducted in 2017 for the Scottish Government and the UK Department for Environment Food & Rural Affairs (DEFRA) suggests an organised illicit market runs in parallel to the legal trade.

This chapter situates the research findings from this UK study (henceforth referred to as the UK IPT study) on the illegal puppy trade within EUROPOL's (2017) conceptualisation of organised crime groups, in order to evaluate their role in the puppy trade. Prior to considering how the changing nature of the puppy trade has opened doors to criminality and organised crime, the chapter briefly outlines the methodology underpinning this UK study. It then considers the defining characteristics of organised crime groups and the evidence supporting their involvement in the illegal puppy trade. To conclude, the implications of responding to sophisticated and organised criminals in the puppy trade are considered.

Methodology

In 2017, the Scottish Government identified a need to determine the nature, extent, and value of the illegal importation and farming of puppies, and to develop recommendations on how to prevent this trade. The research initially focused on Scotland and involved a multi-method and multi-site study. With additional funding, data collection expanded to Great Britain. Five stages of data collection, as detailed in Fig. 1, facilitated data triangulation. Initially, academic and grey literature from the UK, Europe and the US was reviewed. The analysis of economic and trade data, including using the Animal and Plant Health Agency (APHA) TRACES database and Trading Standards Scotland complaints data, facilitated an understanding of the economics of the puppy trade. The former database records the legal commercial movement of dogs and identifies transgressions within this trade. The latter records complaints from puppy purchasers in Scotland. The literature and economic analysis informed the next stages of empirical data collection.

From October 2016, the online advertisements for puppies in Scotland on seven popular platforms were monitored for a period of 12 weeks. Notable information, such as the Local Authority (LA) location, breed, number per litter and sex of the puppies, price per puppy and in total, seller's phone number and name, and Kennel Club and LA registration was recorded where provided. A total of 4074 puppies were offered for sale in 1479 advertisements, costing on average £818 each. This suggested a total market value of £3.3 million during this period. While the data identified many noncompliant advertisements and indicators of

Method & Data	Data	Location
Systematic Literature Review	Academic Peer reviewed literature	Worldwide
Secondary Analysis of Economic & Complaints Data	TRACES (commercial sales) for 2016 Trading Standards (complaints) 2015-2016	Scotland, England & Wales
Expert Semi-structured Interviews	14 participants, 12 interviews	UK & Ireland
Quantitative analysis of Online Advertisements	12 weeks, 7 websites*	Scotland
Stakeholder Questionnaire	53 participants	Scotland
Consumer Focus Groups	160 participants, 40 focus groups (25 face-to-face, 15 phone/online)	Scotland, England & Wales

*Craigslist, Dogs & Puppies UK, Epupz, Freeads, Gumtree, Pets for Homes, and Pets Viva Street

Fig. 1 Overview of the research methodology. (*Craigslist, Dogs & Puppies UK, Epupz, Freeads, Gumtree, Pets for Homes, and Pets Viva Street)

illegality, it was not possible to determine the portion of advertisements involving illegal or irresponsible breeders. 14 experts from statutory agencies, NGOs, veterinarians, and breeding organisations, with specialist knowledge and/or direct experience in responding to the puppy trade (i.e., purposive sampling) engaged in semi-structured interviews. These interviews focused on the nature and prevalence of the illegal puppy trade and the agency responses in place. To capture the experiences of other stakeholders who engage with consumers and puppies from irresponsible or illegal sources, a link to an online questionnaire was sent to relevant organisations in Scotland. 53 professionals, including dog walkers, groomers, trainers, boarders, and veterinarians responded. Data analysis through NVIVO (interviews and surveys) and SPSS (surveys) informed the focus groups with consumers. To facilitate convenience and snowball sampling, LA websites and non-governmental organisations (NGOs) across Great Britain advertised the focus group invite. 160 consumers, who were considering buying or had purchased a puppy over the previous two years, engaged in 40 focus groups. Their data was analysed in NVIVO.

The data collected represents the perspectives of many of the key stakeholders in the puppy trade. Data triangulation was important for validating their experiences and to negate some of the data limitations evident in the study. For example, official statistics are likely to have a large 'dark' figure due to the covert nature of the illegal trade and consumers' uncertainty in reporting it. Furthermore, the focus group sample was limited by selection bias. It is likely that those people concerned about dog welfare would be more likely to participate in the study. The participants were mostly women aged over thirty, which is consistent with other studies on animal abuse involving the public. Those who purposefully engaged with dubious puppy suppliers or who experienced the death of their puppies were less likely to participate due to the sensitive and emotional nature of their experiences. While many consumers did not consider the type of criminality or offenders involved in the trade, experts and professionals were informed and commented on the organised nature of the illegal puppy trade. However, due to the lack of administrative data on organised puppy trade offences and the difficulty in identifying organised offenders, the research must speculate as to their involvement.

The Changing Nature of the Puppy Trade

Numerous reports (IFAW, 2012; RSPCA, 2012; Four Paws International, 2013; IBF International Consulting et al., 2015; Dogs Trust, 2014, 2018, 2020, 2023; Eurogroup for Animals, 2020) have identified a sea-change in the scale and nature of the puppy trade over the past decades. Central to these changes was a cultural shift in viewing dogs as desirable commodities, and the subsequent growth in large-scale industrial commercial breeders, international trade, and the use of the internet as the medium for third-party sales. The commodification of dogs has largely been stimulated by the demand for designer breeds, the desire for which has waned and ebbed like other fashions, creating a fluid and robust marketplace for sellers (Maher & Wyatt, 2019). Consumers, confident in purchasing from online marketplaces, and overwhelmed by the number and cuteness of the puppies advertised, have engaged in impulse buying with little prior research or due diligence (Wyatt et al., 2017). While there are substantial data limitations in determining the UK dog population, longitudinal estimates are provided in commercial data.[4] Between 2010 and 2020, the UK dog population gradually increased by 1.3 million. Between 2020 and 2022, estimates from the industry indicate the dog population increased rapidly from nine to 13 million (Fig. 2) (Pet Food Manufacturers Association [PFMA], 2022). According to these figures, young adults (aged 24–35 years) were largely responsible for the spike in the dog population (PFMA, 2020), with 35% becoming new owners or planning to acquire a dog, compared to 11% among the general population.

Online Marketplaces and International Trade

The role of online marketplaces and social media in facilitating illegal pet trades, was initially raised by CITES in 2010 regarding CITES-listed wildlife species, highlighting the "urgent need for scientific, peer-reviewed

[4] A 2011 study on the UK dog population using Kennel Club registrations, veterinary practice customer archives and pet insurance records suggested the true figure fell between 3.6 and 21.5 million dogs—suggesting PFMA commercial data may be a substantial underestimate (McMillian et al., 2024; PMFA, 2022).

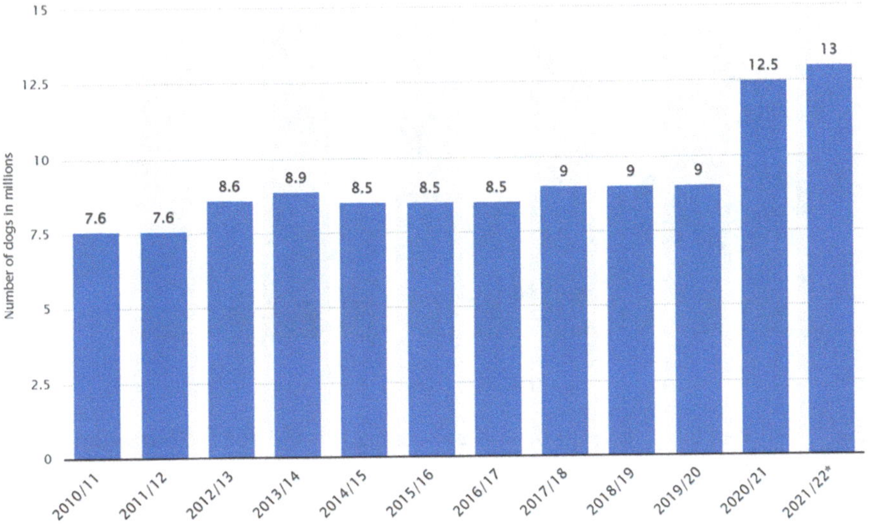

Fig. 2 Estimated UK dog population 2010–2022. (Source: PFMA, 2022)

articles to confirm or refute the claims that the internet fosters the illegal wildlife trade" (CITES, 2010, n.p.). In 2008, a PFMA survey of UK pet owners found 5% of owners acquired their companion animals through online breeders and advertisements (Davenport & Collins, 2016). Over half of pets were acquired from friends, acquaintances, family, or rescues, while the rest were sourced from commercial sellers and breeders (including 21% pet shops, 10% breeders, 10% private advertisement). The report highlighted "an area of concern recently has been the use of the internet as a trade medium" (Davenport & Collins, 2016, p. 9). The Dogs Trust (a dog NGO in the UK and Ireland) compared their survey of UK dog owners in 2018–2019 to that of 2020–2021, reporting a 7.6% increase in people acquiring dogs from pet websites (Anderson et al., 2023). Their study identified 21% of dogs were acquired through pet-selling websites, compared to 40% through family and friends, and charities. Dog owners in 2020/2021 were more likely to be first-time owners, and two-thirds of dogs were under one year of age. Prior to the Covid-19 pandemic, most puppies cost £1000 or less. From 2020, almost half of dogs for sale cost over £1000 (see Fig. 3), reflecting that market supply was unable to meet the demand, and so consumers were willing to

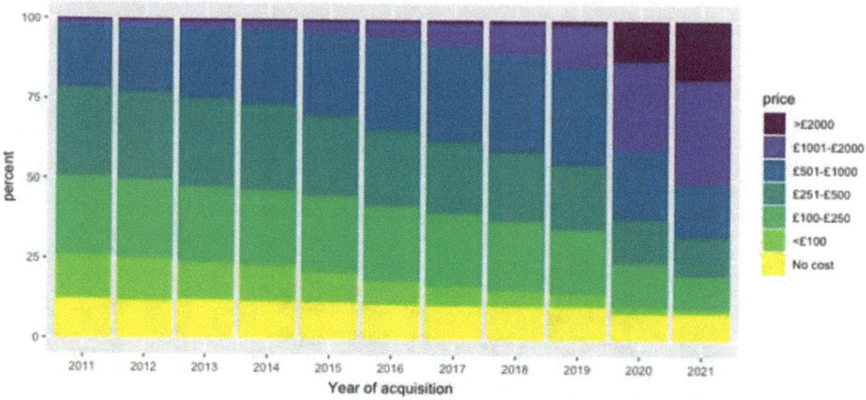

Fig. 3 Price paid for dogs in the UK between 2011 and 2021. (Source: Anderson et al., 2023)

pay more. The Covid-19 pandemic exacerbated consumer demand, resulting in a rise in prices and thereby the profits available to traders. This pattern was not exclusive to the UK, it was evident worldwide in many of the countries with the largest companion dog populations, such as, Europe, the US, China, Russia, Brazil, Australia.

Despite the demand in the UK outstripping the supply (Packer et al., 2021), Shahbandeh (2024) reports that only 3.8% of people surveyed reported their dogs came with a passport from overseas (predominantly Ireland, Romania and then Spain), which is inconsistent with what is known about the capacity of legitimate UK breeders to supply demand. While it is impossible to accurately estimate the scale of the illegal dog trade, Eurogroup for Animals (2020) suggest up to 90% of online ads on eBay, a primary dog trade site, are from illegal breeders and that "The available estimates indicate that Europe's pet trade (cats, dogs, exotics) is the third most profitable illegal trade in the EU, after narcotics and weapons." The UK Pet Advertising Advisory Group [PAAG][5] has set minimum voluntary standards for advertisers; however, their research has exposed the limitations of self-regulation among advertising platforms (PAAG, 2015a,

[5] PAAG set up in 2001 in combat the irresponsible advertising of pets for sale online is made up of animal welfare organisations, trade associations and veterinary bodies and is endorsed by Defra and the devolved administrations.

2015b). They concluded that despite PAAG support few websites consistently met the minimum standards and that they could not be trusted to self-regulate. Their research on social media platforms exposed the sale of companion animals through Facebook (finding 930 closed groups in 24 hours), of which 58% were in breach of PAAG minimum standards.

Regulation of the Trade

By moving online, traders were able to distance their breeding establishments and practices from consumers, thereby making it easier to conceal the harms and the illegal nature of the trade (Maher & Wyatt, 2019). Simultaneously, official regulation of online marketplaces and of breeders and traders has not kept pace with market growth or the expansion online. Rather, austerity resulted in regulators across various agencies (e.g. Local Authorities, Border Force) experiencing a decline in financial and other resources, resulting in little regulation of dog breeders and traders. As the demand for certain designer puppy breeds and crossbreeds outweighed local supply, puppies were increasingly supplied by commercial breeders from abroad with no oversight from the UK authorities. Changes to the EU Pet Travel Scheme in 2012 made it quicker and easier to bring companion animals into the UK. In the first year, the UK experienced a 61% increase in the number of dogs entering from Europe, with the greatest increases from Eastern European countries (Dogs Trust, 2014).

The European Commission (2023) recognises that this freedom of movement has provided a vital modus operandi for the illegal trade. An individual can travel with up to five puppies (from 15 weeks of age) as non-commercial pets, allowing offenders to disguise their commercial activities and evade tax and official scrutiny. During the pandemic, travel restrictions necessitated a change in the modius operandi of the illegal trade. A significant rise was noted across Europe in the commercial moment of dogs (under the Balai Directive). In 2020, the movement of commercial dogs to England and Wales, measured by Intra Trade Animal Health Certificates (ITAHC),[6] rose by 87% (RSPCA, 2020).

[6] Comparing certificates issued between March and October 2020 and those issued certificates between May and September 2019.

Concurrently, there was an 88% decline (when compared to previous years) in the dogs moved to England and Wales under the EU passport scheme (RSPCA, 2020). According to the Dogs Trust (2020), commercially imported dogs are not subject to checks at the border and during lockdown UK post-import checks were suspended. The use of forged documents, dogs concealed in legal consignments and the movement of pregnant bitches across boards in the final weeks of pregnancy have facilitated the illegal trade (EFRA, 2019).

Aligned, these factors have created the proverbial 'perfect storm' for a buoyant legal and illegal trade in dogs, which has attracted the attention of criminal networks across Europe. The value of dogs for our physical and emotional health and in providing a special bond has never been clearer than during the pandemic, especially among the expanding middle class in emerging markets. The monetary value to offenders has similarly been affirmed.

Impact of the Trade

The changing acquisition choices of consumers and breeding practices of sellers have compromised the health and welfare of both dogs and humans. For humans, the changes in the puppy trade have heightened public health risks (e.g. zoonosis, dog bites), costs to owners and economic disadvantage for legitimate sellers, and has enabled millions in undeclared income. Unpaid taxes and loss of government income also applies to those businesses providing services to the illegal puppy trade (e.g. passports, vaccinations), a loss estimated by CAROcat (2015) to be €312 million annually across the EU.

Puppies purchased abroad and during the pandemic are more likely to have been exposed to risk factors associated with problem dog behaviours (Brand et al., 2024). These risk factors include how and when puppies are acquired, early socialisation experiences and owners' experience levels. In the Dogs Trust study, a quarter of puppies were identified as acquired under the recommended minimum age of eight weeks (Kinsman et al., 2020). When compared to puppies who remained in maternal care up to and beyond eight weeks, research shows early separation can have significant impact on their physiological and behaviour development (including destructive and aggressive behaviour, disease susceptibility and mortality) (Kinsman et al., 2020). Dogs purchased during the pandemic are noted to be more at risk from behavioural problems (Christley et al., 2021; Harvey et al., 2022). Furthermore, Tateo et al. (2022, p. 540) cite various studies which document transport as a stressful event for dogs, especially those new to travel (e.g. puppies). A variety of stimuli disrupt the dog's "homeostasis and metabolism", affect their "physiological and emotional state" and may represent a major threat to their welfare and health. Puppies from industrial-scale operations (i.e. puppy farms) are also more likely to have genetic and infection disorders (USPCA, 2021). Parvovirus, for example, can be fatal and cost up to £4000 to treat (Muldoon et al., 2017). Due to dog behavioural and health issues, the UK is now experiencing an increase in the relinquishment and euthanasia of young dogs (Brand et al., 2024). The negative impact of the illegal puppy trade extends well beyond the dogs involved (e.g. puppies and their mothers); it influences the state of health of the nation's dogs.

Animal welfare refers to a duty of care on people to ensure the physical and mental needs of dogs are met. This includes a suitable environment and diet, expressing natural behaviours and not suffering pain, fear, and distress (RSPCA, 2024). The ability of the legal trade to meet these needs is in question (BBC, 2015; Meyer et al., 2022). The harms escalate in the illegal trade, with breeding bitches and puppies experiencing cruelty, illness, unnatural and psychologically damaging rearing conditions and often death. For example, the International Organisation for Animal Protection (2016) estimated that 20% of puppies bought on the internet died before they reached six months of age and 12% developed costly long-term health problems. Another NGO (Surge Campaigning, 2021) identified that four out of five puppies transported to Spain die en route. The nature of the trade prohibits an accurate assessment on the scale of these harms. These studies thereby provide a glimpse of the significant and pervasive harms in the changing puppy trade, especially when facilitated by illegal and irresponsible breeders and traders.

What Makes the Puppy Trade Attractive to Offenders?

The puppy trade is vast and profitable. It is just one example of a much broader issue evident in the sale of cats, rabbits and exotic pets: how the commodification of companion animals has resulted in their appeal to criminals. The UK IPT research (Maher & Wyatt, 2019, 2021; Wyatt et al., 2017) found the illegal puppy trade represents the classic routine activity theory convergence (Cohen & Felson, 1979). For example, most criminal acts require a motivated offender (e.g. high profits and high demand among consumers), a suitable target (e.g. vulnerable puppies, willing consumers) and the absence of capable guardians (e.g. little risk of detection, prosecution or punishment). In particular, the high profits and low risk make the trade attractive to offenders.

High Profit

The shifting puppy marketplace, from local to global and online, the growth in largely unregulated third-party sales through online advertisements, and consumers whims for fashionable breeds—has created a lucrative market which is attractive to criminals. Together, increased demand and easy supply have given rise to a buoyant illegal trade in dogs, whereby the animals involved are simply a lucrative commodity and, accordingly, profit rather than welfare is the priority. A recent European Commission report (2023) estimated that at any one time in the EU around 438,000 dogs were for sale. High profit is evident when you consider that puppies can be bred in Eastern Europe and Ireland for as little as £25, but sold in a Western European country such as the UK for between £1000 and £7000. When the scale of the trade is taken into account, this represents vast profits for all those involved in the supply chain. For example, a Hungarian puppy dealer, reported by Dogs Trust (2018), detailed the weekly exportation of roughly 400 puppies from one village, an annual number of approximately 20,000 puppies, representing an annual turnover of £28 million (given an estimated sale price of around £1400 per puppy). During the COVID-19 pandemic, new markets emerged with lifestyle changes and increased reliance on the internet. The value of the puppy trade increased with growing demand and shortages in designer breeds, resulting in a 125% rise sin adverts online and increased profit for legal and illegal breeders/traders as the price for puppies, in many cases, more than doubled (Anderson et al., 2023).

Low Risk

Despite the potential profits, the illegal dog trade has essentially been under-enforced. The nature and vulnerability of the dog victims, who cannot alert us to their plight, enhances the invisibility of offenders. It has been NGOs advocating for these nonhuman animals that have largely been responsible for uncovering the illegal trade (IFAW, 2012; Four Paws International, 2013; Dogs Trust, 2014, 2018, 2020, 2023; Eurogroup for Animals, 2020). However, their resources and capacity to prevent these illegal practices is very limited. Equally, the future guardians who buy the puppies through online advertisements are distanced from the breeding establishments and harms involved. Consumers often lack the knowledge or capacity to identify whether or not an offence is being committed, and where they are aware they choose to rescue the puppy, thereby funding and encouraging the criminality (Wyatt et al., 2017). Consumer reliance on and trust in technology and social media, has, according to Haenlein and Lord Evans of Weardale (2023), brought radical change to our daily lives, meaning that the public unwittingly use the services of organised crime groups in their consumption. By moving to online marketplaces, traders can act with anonymity and impunity. Offenders effortlessly conceal the illegal trade from consumers and officials amongst the large volume of online advertisements (e.g. 438,000 across Europe at any given time). The risk of offenders being identified or punished is extremely low due to easily exploited loopholes in legislation (such as the EU pet passport), poor enforcement (due to insufficient resources and expertise and prioritisation), and the reliance on market self-regulation at the boarders and online (Maher & Wyatt, 2021). The UK's exit from the EU has highlighted further vulnerabilities associated with the Common Travel Area, which provides a 'back door' for people smuggling puppies into the UK.

Overall, the evidence suggests that as profits have increased alongside the value of dogs, so too has the interest of organised crime groups (OCGs), who have adapted their criminality to take advantage of the marketplace and gaps in regulation.

The Puppy Trade as an Organised Crime

Organised Crime

The UNODC (2000) Convention on Transnational Organised Crime's (UNTOC) defines an organised crime group as three or more persons existing over a period of time, acting in concert with the aim of committing crimes for financial or material benefit. This is consistent with the definition given by the UK National Crime Agency (Home Office, 2015). EUROPOL (2017) extend this definition to reflect the "complex and flexible nature of modern organised crime networks". Their analysis of OCGs found diverse and varied actors, which differ in their structure, level of sophistication and type of criminality. The defining features of organised crime in the UK and Europe are flexibility and adaptability, international reach, the exploitation of technology, and networks (EUROPOL, 2017; HM Government, 2023).

EUROPOL (2017) has observed that "many OCGs are highly flexible and able to shift from one criminal activity to another or to add new criminal activities to their crime portfolio". OCGs are opportunist, moving easily between new markets and modus operandi as profit opportunities emerge and barriers develop. The UK National Strategic Assessment 2023 and Serious and Organised Crime Strategy 2023–2028 highlight the adaptability of OCGs in their response to the changing environment of the UK exit from the EU and COVID-19 pandemic restrictions (NCA, 2023; HM Government, 2023). According to EUROPOL (2017), organised offenders increasingly adapt "the supply chain models of global online retailers". As a result, the benefits for legitimate businesses derived from the internet, low-cost transport and international supply chains has likewise transformed OCG operations. Market advancements have made it easier and less risky for their criminality. Consequently, most serious and organised crime in the UK has an international dimension, making it more difficult to detect and prevent (Haenlein and Lord Evans of Weardale, 2023). The integration of digital systems and online trade is integral to the black market; it has transformed organised crime and the necessary responses (EUROPOL, 2017; HM Government, 2023). Crucially, technology enables scale, facilitating contact with thousands of potential victims and offenders in their networks (Haenlein and Lord Evans of Weardale, 2023).

OCGs vary in scope from loose social networks who offend together to well-organised international operations. A network study of OCGs in one region of the UK identified a majority of small and localised groups, formed by people with the same background (Campana & Varese, 2022). UK OCGs often involve non-core members, including "specialists or professionals", to facilitate their operations (Local Government Association, 2015). The UK SOC strategy 2023–2028 refers to these persons as "enablers", who, either knowingly or unwittingly, are often a key part of the OCGs' ability to commit offences (CPS, 2019). The ability of OCGs to engage with legitimate actors and to hide their activities within the legal trade is also central to their successful engagement with the public. For trafficking offences, OCGs are known to engage with corrupt officials and to use fraudulent documents at the borders (CPS, 2023). While the covert nature of OCGs makes it difficult to assess their nature and prevalence, the National Crime Agency's strategic assessment (2023, p. 12) estimates a minimum of 59,000 known members of serious and organised crime groups, who are estimated to cost the UK over £47 billion per year. Almost half of OCGs are involved in the illegal drugs trade (HM Government, 2017). Alongside the drugs trade, OCGs are most likely to engage the trade in firearms and other physical commodities, organised immigration, slavery or human trafficking, child sexual abuse, and financial crimes (money laundering, fraud, bribery and corruption, illicit finance) and cybercrime.

Evidencing Organised Crime in the Puppy Trade

Prior to the Scottish Government study, in 2015 the UK Border Force (customs) recognised the possible involvement of OCGs in the puppy trade. However, as recorded in the records of an RSPCA and DEFRA-organised workshop meeting, "there are still huge unknowns in how the criminal networks operate" (APHA, 2017: n.p.). Subsequently, the role of OCGs in the illegal puppy trade has been officially recognised in both the UK and the EU. As part of the European illegal pet trade, puppy trafficking was linked to serious and organised crime (European Parliament, 2019) and argued to have become the third-largest illegal trade in Europe (Eurogroup for Animals, 2018). A report based on an international conference organised by Eurogroup for Animals (2020, p. 19) on behalf of the Croatian Presidency of the Council of the European Union maintained "A change of attitude is needed in central and local government to accept that the puppy trade is controlled by criminal gangs in the same way as the drug trade or human trafficking are." The EU (2021) 2021–2025 Strategy to tackle organised crime identified that "companion animals continue to be traded illegally, often on a large scale and sometimes with potential devastating consequences". The recent EU Serious and Organised Crime Threat Assessment (EUROPOL, 2025) also identified the smuggling of pets as part of the changing nature of organised crime in the EU. Within this analysis they note the veterinary, scientific and market expertise within these OGCs in the wildlife pet trade. This expertise and engagement with legitimate businesses and veterinarians is also evident in the puppy trade, as discussed below.

A multi-agency and cross-border Bio-Crime Project in Italy established that 53% of pets (including dogs) traded across the Austrian–Italian border had no documentation (Organised Crime and Corruption Reporting Project, 2021). Their research found that the organised illegal puppy trade was horizontally structured, like the drugs trade, and part of a diverse crime portfolio. For example, they maintained that Italian mafia groups (termed the 'zoo-mafia') were diversifying into the illegal puppy trade, and, thereby, that trafficking offences might include drugs, animals and people. OCGs have also been reported to be expanding into both the illegal (Dalton, 2023) and legal (Mooney, 2020) puppy trades. Research conducted by an NGO in 2022–2023 monitored 330,000 online advertisements across EU websites, identifying organised networks concealing a high volume of puppy trade as rescue operations (European Commission, 2023). The 2022–2023 EU enforcement action identified commonalities in forged paperwork and enablers (professional operators), and indicated the role of OCGs in the puppy trade (European Commission, 2023). According to one case study, TRACES data showed nine suspicious Russian and Belarusian exporters accounted for just under 27,000 of EU imported animals over the previous four years. This represented more than 13% of EU imports of cats and dogs from around the world.

In the UK, Her Majesty's Revenue and Customs (HMRC)'s dedicated puppy trade task force recovered over £5 million in unpaid taxes from 257 cases over a four-year period (HMRC and Stride, 2019). Although further investigation is necessary to firmly link this tax evasion to organised crime, the 2022 Police Scotland (2022) Multi-Agency Strategic Threat Assessment has included the illicit puppy trade as a key threat from what they term serious organised crime groups. The National Crime Agency (2023) recently reported dogs being used to facilitate other organised crimes, such as, for example dogs transported to the UK from Mexico with 10 kilograms of cocaine. During the peak UK demand for dogs, DEFRA (2021) reported the increased value of puppies resulted in OCGs adapting their criminality to pet theft, in order to breed them illegally. In 2010, approximately 2000 dog theft crimes were reported to police in England and Wales (HM Government, 2021). Although further research is required to clarify the role of organised crime groups in dog theft, an increase in multiple dog thefts has also been recorded by another NGO, DogLost (Selby-Fell & Allen, 2021). The Welsh Government Agency Intelligence Network (GAIN) estimated that between 10 to 15% of current OCGs in Wales are involved in dog theft or illegal dog breeding, with a link between these crimes (Tarian Regional Organised Crime Unit cited in HM Government, 2021). Stolen dogs are used for breeding or sold online as puppies, with offenders focusing on dogs in private homes, breeders and boarding kennels (BBC, 2020). Anecdotal evidence suggests that those importing puppies from Eastern European puppy farms are returning with stolen dogs in order to maximise their profits (Butcher, 2022). For example, the dog theft investigator Butcher (2022) describes the theft of Bracken, who was stolen from his home in the UK and found eight months later by an NGO in Poland.

The Role and Nature of Organised Crime in the Puppy Trade in Great Britain

Despite widespread recognition that OCGS are involved in the illegal puppy trade, their covert activities are difficult to expose. Consequently, the nature and scale of their involvement in the UK and Europe is unknown. This limitation was evident in the UK IPT study. Yet many of the research experts and professionals argued an organised illicit market ran in parallel to the legal trade, which was increasingly organised and sophisticated in nature. To explore this further, the defining features of organised crime noted above—international reach, networks, flexibility and adaptability, and exploitation of technology— frame the study findings below (EUROPOL, 2017; HM Government, 2023).

International Reach

The UK demand for puppies is supplied by domestic illegal commercial operations and the illegal trade from Ireland and Eastern European countries. The origin of puppies from abroad diverged by destination country and modus operandi. Wales and Scotland were predominantly the destination for puppies from Ireland, while England received trade from both mainland Europe and Ireland. At the time of the study, puppies were sourced largely from Poland, Lithuania, Hungary and Romania for the UK. The trade in these countries was poorly regulated and supported by poor rural communities who benefited significantly from diversification into the puppy trade. While these remain the major countries of concern, the 2023 European Commission study highlighted the illegal movement of puppies originating from outside the EU (Turkey, Serbia, Russia and Belarus). The international scope of the puppy trade is consistent with most UK serious and organised crime, making it more difficult to detect and prevent (Haenlein and Lord Evans of Weardale, 2023).

Extensive Networks

Expert participants described the illegal actors in the trade as non-compliant individuals, OCGs and ancillaries. Likewise, stakeholder survey respondents (n = 53) identified illegal puppy suppliers principally as follows: an organised network (n = 43), opportunistic offenders (n = 30) and a part of a legitimate business (n = 20). OCGs were identified as: (i) groups who developed around puppy trafficking; (ii) groups who diversified into puppy trafficking; and (iii) groups using the legal puppy trade to launder illegal finance. For example, experts explained:

(i) "So puppies are basically coming in from Central and Eastern Europe, but that's your starting point, but these puppies will be distributed right across the country. The individuals involved are becoming very clever…Certainly we're aware that some of the service stations down in Kent are being used for the transfer of puppies, not necessarily to owners…they're effectively using a distribution network to get them across the country" [EI2].

(ii) Despite the Irish government trade database indicating that approximately 3000 puppies were being transported commercially to the UK, he was able to "directly follow over 52,000 pups to the UK every year and that was the ones we knew about" [EI5].

(iii) "And the trade has moved on…you're getting serious and organised criminals [in the UK] that are using the puppy trade as a way to launder the proceeds of conventional criminality. So, if you're importing drugs or selling guns or doing whatever and you've got all these, what used to be tanning studios or nail bars or whatever, they're now using the front of selling pups. So, when they do get detected and they've got X amount of hundreds and thousands in the bank they say that they've earned that money from puppy trading,[7] but in actual fact it's from the conventional criminality" [EI4].

---

[7] The latter was evidenced in the breeding of 'pocket bullies', for example, with puppies being priced at £6000–7000 each.

The lengthy international and cross-border movement of live 'contraband' and the fluid movement within countries requires connections, coordination, and funding. A network is required as there are multiple stages to the trade—planning, breeding, rearing, transporting, advertising, and sale—which require diverse skills. A variety of legal and illegal actors working as a network make the illegal trade possible. The level of sophistication in some areas of the illegal puppy trade implies that OCGs are, at a minimum, involved in the smuggling of puppies.

Puppy farmers who plan, breed and rear dogs in Eastern Europe, for example, may be legal (e.g. licensed) or non-compliant (e.g. unlicensed and not declaring profits). For these puppy farmers, a link to OCGs permits them to increase the scale and sophistication of their operation and to connect directly to lucrative marketplaces. The OCGs benefit by mass-producing puppies in locations that are relatively cheap and poorly regulated. Criminality may be engaged in by these farmers as a result of them declaring their business or facilitating necessary inspections and health requirements. EU regulations require the premises from which commercial puppies originate to be a registered business. Similarly, local agencies and professionals may fail to conduct inspections or fulfil duties in regulating these farms. The IBF (2015) study highlighted licensing inconsistencies and poor compliance was documented across EU member states, suggesting limited regulation and enforcement is central to the success of the illegal trade. This may be facilitated by corruption and/or poor resourcing and prioritisation among officials.

Puppies intended for international travel, either commercially or as companion animals, must be accompanied by a health certificate (ITAHC) or a pet passport (Pet Travel Scheme [PTS]—detailing microchip number and rabies vaccine) completed by a veterinarian. In the illegal trade, this may be facilitated by 'enabler' veterinarians who are paid to forge paperwork (e.g. the dogs age and vaccination status) or using counterfeit veterinary materials. For example, a UK NGO provided details of Hungarian breeders who claimed to move 400 underage puppies a week to Spain, and Hungarian, Serbian and Lithuanian vets who sold undercover investigators fraudulent passports and sedatives to help conceal puppies. Puppies are commonly transported to Great Britain by land, then either ferry or train (Eurotunnel). Legitimate companies may be used to transport the dogs either covertly or overtly—they may be concealed in smugglers' vehicles, or transported by their fictitious 'owner' using a pet passport. In previous research, trade companies involved with Eastern European breeders were associated with the mafia and corrupt police (Four Paws International, 2013). These actors have direct contact with border enforcement and the transport companies who regulate the trade. The UK IPT research did not identify UK officials who permitted illegal movement; rather, it was the limited resources and lack of specific expertise at the border facilitated the illegal trade. The IBF study identified that the UK, like most of the countries reviewed, did not have a national framework specific to the transport of dogs, and those that did reported high levels of noncompliance among national competent authorities. In some instances, enforcement is delegated to private organisations or based on standard tools, such as checklists and guidelines. Dogs travelling under the pet passport are not registered or traceable once they reach the UK, permitting easy movement across the UK. Likewise, the use of forged ITAHC paperwork ensures the movement of commercial dogs are untraceable. Consequently, the international trade seamlessly merges with the UK domestic trade. At this point, puppies may be passed directly to their new owners or to enabling 'stooge' breeders or sellers who provide a legitimate front to engage with consumers. The use of intimidation by sellers was also noted by some experts and members of the public. This is an effective control strategy commonly used by OCGs (HM Government, 2023).

Online traders who advertise the puppies are central to the illegal network. These may be legitimate businesses or criminals in the UK or abroad. The UK IPT study analysis of online advertisements, for example, found that advertisements with local addresses in Scotland could be for puppies bred abroad and sold by local third-party traders. In response to UK awareness campaigns, local sellers used fake houses and bitches, or imported pregnant dogs to allow consumers to view puppies in their 'home' with their 'mother'.

Flexibility and Adaptability

To breed and sell puppies, offenders must plan which dogs to breed, how to move them to lucrative marketplaces, how to engage consumers, and where to house the puppies to meet or be passed on to their new owners. This is further complicated by a rapidly changing market driven by the desire for trendy breeds. In recent years, the UK exit from the EU, the COVID-19 pandemic and changing enforcement responses have required considerable adaptability. The UK IPT research found those involved in the illegal trade appear to have the flexibility and infrastructure to manoeuvre through these market, social and enforcement changes. Although the number of puppy breeding operations in Eastern Europe and Ireland is unknown, the ability to swiftly supply new fashionable breeds to international marketplaces indicates that there must be a considerable number (pregnancies in dogs last about nine weeks).

The study identified new approaches aimed at reducing costs and vulnerabilities. Offenders have advertised for sale puppies not yet born, allowing them to develop a list of desired dogs which would be passed to breeders. A consignment of the desired puppy breeds would be moved, underage, from the breeder directly to the buyer. This approach avoided enforcement in the UK, as puppies were not held by UK sellers or 'stooge' families.[8] Experts noted the offenders' ability to avoid changes in the enforcement response, including moving from vans to cars, to trafficking puppies as foot passengers. During an enforcement operation (Operation Delphin), experts noted that illegal transporters who were stopped, would promptly message other smugglers, advising them to change the point and day of entry to the UK. Likewise, by diversifying from pet passports to commercial movement during the pandemic, offenders utilised the least-monitored and most straightforward method. This flexibility and sophistication in adapting to changes to the supply chain model to circumvent regulations and enforcement makes the organised nature of the illegal puppy trade quite clear.

[8] 'Stooge' families are set up in rental accommodation or in the home of owners of female dogs. Puppies are placed with these 'families' in order to show the puppy with its mother (which is required by law) and to indicate the puppy was bred locally rather than in large-scale breeding facilities, either in the UK or abroad.

Exploitation of Technology

The UK IPT research found technical advancements have transformed the landscape of the puppy trade from physical markets to virtual platforms. Focus group participants highlighted that the internet was a trusted marketplace, and emphasised the assumption that it was regulated, and the credentials of the puppy sellers verified. By exploiting this technology and trust, offenders have engaged at scale with an extensive pool of victims. Crucially, experts how difficult it is to regulate the internet; as there are few enforcers, and those who are monitoring the trade find it difficult to distinguish between legal and illegal trades. However, due to resource limitations and a reliance on industry self-regulation, sellers' details cannot be easily and reliably checked. Experts cautioned that the over-regulation of established online sales platforms may force them to close or move abroad, encouraging offenders to migrate to other platforms. Offenders, for example, diversified into advertising puppies for sale on social media groups, where sale is prohibited and thereby unregulated. Where legislation and regulation has become more robust, offenders expertly navigate the process by providing legitimate but fraudulent information (e.g. microchip, license number) and utilising multiple email addresses and mobile phone numbers to conceal their identity and the scale of their online presence.

Offenders in the trade vary in scale: from opportunistic and small-scale offenders to noncompliant legitimate operations to complex illegal international operations. The latter involve significant geographical and financial scale and infrastructure (people and transport), and sophistication to successfully navigate the market and enforcement. Considering the EUROPOL (2017) and HM Government (2023) analysis of OCGs as involving diverse and varied actors, focused on profit, who are flexible, adaptable and exploit international borders and technology, some offenders in the puppy trade undoubtedly fit this description. Without considerable enforcement engagement and research, the number of OCGs and level of involvement can only be speculated. Nonetheless, it is crucial to consider if the responses in place are adequate for protecting dogs and consumers from OCGs.

Responding to the Illegal Puppy Trade

The UK IPT research identified significant limitations in the regulation of the puppy trade. Specifically, gaps in legislation facilitated the misuse of the pet passport for commercial trade, and the use of online advertisement and social media platforms for the illegal sale in dogs. The enforcement approach was fragmented and inadequate due to limited resources and prioritisation at the boarders and within the UK. Expert interviewees noted that enforcement efforts to target pervasive offenders in the trade were constrained due to the significant resources available to some offenders to avoid detection, and to contest any legal actions if caught. Law enforcement lacked the necessary intelligence and resources to respond to the scale, capacity, fluidity and sophistication of OCG offenders involved in the trade. Since the UK IPT study, changes in UK regulation and enforcement have enhanced barriers for offenders. Furthermore, a bill currently (2024) proceeding through Parliament has the intention of reducing the misuse of the pet passport by increasing the age at which puppies can travel to six months, reducing the number of dogs permitted to five per vehicle rather than per person, and imposing a ban on importing heavily pregnant dogs. Developments in the EU suggest there are additional measures which would support a more robust response to OCGs involved in the illegal puppy trade.

In other illegal pet trades, Keatinge et al. (2018) note, finance-based responses must be deployed more systematically, to tackle the 'financial' appeal to offenders. Both the UK government and the EU Bio Crime Project have utilised a finance-based response for illegal traders. For example, in 2019, the UK HMRC task force successfully targeted tax evasion in the puppy trade (HMRC and Stride, 2019). No further investigations have been reported. Given the noted decline in the use of pet passports during the COVID-19 pandemic, financial investigations into the commercial trade during this period would be critical to understanding the nature of OCGs in the illegal trade.

Multi-agency partnerships involving intelligence generation and sharing have been central to responding to organised crime in the UK. Since 2021, the illegal puppy trade has been included in the strategic approach to the UK Serious and Organised Crime Strategy. Intelligence relating to

OCGs and illegal breeding is fed into Operation OPAL in order to improved national intelligence (HM Government, 2021). Consequently, other financial instruments effectively used to tackle organised crime, such as the process for the seizure of criminal assets, could also be adopted in responding to the most pervasive offenders in the illegal puppy trade. Lessons can also be taken from the Bio-crime Model (Zucca et al., 2020) which has been operating on the Italian–Austrian border since 2017. This EU-funded project, recognised as best practice, facilitates multi-agency cross-border collaboration between all public agencies involved in tackling transnational organised crime related to the illegal animal trade. Importantly, this includes law enforcement and customs, the justice system, and veterinary health. To ensure close collaboration, the Bio-crime Veterinary Medical Intelligence Centre is based in the International Police and Custom Cooperation Centre (IPCCC). The project recognises that tackling OCGs is central to preventing the spread of zoonotic diseases which significantly impact human and nonhuman animal health and wellbeing.

As dogs pass through a complex supply chain to reach UK consumers, it is difficult to establish their real origins and identify offenders. In response, since 2020 a ban on the commercial third-party sales of dogs (also known as Lucy's Law) has been adopted across Great Britain (with moves also made to introduce it also to Northern Ireland). Consumers looking to purchase a puppy must do so directly from the breeder, including visiting the breeding establishment and seeing the puppy with its mother (DEFRA, 2020). Effectively, the law aimed to remove the 'middleman' and the anonymity of illegal breeders. As all puppies must move directly from the breeder's establishment to the owner, traceability is enhanced, and consumers become more capable guardians. This creates barriers for illegal traders. These can be circumvented if consumers do not comply, and due to the loophole which does permit the third-party sale of imported dogs. For example, online advertisers who identify puppies as bred outside the UK can continue to provide forged paperwork to buyers.

While technology is effectively employed to facilitate illegal pet trades, it can also be harnessed to improve monitoring and disruption techniques (Global initiative 2023). Microchipping is mandatory for UK dogs (since

2015); breeders are required to microchip puppies prior to sale, in order to provide traceability. However, the fragmented microchipping databases within the UK (a total of 16 commercial UK databases according to the HM Government, 2021) and abroad makes traceability difficult and incompatible with the desire for online advertising platforms to validate advertisers. A single point of access to these databases is required to ensure traceability for consumers, online platforms and enforcement agencies seeking to verify the origin of the dog. The EU Commission (2024) recently proposed new rules to enhance the welfare and traceability of dogs (and other companion animals) in EU transport and trade. These measures will include strict traceability requirements (microchip and authorised establishments) and automated checks for online sales (identification and registration of advertisers, which can be checked), to enhance enforcement by statutory agencies, NGOs and consumers. This will be facilitated through the EU Animal Health Law (2021), which requires all dog breeders to register with relevant national authorities. In addition, the EU Digital Services Act (2024) puts pressure on large online platforms to regulate online advertisers. This approach is critical, according to the UK advisory group PAAG, for facilitating the traceability and regulation of online marketplaces in the UK (DEFRA, 2021). For example, between March and November 2020, one popular online advertising platform (Pets4Homes) reported blocking more than 120,000 adverts from appearing on their website because advertisers did not pass the prepublication approval process created to implement PAAG guidelines. Where advertising platforms demand robust data from advertisers, it creates barriers for offenders. However, the regulation of these online platforms is not legislated for in the UK. European NGO Four Paws (2021) has proposed a scheme known as VeriPet (also known as PetSAFE), a two-factor authentication technical solution for online advertising platforms to ensure that only registered dogs can be advertised online. The system was launched in Ireland in 2021 on classified advertising site Dogs.ie, and is proposed as a pan-Europe database, which will hinder the anonymity of offenders and their effortless access to consumers.

In response to the misuse of the pet passport to facilitate the illegal commercial trade, since 2019 animal health and welfare have been included in the EU iRASFF (Rapid Alert System for Food and Feed) platform. Under the Regulation on Official Controls (EU) No 2017/625, Member States are required to exchange information and conduct investigations via the iRASFF on "encountered irregularities, non-compliances and suspicions of fraud" related to the illegal puppy trade. A recent evaluation of iRASFF (2022–2023) identified 467 notifications relating to dogs and cats, predominantly relating to severe violations and including intelligence on established organised crime networks. This data provides valuable intelligence, facilitates the sharing of enforcement resources and more efficient and effective targeted responses across the EU. Since the UK's exit from the EU, it has been unable to benefit from this cross-border collaboration and intelligence sharing in their response to puppy trafficking. In 2021, under the UK–EU Trade and Cooperation Agreement, the UK National Crime Agency and Europol signed a working agreement to support collaboration post-EU exit (NCA, 2021). The agreement allows the UK to make use of EUROPOL's national units, EIS (information system) and SIENA (information exchange network), to respond to cross border OCGs. This partnership would be vital in the response to OCGs involved in the illegal puppy trade.

Current UK and EU responses have arguably enhanced capable guardians, improved monitoring and disruption techniques, and facilitated more robust enforcement. Further research and administrative data are required to identify their impact on OCGs involved in the puppy trade. Clarity on the scale and nature of OCG involvement will be central to developing and evaluating a targeted response to their offences in the puppy trade. More broadly, however, responses to the IPT require appropriate consideration of the victims involved.

IPT offences encompass a wide range of victims, the breeding dogs and puppies are the first casualties, with humans mostly secondary victims. Green criminologists stress the importance of species justice, and the need to respond directly to the harms to nonhuman animals (White, 2014). The commodification of dogs and their status as property (Maher & Wyatt, 2019) is characteristic of the widespread speciesism which drives the neoliberal free-market trade in animals (Byrne 2017; Horta 2010; Nurse and Wyatt 2021; Beirne et al. 2018). The legal and illegal pet trades define puppies as an exploitable resource, there to serve human desires. Consumers' desire for, and impulsive purchasing of, aesthetically pleasing and fashionable dogs makes the illegal trade profitable for OCGs. Recognising the sentience and intrinsic value of these dogs is crucial for challenging the harms implicit in the trade and in responding to offenders in the trade. It also challenges the glaring paradox that the desire for an emotional bond with a dog drives their material objectification and abuse. This incongruency has been recognised in the Pet Abduction Act 2024 (England and Northern Ireland), which was introduced in response to increased dog theft. It recognises that "dogs are not inanimate objects but sentient beings capable of experiencing distress and other emotional trauma when they are stolen from their owners or keepers" (HM Government, 2024). MacBeth and Whitfort (2024) report the use of species victim impact statements have increased sentences for criminals elsewhere. These could present another step towards species justice by giving a voice to dog victims, and an opportunity to enhance the response to offenders.

Conclusion

The UK legal trade in both domestic and international puppies has provided opportunities for widescale illegal and irresponsible trade. The illegal puppy trade is a buoyant and lucrative enterprise, multifarious in its nature, harms, and the complexities it poses to enforcement agencies. The UK IPT study supports more recent acknowledgements in the UK and EU, that the trade is a serious crime and harm, which is facilitated in part by OCGs. Using the EUROPOL OCG definition, the factors shaping organised crime are evident in the UK illegal puppy trade, particularly when the international reach, flexibility and adaptability, exploitation of technology, and networks involved are taken into consideration. The illegal trade is structured around numerous offenders engaging across borders and intertwined inextricably with the legal trade. Offenders in the trade respond fluidly to the changing market, and exploit legislative loopholes, enforcement capabilities, and consumer detachment. While the scale and nature of OCG involvement remains unclear, this research, and subsequent studies, evidence the sophisticated nature of the offences and offenders involved.

The scale and severity of animal abuse, potential health risks to the public and general dog population, and the negative financial and social consequences that extend to the government, civil society, and the public are clearly evidenced in the illegal trade. Likewise, growing concerns among NGOs, enforcement agencies, and politicians (EFRA, 2019) have been voiced. Despite recognition of the illegal puppy trade within official serious and organised crime strategies, significant gaps remain in the enforcement response.

Recent UK and EU developments have resulted in enhanced legislation, traceability, and disruption techniques, which will interrupt offenders. It is not clear; however, how effective such strategies are in disrupting sophisticated offenders, who often easily circumvent enforcement efforts. Understanding the role and modus operandi of OCGs has important implications for developing targeted prevention and detection strategies. Focusing on illegal finance and disruption techniques through enhanced technology and traceability, and cross-border collaboration and intelligence is required to respond to OCGs involved in the puppy trade. This would require the prioritisation of the illegal puppy trade by UK enforcement agencies, including those responsible for responding to OCGs. EU developments should be monitored for their effectiveness and applied to the UK, where viable. More generally, removing loopholes for offenders is crucial, for example, by improving physical/visual scrutiny at borders, increasing the age at which puppies can be moved across borders, and improving monitoring and traceability online.

The commodification of dogs is central to the ability of OCGs to engage in the illegal trade. Recent studies of UK dog owners highlight that consumers are both willingly and blindly engaging in the illegal trade due to the emotion implicit in sourcing a puppy. The material objectification inherent in the trade leaves an increasing number of dogs vulnerable to abuse and exploitation. By failing to recognise the dog victims, the illegal trade may appear a victimless crime and will remain a lower priority when responding to OCGs. Consequently, it will continue to be a low-risk offence with minimal prosecution and punishment. Dogs are generally recognised as legal property, and so not considered in prosecutions. This has changed somewhat with the UK Pet Theft Act 2024, which legally recognises the sentience of dogs and their distress as victims of abduction, with punishments of up to five years' imprisonment for offenders. This recognition must be extended to dog victims of the illegal puppy trade, alongside the use of species victim impact statements to emphasise the intrinsic worth of the victims and the severity of the harm inflicted. These steps will enhance protection for our dearest companions while providing more effective enforcement responses to OCGs. From a green criminology perspective, the benefit of challenging enduring speciesist tendencies is apparent, in spite of the complexity of doing so.

References

Anderson, K. L., Casey, R. A., Cooper, B., Upjohn, M. M., & Christley, R. M. (2023, March 16). National Dog Survey: Describing UK Dog and Ownership Demographics. *Animals, 13*(6), 1072.

Animal and Plant Health. (2017). *Response to a freedom of information 2000 request*. Retrieved November 2, 2023, from https://assets.publishing.service.gov.uk/government/uploads/system/uploads/attachment_data/file/643330/ATIC1140.pdf

BBC. (2015). *The dog factory*. Available at: The Dog Factory a Shocking BBC Documentary—YouTube. Retrieved March 20, 2024.

BBC. (2020). Dog theft: Organised crime driving 'epidemic' of dog snatching. Available at: Dog theft: Organised crime driving 'epidemic' of dog snatching—*BBC News*. Retrieved October 12, 2024.

Beirne, P., O'Donnell, I., & Janssen, J. (2018). *Murdering animals. Palgrave Studies in Green Criminology*. Palgrave.

Beirne, P., & South, N. (2007). *Issues in green criminology*. Routledge.

Brand, C. L., O'Neill, D. G., Belshaw, Z., Dale, F. C., Merritt, B. L., Clover, K. N., Tay, M. M., Pegram, C. L., & Packer, R. M. (2024, January 22). Impacts of puppy early life experiences, puppy-purchasing practices, and owner characteristics on owner-reported problem behaviours in a UK Pandemic Puppies Cohort at 21 Months of Age. *Animals, 14*(2), 336.

Butcher, C. (2022). *How is organized crime impacting on dog theft & illegal breeding*? Available at: How is Organized Crime Impacting on Dog Theft & Illegal Breeding?—Colin Butcher Author. Retrieved October 12, 2024.

Campana, P., & Varese, F. (2022). The determinants of group membership in organized crime in the UK: A network study. *Global Crime, 23*(1), 5–22. https://doi.org/10.1080/17440572.2022.2042261

CAROcat. (2015). *France: A new law to limit private breeding of dogs and cats and internet trade in these animals.* Retrieved February 22, 2022, from http://carocat.eu/france-a-new-law-to-limit-private-breeding-of-dogs-and-cats-and-internet-trade-of-these-animals/

Christley, R. M., Murray, J. K., Anderson, K. L., Buckland, E. L., Casey, R. A., Harvey, N. D., Harris, L., Holland, K. E., McMillan, K. M., Mead, R., et al. (2021). Impact of the first COVID-19 lockdown on management of pet dogs in the UK. *Animals, 11*, 5.

CITES. (2010). *E-commerce of CITES-listed species.* Retrieved March 19, 2024, from https://cites.org/sites/default/files/eng/news/pr/2010/20100316_e-commerce_kit.pdf

Cohen, L., & Felson, M. (1979). Social change and crime rate trends: A routine activity approach. *American Sociological Review, 44*, 588–608.

Crime Prosecution Service. (2019). *Organised Crime Group participating in activities of an.* Retrieved February 23, 2024, from https://www.cps.gov.uk/legal-guidance/organised-crime-group-participating-activities

Crown Prosecution Service. (2023). No place to hide: Serious and organised crime Strtagey 2023-2028. Retrieved July 19, 2025, from https://assets.publishing.service.gov.uk/media/65798633254aaa0010050bdc/SOC_Strategy_23-28_V9_Web_Accessible.pdf

Dalton, J. (2023). The illegal puppy trade. *Independent.* Available at: Thriller writer Peter James exposes puppy-smuggling gangs risking bringing rabies to UK | The Independent. Retrieved October 12, 2024.

Davenport, K., & Collins, J. (2016). *European code of conduct on pets and invasive Aline species.* Council of Europe. ES139315_PREMS 032115 GBR 2585 Code of conduct on pets and invasive aliens species Web 16 × 24.pdf (europa.eu). Retrieved March 19, 2024.

DEFRA. (2021). *Pet theft taskforce policy paper.* Available at: Pet Theft Taskforce (publishing.service.gov.uk). Retrieved February 21, 2024.

Department for Environment Food and Rural Affairs [DEFRA]. (2020). *Lucy's Law spells the beginning of the end for puppy farming.* Available at: Lucy's Law spells the beginning of the end for puppy farming—GOV.UK (www.gov.uk). Retrieved March 19, 2024.

Dogs Trust. (2014). *The Puppy Smuggling Scandal: An investigation into the illegal entry of dogs into Great Britain under the Pet Travel Scheme.* Retrieved February 23, 2024, from https://www.dogstrust.org.uk/press-materials/dt_puppy_smuggling_report_v12_web(1).pdf

Dogs Trust. (2018). *Puppy smuggling: When will this cruel trade end?* Retrieved February 2, 2024, from https://www.dogstrust.org.uk/puppy-smuggling/111018_puppy%20smuggling%202018_final.pdf

Dogs Trust. (2020). *Puppy smuggling: Puppies still paying as Government delays.* Available at: 2020 Puppy smuggling report.pdf (dogstrust.org.uk). Retrieved February 2, 2024.

Dogs Trust. (2023). *National Dog Survey.* Available at: National-Dog-survey-Results-23_1.pdf (dogstrust.org.uk). Retrieved March 19, 2024.

EC. (2024). *New rules to improve animal welfare.* Retrieved April 5, 2024, from https://ec.europa.eu/commission/presscorner/detail/en/ip_23_6251?_jtsuid=85135171930609642050962

Environment, Food and Rural Affairs Committee [EFRA]. (2019). *Puppy smuggling inquiry launched.* Available at: Puppy smuggling inquiry launched—Committees—UK Parliament. Retrieved February 2, 2024.

Eurogroup for Animals. (2018). *Booming online puppy trade: Illegal, fraudulent and heart-breaking for dogs and buyers.* Retrieved December 2, 2023, from https://www.eurogroupforanimals.org/news/booming-online-puppy-trade-illegalfraudulent-and-heartbreaking-dogs-and-buyers

Eurogroup for Animals. (2020). *Illegal pet trade: Game over.* Retrieved February 2, 2024, from https://www.eurogroupforanimals.org/sites/eurogroup/files/2020-06/Report_Illegal%20Pet%20Trade_%20Game%20Over_2020.pdf

European Commission [EC]. (2023). *Illegal Trade of cats and dogs EU enforcement action.* Available at: Report—Illegal trade of cats & dogs (europa.eu). Retrieved April 5, 2024.

European Parliament. (2019). *Texts adopted.* Retrieved March 3, 2024, from https://www.europarl.europa.eu/doceo/document/TA-8-2016-0065_EN.pdf?redirect

European Union. (2021). *EU strategy to tackle organised crime 2021–2025*: Annexes to COM(2021)170. Available at: Annexes to COM(2021)170—EU Strategy to tackle Organised Crime 2021–2025—EU monitor. Retrieved May 4, 2024.

EUROPOL. (2017). *Defining serious and organised crime.* Retrieved March 22, 2024, from https://www.europol.europa.eu/socta/2017/defining-serious-and-organised-crime.html#top

Europol. (2025). European Union Serious and Organised Crime Threat Assessment: The Changing DNA of Serious and Organised Crime. The Hague: Europol. Retrieved July 19, 2025, from https://www.europol.europa.eu/publications-events/main-reports/socta-report

FEDIAF. (2023). *Facts & Figures 2022.* Available at: FEDIAF_Annual-Report_2023_Facts-Figures.pdf (comingsoon.site). Retrieved March 19, 2024.

Four Paws International. (2013). *Puppy Trade in Europe: Research on the impact of illegal businesses on the market, on consumers, on the one-health concept and on animal welfare*. Four Paws International.

Four Paws UK. (2021). New solution to block unscrupulous online traders. Available at: New solution to block unscrupulous online traders—Animal Charity—Animal Welfare Organisation—FOUR PAWS UK (four-paws.org.uk). Retrieved March 19, 2024.

Haenlein and Lord Evans of Weardale. (2023). A boundless threat? The rise of organised crime in the UK. RUSI. Available at: A Boundless Threat? The Rise of Organised Crime in the UK | Royal United Services Institute (rusi.org). Retrieved March 25, 2024.

Harvey, N. D., Christley, R. M., Giragosian, K., Mead, R., Murray, J. K., Samet, L., Upjohn, M. M., & Casey, R. A. (2022). Impact of changes in time left alone on separation-related behaviour in UK Pet Dogs. *Animals, 12*, 482.

Home Office. (2015). Serious Crime Act 2015. Retrieved July 19, 2025, from https://assets.publishing.service.gov.uk/government/uploads/system/uploads/attachment_data/file/415967/Fact_Sheet_-_Participation_Offence_-_Act.pdf

HM Government. (2017). *Drug strategy*. Home Office. Retrieved October 12, 2024, from https://www.gov.uk/government/uploads/system/uploads/attachment_data/file/628148/Drug_strategy_2017.PDF

HM Government. (2021). *Pet theft taskforce policy paper*. Available at: Pet Theft Taskforce (publishing.service.gov.uk). Retrieved March 23, 2024.

HM Government. (2023). *No place to hide: Serious and organised crime strategy 2023–2028*. CP992. Retrieved February 2, 2024, from https://assets.publishing.service.gov.uk/media/65798633254aaa0010050bdc/SOC_Strategy_23-28_V9_Web_Accessible.pdf

HM Government. (2024). *Pet abduction bill becomes law*. Available at: Pet Abduction Bill becomes law—GOV.UK (www.gov.uk). Retrieved October 12, 2024.

HM Revenue & Customs and The Rt Hon Mel Stride MP. (2019). Millions raised from tax evading dog breeders. Retrieved July 19, 2025, from https://www.gov.uk/government/news/millions-raised-from-tax-evading-dog-breeders

IBF International Consulting, VetEffect, Wageningen University and Research Centre, and Institute Zooprofilattico Sperimentale dell Abruzzo e del Molise G. Caporale. (2015). *Study on the welfare of dogs and cats in commercial practices*. Specific Contract SANCO 2013/12364: Final Report. European Commission. Retrieved November 2, 2020, from https://ec.europa.eu/food/

sites/food/files/animals/docs/aw_eu-strategy_study_dogs-cats-commercial-practices_en.pdf

International Fund for Animal Welfare. (2012). *How much is that Doggie on my browser: The truth behind online puppy sales.* IFAW.

International Organisation for Animal Protection. (2016). *The Illegal Trade of Puppies in Europe.* Available at: OIPA—International Organization for Animal Protection. Retrieved March 23, 2024.

Keatinge, T., Haenlein, C, & Reid, A. (2018). A Missing Page: Strengthening the Response to the Illegal Wildlife Trade. [online] Royal United Services Institute. Retrieved July 19, 2025, from https://www.rusi.org/explore-our-research/publications/commentary/missing-page-strengthening-response-illegal-wildlife-trade

Kinsman, R. H., Casey, R. A., Knowles, T. G., Tasker, S., Lord, M. S., Da Costa, R. E., Woodward, J. L., & Murray, J. K. (2020, August 8). Puppy acquisition: factors associated with acquiring a puppy under eight weeks of age and without viewing the mother. *Vet Rec, 187*(3), 112.

Local Government Association. (2015). *Tackling serious and organised crime a local response.* Retrieved February 22, 2024, from https://www.local.gov.uk/sites/default/files/documents/tackling-serious-and-orga-44a.pdf

MacBeth, A., & Whitfort, A. (2024). Species Victim Impact statements: Giving a voice to the unheard victims of environmental crime. *Global Initiative Against Transnational Organized Crime.* Retrieved March 13, 2024, from Alastair-MacBeath-et-al-Species-victim-impact-statements-GI-TOC-January-2024_2.pdf (globalinitiative.net).

Maher, J., & Wyatt, T. (2019). Rural-urban dynamics in the UK Illegal Puppy Trade: Trafficking and Trade in 'man's best friend'. *International Journal of Rural Law and Policy, 9*(2). https://doi.org/10.5130/ijrlp.2.2019.6266

Maher, J., & Wyatt, T. (2021). Organised crime and the puppy trade. Trends in Organized Crime Special Journal Issue. Springer. Available at: European illegal puppy trade and organised crime—PMC (nih.gov).

McMillian, K., Christley, R., Upjohn, M., Harrison, X., & Casey, R. (2024). *UK pet dog population project: Research.* Dogs Trust. Retrieved May 5, 2024, from https://www.dogstrust.org.uk/how-we-help/professionals/research/ukpdpp

Meyer, I., Forkman, B., Fredholm, M., Glanville, C., Guldbrandtsen, B., Izaguirre, E., Palmer, C., & Sandøe, P. (2022). Pampered pets or poor bastards? The welfare of dogs kept as companion animals. *Applied Animal Behaviour.* Available at: Pampered pets or poor bastards? The welfare of dogs kept as companion animals (sciencedirectassets.com). Retrieved February 22, 2024.

Mooney, J. (2020). Irish criminal gangs go to the dogs to launder drugs cash. *The Sunday Times*. Available at: Irish criminal gangs go to the dogs to launder drugs cash (thetimes.com). Retrieved October 12, 2024.

Muldoon, J., Williams, J., & Scottish SPCA. (2017). *The impact of the dog trade on canine welfare: The Current Situation and Recommendations for Policy and Practice*. Available at: illegal_dog_trade_briefing_paper.pdf (ed.ac.uk). Retrieved February 12, 2024.

National Crime Agency. (2021). *NCA and EUROPOL sign up to a new working arrangement*. Available at: NCA and EUROPOL sign up to a new working arrangement—National Crime Agency. Retrieved February 4, 2024.

National Crime Agency. (2023). *Dog Rescued from a life of Organised Crime*. Available at: Dog rescued from a life of organised crime—Happy ending for "unwitting accomplice" Camila—National Crime Agency. Retrieved February 23, 2024.

Nurse, A., & Wyatt, T. (2021). *Wildlife crime*. New Horizons in Criminology. Bristol University Press.

Organised Crime and Corruption Reporting Project. (2021). *The Canine connection: Europe's illegal dog trade*. Available at: The Canine Connection: Europe's Illegal Dog Trade (occrp.org). Retrieved March 22, 2024.

Pets Advertising Advisory Group [PAAG]. (2015a). *Gumtree PAAG Report*. PAAG.

Pets Advertising Advisory Group [PAAG]. (2015b). *Facebook Animal Group Moderation*. PAAG.

Packer, R. M., Brand, C. L., Belshaw, Z., Pegram, C. L., Stevens, K. B., & O'Neill, D. G. (2021). Pandemic puppies: Characterising motivations and behaviours of UK owners who purchased puppies during the 2020 COVID-19 pandemic. *Animals, 11*, 2500.

Pet Food Manufacturers Association [PFMA]. (2020). *PFMA confirms dramatic rise in pet acquisition among millennials*. Retrieved February 23, 2024, from https://www.pfma.org.uk/news/pfma-confirms-dramatic-rise-in-pet-acquisition-among-millennials-#:~:text=acquisition%20among%20Millennials-,Today%2C%20the%20Pet%20Food%20Manufacturers'%20Association%20releases%20its%20brand%20NEW,a%20pet%20to%20their%20families

PFMA. (2022). *2022 Annual Report*. Available at: PFMA 2022 Annual Report | Pet Food Manufacturers Association (ukpetfood-reports.co.uk). Retrieved February 23, 2024.

Police Scotland. (2022). *Scottish multi-agency strategic threat assessment*. Available at: Scottish Multi-Agency Strategic Threat Assessment—Police Scotland. Retrieved February 22, 2024.

Royal Society for the Prevention of Cruelty to Animals [RSPCA]. (2012). *Do puppies have secrete powers: Understanding the irrational behaviour of the puppy buying publics*. RSPCA.

RSPCA. (2020). *Stop puppy imports*. Available at: Stop Puppy Imports | #ActNowForAnimals | RSPCA—RSPCA. Retrieved February 23, 2024, from rspca.org.uk

RSPCA. (2024). Animal Welfare Act. Retrieved April 14, 2024, from www.rspca.org.uk

Rubio, D. P., & Gupta, M. (2023). Analysis: European pet population grew by 11% in 2022. *GlobalPETS*. Available at: Analysis: European pet population grew by 11% in 2022—GlobalPETS (globalpetindustry.com). Retrieved March 19, 2024.

Selby-Fell, H., & Allen, D. (2021). *Dog Theft: What can we infer from the evidence so far?* Retrieved October 12, 2024, from http://www.pettheftreform.com/research

Shahbandeh, J. (2024). Where owners got their pet dogs from in the United Kingdom (UK) 2023. *Statista*. Available from: UK: main sources for getting a dog 2023 | Statista. Retrieved March 19, 2024.

Surge Campaigning. (2021). *Pet trade death trade: Europe's forgotten victims of trafficking*. Available at: Pet trade death trade: Europe's forgotten victims of trafficking—Surge | Creative Non-Profit for Animal Rights (surgeactivism.org). Retrieved February 22, 2024.

Tateo, A., Nanni Costa, L., & Padalino, B. (2022). The welfare of dogs and cats during transport in Europe: A literature review. *Italian Journal of Animal Science, 21*(1), 539–550.

Ulster Society for the Prevention of Cruelty to Animals [USPCA]. (2021). *Puppy dog fortunes: A report on the puppy trade in Northern Ireland*. https://www.uspca.co.uk/media/uploads/uspca-puppy-dog-fortunes.pdf

United Nations Office on Drugs and Crime [UNODC]. (2000). *United Nations Convention against Transnational Organized Crime definition*. Retrieved March 22, 2024, from https://www.unodc.org/unodc/en/organized-crime/intro/UNTOC.html

White, R. (2014). Eco-justice and Problem-solving approaches to Environmental Crime and Victimisation. In T. Spapens, R. White, & M. Kluin (Eds.), *Environmental crime and its victims*. Ashgate.

World Organisation for Animal Health. (2023). *Stepping up dog population management to achieve rabies elimination*. Available at: Stepping up dog population management to achieve rabies elimination (woah.org). Retrieved March 19, 2024.

Wyatt, T., Maher, J., & Biddle, P. (2017). *Scoping research on the sourcing of pet dogs from illegal importation and puppy farms 2016–2017*. Retrieved February 23, 2024, from https://www.gov.scot/publications/scoping-research-sourcing-pet-dogs-illegal-importation-puppy-farms-2016/pages/16/

Zucca, P., Rossmann, M. C., Osorio, J., Karem, K., et al. (2020). The "Bio-Crime Model" of cross-border cooperation among veterinary public health, justice, law enforcements, and customs to tackle the illegal animal trade/bio-terrorism and to prevent the spread of zoonotic diseases among human population. *Frontiers in Veterinary Science, 7*, 855.

Wildlife Trafficking in Spain: Causes, Consequences and Future Considerations

Mònica Pons-Hernández

Introduction to the Illegal Wildlife Trade

We are facing an unprecedented biodiversity crisis. The Intergovernmental Science-Policy Platform on Biodiversity and Ecosystem Services (IPBES) Global Assessment of Biodiversity and Ecosystem Services in 2019 projected that approximately 1 million animal and plant species are in danger of disappearing. Among the threats animal and plant species are facing, IPBES (2019) highlights the role of habitat degradation, overexploitation, and illicit trade. Although all factors have an impact on species survival, this chapter focuses on the illegal wildlife trade. "The illegal

This work has been carried out as a member of CEDAT, IU-RESCAT, and the Research Group of Universitat Rovira i Virgili "Territory, Citizenship, and Sustainability," recognized as a consolidated research group supported by the Department of Research and Universities of the Government of Catalonia (2021 SGR 00162). The author also acknowledges funding from the European Union's Horizon 2020 research and innovation programme under the Marie Skłodowska-Curie grant agreement No 101034309 Shaping European Research Leaders for Marine Sustainability (SEAS).

M. Pons-Hernández (✉)
University of Bergen, Bergen, Norway
e-mail: monica.pons@urv.cat

D. van Uhm, D. Siegel (eds.), *Global Green Crime and Ecojustice*, Palgrave Studies in Green Criminology, https://doi.org/10.1007/978-3-031-91557-4_7

wildlife trade or wildlife trafficking is the complete process from killing or kidnapping wildlife, through alteration into products if necessary, then smuggling within or between countries, and selling to the final buyer in person or online" (Wyatt, 2022, p. 9).

The instrument for regulating the trade of wild species is the Convention on International Trade in Endangered Species of Wild Fauna and Flora (CITES). CITES lists species into three appendices according to the degree of protection they need[1] (CITES, n.d.-a). So, depending on their classification, a different set of permits is required for legal trade (Wyatt, 2021). At the time of writing, the trade of more than 40,900 species—including approximately 6610 species of animals and 34,310 species of plants—are protected by CITES against overexploitation through international trade (CITES, n.d.-b). Currently, the implementation of CITES in the EU, and therefore in Spain, is carried out through *Council Regulation (EC) No 338/97 of 9 December 1996 on the protection of species of wild fauna and flora by regulating trade therein*, and through the implementing Regulation, *Council Regulation (EC) No 865/2006 of 4 May 2006, which lays down detailed rules for the implementation of Regulation (EC) 338/97* (Gobierno de España, n.d.). Following *Council Regulation (EC) No 338/97*, Spain incorporates CITES in its legal framework through the *Royal Decree 986/2021, of 16 November on measures to implement the Convention on International Trade in Endangered Species of Wild Fauna and Flora (CITES), made in Washington on 3 March 1973* and the *Regulation (EC) 338/97 of the Council of 9 December 1996, concerning the protection of species of wild fauna and flora through the control of their trade.*

Despite the adoption and implementation of CITES, the illegal wildlife trade is on the rise (TRAFFIC, 2024; WWF, 2018). As Fajardo (2016a) explains, this might respond to the fact that this is one of the most lucrative illicit markets because the penalties it carries are very low compared to other market crimes. Currently, it is estimated to generate between US$10 and 20 trillion annually, not including timber trafficking

[1] CITES manages trade solely for species listed in one of its three Appendices, which are categorized based on evaluations of extinction risk and the impact of trade (CITES, n.d.-a). Appendix I covers species that are at risk of extinction, permitting trade only in exceptional cases; Appendix II includes species that require regulated trade to avoid threats to their survival; and Appendix III consists of species that are protected in at least one country (CITES, n.d.-a).

and illegal fishing (Wyatt, 2022). If illegal logging and fishing are included, it is estimated to generate between US$48 and 216 trillion annually (Wyatt, 2022). Nevertheless, it should be noted that the figures could be higher, as there are fewer resources available, and personnel involved in detecting the illegal trade of wildlife (Wyatt, 2021). Consequently, the "dark figure" may be larger as it is considered a low-priority crime (Wyatt, 2022; Fajardo, 2016a; WWF, 2018). Overall, experts refer to it as one of the largest illicit markets (Beirne, 2018; Wyatt, 2022). Although there is no concrete evidence of the exact ranking, the World Bank (2019) equates the economic value of wildlife trafficking to that of drugs and human trafficking.

The Rationale of This Research[2]

The illegal trade in flora and fauna affects a wide variety of species and regions of the planet. However, while criminological research on this illegal market is growing, the study of wildlife trafficking has largely focused on charismatic megafauna in Africa and Asia (see Hübschle, 2017; Moreto, 2015). Thus, with some exceptions (see for example Petrossian et al., 2016; Pires & Marteache, 2024), neglecting the role of the European Union and other countries in the Global North and ignoring those less-charismatic species and spaces (Wyatt, 2022; Wyatt & Arroyo Quiroz, 2018; Arroyo Quiroz & Wyatt, 2019). Still, Fajardo (2016b), WWF (2018), and, more recently, TRAFFIC (2024) highlight Spain's prominent role in the global market. As WWF (2018) claims, Spain plays a fundamental role due to its strategic location, bordering North Africa, its commercial position with numerous important ports and airports, and its historical ties with North Africa and Latin America linked to colonialism and post-colonialism (see also Goyes, 2019).

In Spain, empirical research providing an overview of wildlife trafficking is limited. Fajardo (2016b) and WWF (2018) released the first reports

[2] This contribution is based on a paper previously published in Spanish at the Revista Española de Investigación Criminológica (see Pons-Hernandez, 2023). It has been translated and adapted for inclusion in this volume.

of this nature, but these do not incorporate data after 2015; Arroyo Quiroz et al. (2021) explore wildlife trafficking between Mexico and Spain from an eco-global criminology perspective; and León Garrido (2022) approaches this phenomenon at a global, European, and Spanish level, evaluating its security implications. More research has been done on the illegal trade of specific species in Spain. Salinas et al. (2011) provide evidence of illegal trade of the Mediterranean spur-thighed tortoise in Southern Spain and Northern Africa; Auliya et al. (2016) examine the role of the European market in the trade in live reptiles and highlight Spain as one key entry port; van Uhm (2016b) evidences the high number of seizures in Spain of Barbary macaques coming from Morocco and, to a lesser extent, Algeria; and Alonso and van Uhm (2023) show the interrelation between the legal and illegal market for European eel.

Although the importance of Spain in this transnational crime has been emphasized and the existence of this illicit market in Spanish territory has been evidenced, updated information on its causes, consequences, and future considerations is limited. Accordingly, this chapter aims to explore wildlife trafficking in Spain. With this goal in mind, the text aims to answer the questions: *What is Spain's role in the supply chain? What are the most trafficked species in Spain and what is their purpose?* To answer the research questions, the next pages present the research methods and results and offer a conclusion to highlight future research avenues.

Research Methods

Responding to the exploratory nature of this research, I conducted a qualitative content analysis to provide an overview of wildlife trafficking in Spain. The data upon which this analysis is based primarily come from press releases detailing the operations carried out by the Nature Protection Service (SEPRONA in its Spanish acronym) of the Guardia Civil from 2017 to 2021. The information was provided by the institution itself and included the operations related to offences against flora and fauna. To offer more detailed insights into species trafficking in Spain, however, has been taken into consideration data from the Wildlife Trade Portal (WTP) and information extracted from governmental and non-governmental reports, academic articles, and journalistic reports. WTP is an open-source tool by TRAFFIC that provides information about incidents and seizures. Access to the database was requested on February 28, 2022, and access was granted by the organization on March 3, 2022. The data from WTP covers incidents in Spain from January 1, 2003 (the start date of records) to January 1, 2023.[3] The search returned 103 results, of which 96 were qualitatively analysed. The seven results that haven't been included in the analysis are cases related to wildlife crimes but not specifically about the illegal trade. Most of them were related to human–wildlife conflicts in rural areas in Spain.

Illegal Wildlife Trade in Spain: A Country of Source, Transit, and Demand of Wildlife

Research on wildlife trafficking, similar to that related to other illegal markets, is commonly divided into the source, transit, and destination countries, known as the supply chain (Mackenzie, 2020). Traditionally, the origin country is often a developing nation that, unable to compete in the global economy through legal means primarily occupied by multinational corporations, participates by fulfilling the demand for illegal 'goods', whether they be humans, drugs, weapons, or wildlife species (Shelley, 2019). Transit countries serve as crucial trade hubs, with places like Hong Kong playing significant roles. In terms of demand, it's established that North America, Europe, and Asia are leading the consumption of wildlife and wildlife products (McMurray, 2009; Wyatt, 2022).

WWF (2018) categorized the data based on the trafficked species; here, however, the results are presented in terms of Spain's role in the supply chain. In other words, first are discussed the operations to hinder wildlife trafficking in which Spain was the source country. This is followed by the cases in which Spain was a transit country, and, finally, we present those in which Spain was the destination of the trafficked

[3] WTP provides valuable information on wildlife seizures, but it has limitations. One significant issue is the incomplete coverage, as not all seizures are reported or included in the platform. Consequently, relying solely on the data provided by the WTP may not fully capture the true scale of the illegal wildlife trade, potentially hindering comprehensive understanding.

wildlife. This approach underscores that, as indicated by data from SEPRONA and WTP, Spain simultaneously functions as a country of origin, transit, and destination for trafficked flora and fauna species. In this vein, Spain's endemic species are victims of trafficking, Spain serves as a hub for wildlife trafficking, and it's also a country with a high demand for trafficked species (Table 1).

Table 1 Main seizures of wildlife in Spain

Date of Incident	Species	Discovery location	Method of concealment
01/04/2022	1090 taxidermy animals[a]	Valencia	Found in a private warehouse
01/11/2020	100 kg European eels	Madrid (airport)	In freight
19/07/2020	280 exotic birds	Murcia, Alicante, Granada, Asturias	–
30/04/2020	722 kg European eels	Murcia, Castilla y León, Barcelona	Passenger luggage (airport) & vehicle
23/03/2020	20 kg gooseneck barnacles	A Coruña (Galicia)	Found in a bag hidden in bushes
01/03/2020	25 Testudo graeca	Puzol (Valencia)	Private house
01/12/2019	195 birds	Teruel (Aragon)	Concealed in the trunk of a vehicle
01/12/2019	80 kg European eel	Irun (Basque Country)	Found in dozen suitcases
30/10/2019	26 kg European eel	Barcelona (airport) destined to Moscow	12 plastic bags in personal baggage
01/02/2019	200 stuffed animals	Alicante	Warehouse
22/08/2018	1100 tortoises	Llucmajor (Majorca)	Turtle farm
06/04/2018	460 kg European eel	Southern Spain	Air cargo and suitcases
01/03/2018	23 bluefin tuna	Port of Tarifa	Fishing vessel routine inspection
11/02/2018	310 kg European eel	Madrid (airport)	In freight
20/11/2017	70 birds (32 dead)	Pamplona (airport)	Inside conical plastic containers
21/10/2017	Four spur-thighed tortoises	Port of Almería	Inside a cardboard box found in the trunk of the car
01/09/2017	153 birds	Elche (Alicante)	9 exotic bird farms

(*continued*)

Table 1 (continued)

Date of Incident	Species	Discovery location	Method of concealment
24/08/2017	66 spur-thighed tortoises	Melilla (coming from Morocco)	Inside two boxes inside the car that was being transported by ferry
01/03/2017	200 wildlife items including products from white, rhino, elephants, lion, hippos and giraffes	Murcia	In 2 private properties
01/12/2016	1424 dry sea horses	Marbella bus station	-
01/01/2016	700 kg European eel	Madrid	Suitcases
15/12/2014	21 Tenerife blue chaffinch	Tenerife South Airport	Suitcase
28/05/2013	512,000 pieces of coral	Madrid	Jewellery stores and warehouses
01/03/2013	111 ivory carvings & carved mammoth tusk	Barcelona	Auction house
17/10/2012	85 birds (40 dead)	Madrid (airport)	Personal baggage
01/02/2012	959 caiman tails	Spain	Incorrect documents
01/03/2010	24 tusks	Barcelona	In premises
02/07/2004	2860 kg of ivory	Madrid	-

Source: WTP (2003–2023)

[a]Not all listed: elephant, rhinoceros, polar bear, cheetah, panther, lion, lynx, crocodile, sable antelope, Bengal tiger, elephant tusks (198), oryx

Spain's Native Species in Global Wildlife Trafficking

Spain is rarely conceived as an origin country for wildlife trafficking (see TRAFFIC, 2024). As Hutchinson (2023) explains, this could be attributed to the lack of research stemming from society's limited interest in conserving uncharismatic species. Nevertheless, Spain does serve as an origin country for certain trafficked species (see, for example, UNODC, 2020). According to data from WTP, Spain stands out as the origin country for European eels (*Anguilla anguilla*), and, to a lesser extent or sporadically, gooseneck barnacles (*Pollicipes pollicipes*), spur-thighed tortoises (*Testudo graeca*), noble pen shells (*Pinna nobilis*), and Tenerife blue chaffinches (*Fringilla teydea*).

Although not explicitly reflected in the WTP data, WWF (2018), Fajardo (2016b), Arroyo Quiroz and colleagues (2021), and the Guardia Civil (2017) also point to Spain's role as the origin country for illegally captured and traded birds of prey (Image 1). These birds of prey victims of illegal trade are used in falconry, by breeders, or to certify the legal origin of birds traded illegally. Moreover, media outlets reported on this issue. As reported in the media (see, for example, Gazdagi, 2023; La Vanguardia, 2023; Segre, 2023; Nació Digital, 2023), between 2011 and 2014, employees of Roc Falcon stole falcon eggs during the breeding season to sell them online. To sell these illegally sourced specimens, the accused individuals forged administrative certificates, rings to identify the species, and individualized bird documentation. As reported, the

Image 1 A baby falcon, a victim of illegal bird of prey trade in Spain. (Source: M.A.P. (SEPRONA), permission to use granted on the 26th of January 2024)

accused have also been using certificates that belonged to deceased birds to launder the birds of illegal origin (La Vanguardia, 2023). These highly desirable specimens can fetch up to 10,000 euros on the market. Nowadays, due to the increasing use of breeding centres to mate birds of prey, the illegal abduction of specimens from the wild is largely carried out by breeders using these birds to avoid inbreeding, or, to put it differently, to renew their breeding lines.

However, at the outset, it is important to highlight the trade of European eel (*Anguilla anguilla*) (Image 2). The trade in European eels poses a key environmental threat in the European Union (EUROPOL, 2022). Although WWF (2018) did not find substantial evidence of European eel trafficking in Spain due to a lack of data during the period covered by its study, currently available data from press releases by the SEPRONA and WTP demonstrate that Spain plays a fundamental role in this traffic. The United Nations Office on Drugs and Crime (UNODC, 2020) points out Spain as the country with the most European glass eel seizures.

As SEPRONA informed, in 2017, 23 individuals were arrested, and around 1200 kg of eels were confiscated as part of *Operation Abaia*. The final destination for these eels was Hong Kong, and part of the

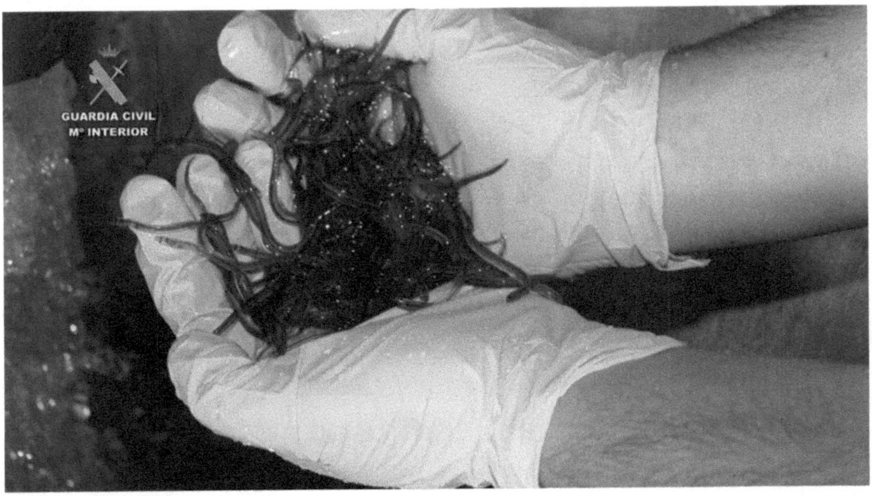

Image 2 European eels (*Anguilla anguilla*). (Source: Guardia Civil, permission to use granted on the 21st of October 2023)

transportation was carried out by air. Similarly, in 2018, the issue of eel trafficking was also highlighted. As reported by the SEPRONA, under *Operation Elver*—linked to Operation Lake coordinated by EUROPOL—a criminal network established in Spain with branches in Portugal and Morocco was dismantled. This operation resulted in 10 arrests and 500 kg of eels seized, destined for Asian markets. Similarly, in 2019, the issue of eel trafficking remained relevant. Under *Operation Fame*, 16 individuals were arrested, and 400 kg of eels valued at over 600,000 euros in the Asian market were seized. In 2020, as part of *Operation Askea*, 20 individuals were arrested, and 722 kg of eels, with a value of 1.5 million euros, were confiscated. And in 2021, under *Operation Askea 2*, the SEPRONA arrested 27 individuals and seized up to 320 kg of eels with a market value that could reach up to 650,000 euros. This marked the first occasion in which one of the detainees was imprisoned for eel trafficking.

As reported by UNODC (2020) and EUROPOL (2022), the smuggling of eels is always carried out by a criminal organization, but it can utilize two methods. On the one hand, the smuggling of live eels can be done using refrigerated containers through concealed or mislabelled air cargo, hidden as other products. This method is typically employed when the sender is a company owned by a Spanish (or European) citizen. On the other hand, when the smuggling is conducted by Chinese nationals, suitcases checked in by 'mules' on commercial flights are used—these individuals acting as mules are usually from Southeast Asia (see EUROPOL, 2022). In this case, the luggage is packed in bags with a bit of water, pressurized oxygen, and some form of cooling, typically frozen water bottles. This method proves to be very harmful to the trafficked eels. In 2018, an infrastructure for shipping over 5000 kg of eels in 364 suitcases was uncovered. It is estimated that the criminal organization's profits could have reached over 37,000,000 euros (Guardia Civil, 2018). This is, however, only the tip of the iceberg of the illegal trade in European eels.

Spain—a Hub for Wildlife Trafficking

Based on data extracted from the Guardia Civil (2019) and WTP, as well as the reports from UNODC (2020) and TRAFFIC (2019), Spain seems to act not only as a country of origin for eels but also as a transit point within the EU before they are illegally exported to a country in North Africa or Asia. As reported, eels illicitly caught in Spain are mixed with those illegally sourced in southern France or Portugal. Once gathered, these eels may be collected in Spain and then sent outside the European Union, for example, to Morocco by plane, car, or conditioned trucks (TRAFFIC, 2019; UNODC, 2020).

In addition to being an eel sourcing and exporting point, Spain is also a source and *entry* point of spur-thighed tortoises (*Testudo graeca*) (Image 3). According to WTP data and Salinas et al. (2011), there is illegal trade of *Testudo graeca* between Africa and Europe through the Gibraltar Strait. But Spain is not only an entry point for spur-thighed tortoises, as reported by the SEPRONA and reflected in WTP data, but also serves as one of the key entry points into the EU for birds, reptiles, mammals, amphibians, and their parts or derivatives. These are primarily destined for the

Image 3 Spur-thighed tortoises (*Testudo graeca*). (Source: Victor Ortega Horcajo (FIEB Foundation-CITES Rescue Center), permission to use granted on the 1st of July 2024)

(exotic) pet trade and rare and exclusive specimens are also to be sold as collectors' items. As mentioned above, WWF (2018) indicated that Spain plays a significant role in the EU due to its strategic location on the border with North Africa, its commercial situation with numerous important ports and airports, as well as its historical relationships with North Africa and Latin America linked to colonialism and post-colonialism (see Goyes, 2019, 2021, 2023; Sollund & Runhovde, 2020).

From the data available in the press releases of operations carried out by the SEPRONA in coordination with INTERPOL and EUROPOL, it can be gathered that in 2018, under *Operation Parum*, 11 individuals were investigated for primate trafficking. Specifically, 5 titi monkeys and one Midas tamarin with a total value of 12,500 euros were seized. Additionally, under *Operation Coahuila*, conducted within the framework of EUROPOL's Operation Nautilus in collaboration with Germany, France, Italy, and Austria, 1100 turtles belonging to 62 different species and over 700 eggs were seized, dismantling Europe's largest turtle breeding centre. In Spain, during *Operation Jungla V* targeting the trafficking of protected reptiles, over 600 reptiles from South America, Asia, and Oceania were seized. Among the confiscated specimens were spotted box turtles, Fijian iguanas, Australian and New Zealand monitor lizards, and chameleons. As reported by the SEPRONA, the network operated from Spain. The individuals involved conducted smuggling through 'mules', and, once in Spanish territory, they transferred the specimens to other European countries, where many were sold at animal fairs such as Terraristik in Hamm, Germany. In the context of *Operation Oratrix* in 2020, a network operating from southern Spain was dismantled. Protected birds from South American countries were imported to Spain and then transferred to Morocco, to be subsequently trafficked to Asia. As reported by the SEPRONA, the significance of this operation lies in revealing the reverse flow. In other words, while exotic species are generally introduced to the EU through Spain, in this case, species were destined for North African and Asian countries through Spain.

Thus, in accordance with WWF (2018) and the SEPRONA, Spain serves as both an entry and an exit point for species within the EU—destined among other purposes to be consumed as luxury foods (as is the case of eels) or to meet the demand for exotic species intended for collecting or the pet trade (as happens with, for example, reptiles). As established in the SEPRONA's reports, species intended for collecting and the pet trade are predominantly reptiles—in 2017, 41% of the more than 883 seized specimens were reptiles—although this can vary based on market trends. Similar to the eel trade, trafficked individuals can be introduced to the destination country or region through two methods: either utilizing legal commercial channels employing a laundering process or through smuggling (WWF, 2018).

For example, in 2019, 76 baby giant tortoises (*Aldabrachelys gigantea*) were intercepted at Barcelona Airport in a suitcase from Seychelles. The animals were transported in a container with holes and were found to be in critical health conditions, so they were transferred to the *Centre de Recuperació d'Amfibis i Rèptils de Catalunya* (Hernández, 2019). Similarly, WTP data reports that in 2017, 70 birds were seized, hidden in cylindrical plastic containers at Pamplona Airport—32 of them were dead—on a flight from Montevideo, Uruguay. As described in the SEPRONA reports and WTP, the trafficking of live birds and reptiles is very harmful as the majority of the specimens die from suffocation during the transport phase. Although these specimens arrived in Spain with the intention of being sent to other EU countries, the country also serves as a destination for these specimens.

Spain—a Destination Country of Trafficked Wildlife

Like other countries in the EU (see McMurray, 2009), Spain has a significant demand for species victim to trafficking (Arroyo Quiroz et al., 2021). As reported by the SEPRONA, in Spain there is a demand for birds, reptiles, mammals, and amphibians, as well as parts or derivatives of these species. In other words, wildlife destined for Spain can be intended for various uses, including pets, collectors' items, and even ornamental purposes. Similarly, data obtained from the WTP platform align with this focus, particularly regarding carved ivory pieces and jewellery made from coral.

As reported by the SEPRONA, in 2017, over 883 different specimens were seized, including birds, reptiles, mammals, amphibians, and 111 pieces of parts or derivatives (elephant hair, ivory, limbs, heads, skins, trophies). For example, under *Operation Suzaku*, an organization transporting birds to Spain from Africa and Latin America was dismantled (Image 4). In 2018, over 4417 different specimens were seized, including birds, reptiles, mammals, amphibians, and 681 pieces of parts or derivatives. In 2019, *Operation Celacanto* resulted in the apprehension of 2296 individuals from 70 species of protected fauna. Inspections took place at ports and airports, as well as at animal sales establishments, wholesalers, private collectors, and antique shops. In the same year, under *Operation Illuminatis*, a clandestine zoo housing African Crested Porcupines, Crowned Cranes, Rheas, Great Horned Owls, Scarlet Ibises, African Spurred Tortoises, and Livingstone's Turacos was uncovered. In 2020, 750 specimens of birds, reptiles, mammals, amphibians, and 390 pieces of their parts or derivatives were recovered. This figure is considerably lower than in previous years, which could be attributed to the expansion of the coronavirus and the mobility restrictions imposed by various governments.

Therefore, in addition to being a transit and entry point to the EU, Spain is also a destination country for trafficked species. It's worth remarking that aside from traditional methods of selling these specimens, such as fairs and pet stores, WTP data, as well as Lavorgna (2015), Fajardo (2016b), Wyatt et al. (2018), and Pacheco (2022), point to the exponential growth of online sales of these specimens. As Lavorgna (2015) states,

Image 4 Blue-and-yellow macaw (*Ara ararauna*). (Source: M.A.P. (SEPRONA), permission to use granted on the 26th of January 2024)

the utilization of the Internet has significantly impacted the pet trade, transforming it into a hybrid market that integrates both conventional and contemporary social and economic opportunities. Similarly, Wyatt et al. (2018) assert not only that live reptiles are sold at expos and pet stores but also that illegal breeders use online structures and magazines to advertise and sell species.

While most operations pertain to the trafficking of animal species, there is also evidence of cases involving plant species in wildlife trafficking. WTP does not report any cases, but SEPRONA's press releases report an operation in 2018. As indicated, under *Operation Thunderstorm*, a total of 104 individuals were investigated, and over 3852 protected cactus specimens were seized (Image 5)—along with various animal specimens. WWF (2018) also reported the existence of live plant trafficking in Spain, highlighting cacti and cycads. While orchid trafficking has been

Image 5 Protected cactus seized in Spain during Operation Thunderstorm. (Source: Guardia Civil, permission to use granted on the 2nd of February 2024)

highlighted in some parts of Latin America (see López-Tarabochia, 2017), no data on such trafficking in Spain were found in the consulted sources. Despite the limited number of cases in Spain, the academic literature suggests that plant smuggling is on the rise (see Goettsch et al., 2015; Hinsley et al., 2018; Margulies, 2023; Margulies & Bullough et al., 2019a). And, as found, similar to the case of animal species, market trends for plant species have an impact in determining which are most trafficked. Brown and Thornycroft (2021) report that the *millennial* trend of having indoor plants and sharing photos of them on social media platforms like Instagram is contributing to a growing black market for succulents from Africa, posing a threat to the survival of these species in the wild. As stated in the Botanic Gardens Conservation International (BGCI) newsletter (BGCI, n.d.), this market has been exponentially growing since 2019.

Wildlife Trafficking in Spain and Future Considerations

As demonstrated through the data collected, in Spain, the activity is present throughout the supply chain, thus at source, in transit, and at the demand end. This means that a particularity of Spain is that it plays a key role in the entire supply chain. As we have seen, at the source, the trafficking of eels destined for Asia stands out. In transit, Spain plays a crucial role in the entry of exotic species into the EU, particularly those destined for the exotic pet market and those sought after by private collectors for their rarity and exclusivity. The dynamic nature of market trends, ranging from reptiles to exotic birds and small primates, further underscores the complexity of Spain's involvement. Additionally, the high demand for exotic pet species and collector's items, including luxury foods, ivory skins, and ornaments, aligns Spain with other EU countries. If looking at the cases reported in WTP that took place during 2023 (after the data collection), one can see that the illegal trade in European eels remains rampant; the trafficking of songbirds is present, as well as of turtles. A case of turtle trafficking detected on the border between Spain and France reveals a new route to move across EU countries—via Latour-de-Carol (through the Catalan Pyrenees). In 2023, also resonated a case of monkey trafficking destined as laboratory animals, as it had been detected in a town near Tarragona (see Suazo Jiménez, 2023). Overall, Spain's distinctive role in the entire supply chain emphasizes the need for further research in the country.

On the one hand, further research about Spain's role in the global wildlife trade is essential to gain a comprehensive understanding of the ecological, socio-economic, and regulatory aspects associated with these activities. Investigation into the ecological impact is crucial to assess the consequences of trafficking on local ecosystems, biodiversity, and endangered species (van Uhm, 2016a; Wyatt, 2022). Additionally, understanding the socio-economic and cultural implications of Spain's involvement can inform policies for wildlife management and for disrupting the supply of and the continuous demand for wildlife (see van Uhm, 2018). Investigating public awareness and attitudes towards wildlife conservation may provide insights into potential strategies for reducing demand and promoting responsible consumer behaviour (see Margulies et al., 2019b). Moreover, examining the regulatory framework is imperative to identify gaps and weaknesses in current legislation and enforcement mechanisms. This includes scrutinizing Spain's measures to combat illegal wildlife trafficking and evaluating the effectiveness of existing regulations (see Alfaro Moreno, 2023). Improved regulations and enforcement strategies can contribute to mitigating the negative effects of the wildlife trade and enhancing conservation efforts. In sum, comprehensive research should delve into ecological, economic, socio-cultural, and regulatory dimensions to formulate evidence-based policies and interventions. This knowledge is crucial for developing effective strategies to address Spain's intricate role in the global wildlife trade.

On the other hand, building upon this research, and as elaborated in greater detail in Pons-Hernandez (2023), it becomes evident that the illegal trade of wildlife inflicts severe harm on trafficked species, subjecting them to instances of violence throughout the entire supply chain. The urgency to conduct further research employing a species justice approach lies in uncovering the specific and often-brutal harms these animals endure (see Fuller et al., 2018). By using a species justice approach, research would regard humans as just another species, deserving the same moral consideration as other animals and attention would be paid to animal rights and animal welfare (White, 2013). Understanding the extent and nature of the physical and psychological suffering of the victims of this crime is crucial to developing targeted interventions that address their welfare. Fuller (2018), Maher and Sollund (2016), Pons-Hernandez (2023, 2024), and Wyatt et al. (2021), among others, evidence the necessity of incorporating animal welfare protection mandates into legislation aimed at preventing and deterring this type of crime (Fajardo, 2024). By delving deeper into the intricacies of the harm suffered by trafficked wildlife, researchers can provide empirical evidence that strengthens the case for robust legal frameworks. These frameworks should not only focus on penal measures for traffickers but also integrate proactive measures to safeguard the well-being of the animals involved (see also, Pons-Hernandez, 2024; Sollund, 2022; Wyatt et al., 2021).

Detailed research can shed light on the various stages of the supply chain where animals face violence, whether during capture, transportation, or sale (Pons-Hernandez, 2023). This nuanced understanding is essential for crafting legislation that not only addresses the criminal aspects but also prioritizes the ethical treatment and protection of wildlife. In conclusion, further research is imperative to comprehensively document and understand the specific harms inflicted on trafficked wildlife. This knowledge forms the foundation for the development of legislation that goes beyond mere deterrence and actively works towards the welfare and protection of these vulnerable creatures. By incorporating such insights into legal frameworks, we can strive for more effective and humane measures to combat the illegal wildlife trade.

Conclusion

The intricate role of Spain in the global wildlife trade, as revealed through this comprehensive analysis, underscores the urgency for further research in various dimensions. Spain's unique position in the supply chain necessitates a deeper exploration of ecological, socio-economic, and regulatory aspects. Understanding the impact of wildlife trafficking on local ecosystems, biodiversity, and endangered species is vital for informed conservation efforts. Additionally, delving into the socio-economic and cultural implications can inform policies aimed at disrupting both the supply and demand for wildlife. Scrutinizing the regulatory framework is imperative to identify and address weaknesses, ensuring more effective measures to combat illegal trafficking.

Furthermore, the findings presented in Pons-Hernandez (2023) shed light on the alarming violence inflicted upon trafficked wildlife throughout the supply chain. This highlights the critical need for research to uncover the specific harms and develop targeted interventions. Incorporating animal welfare protection mandates into legislation becomes imperative, transcending punitive measures to actively prioritize the ethical treatment and protection of these vulnerable beings. In essence, this comprehensive research is pivotal for shaping evidence-based policies that combat the illegal wildlife trade while promoting the welfare and protection of wildlife.

References

Alfaro Moreno, J. A. (2023). Investigaciones financieras en delitos medioambientales, una asignatura pendiente. *Revista oficial de la Guardia Civil, 946*, 48–51.

Alonso, A., & van Uhm, D. (2023). The illegal trade in European eels: Outsourcing, funding, and complex symbiotic-antithetical relationships. *Trends in Organized Crime, 26*, 293–307. https://doi.org/10.1007/s12117-023-09490-5

Arroyo Quiroz, I., Díaz Rentería, P., & Puente Guerrero, P. (2021). Comercio transatlántico de vida silvestre: un estudio de caso entre México y España. In L. Zúñiga Rodríguez & J. Ballesteros Sánchez (Eds.), *Nuevos desafíos frente a la criminalidad organizada transnacional y el terrorismo*. Dykinson.

Arroyo Quiroz, I., & Wyatt, T. (2019). Wildlife Trafficking between the European Union and Mexico. *International Journal for Crime, Justice and Social Democracy, 8*(3), 23–37. https://doi.org/10.5204/ijcjsd.v8i3.1243

Auliya, M., Altherr, S., Ariano-Sanchez, D., Baard, E. H., et al. (2016). Trade in live reptiles, its impact on wild populations, and the role of the European market. *Biological Conservation, 204*(A), 103–119.

Beirne, P. (2018). *Murdering animals: Writings on theriocide, homicide and non-speciesist criminology*. Palgrave Macmillan.

BGCI. (n.d.). *Plant poaching is on the rise. What can we do?* Retrieved January 26, 2024, from https://www.bgci.org/news-events/plant-poaching-is-on-the-rise-what-can-we-do/

Brown, W., & Thornycroft, P. (2021). *Social media houseplant craze blamed for devastating succulent poaching across Africa*. Retrieved January 26, 2024, from https://www.telegraph.co.uk/world-news/2021/08/08/social-media-houseplant-craze-blamed-devastating-succulent-poaching/

CITES. (n.d.-a). *How CITES works*. Retrieved January 25, 2024, from https://cites.org/eng/disc/how.php

CITES. (n.d.-b). *The CITES species*. Retrieved January 25, 2024, from https://cites.org/eng/disc/species.php

EUROPOL. (2022). *Environmental crime in the age of climate change: Threat assessment 2022*. Retrieved January 26, 2024, from https://www.europol.europa.eu/publications-events/publications/environmental-crime-in-age-of-climate-change-2022-threat-assessment

Fajardo, T. (2016a). The European Union approach in the fight against wildlife trafficking: Challenges ahead. *Journal of International Wildlife Law and Policy, 19*(1), 1–21. https://doi.org/10.1080/13880292_2016.1131516

Fajardo, T. (2016b). *Wildlife Crime in Spain: In-depth analysis*. Retrieved January 26, 2024, from https://www.europarl.europa.eu/RegData/etudes/IDAN/2016/578962/IPOL_IDA(2016)578962_EN.pdf

Fajardo, T. (2024). CITES in Spain: Blueprints and challenges of Spanish practice on CITES and welfare of trafficked victims. In R. Sollund & M. Lie (Eds.), *Criminal justice, wildlife conservation and animal rights in the anthropocene*. Bristol University Press.

Fuller, G., Eggen, W. F., Wirdateti, W., & Nekaris, K. A. I. (2018). Welfare impacts of the illegal wildlife trade in a cohort of confiscated greater slow Lorises, Nycticebus Coucang. *Journal of Applied Animal Welfare Science, 21*(3), 224–238.

Gazdagi, O. (2023). *Robaban crías de halcón en Odèn y las vendían por internet a precios desorbitados: hasta 13 años de prisión*. Retrieved January 26, 2024, from https://elcaso.elnacional.cat/es/noticias/robaban-crias-halcon-oden-vendian-internet-precios-desorbitados-13-anos-prision_952500102.html

Gobierno de España. (n.d.). Convenio CITES. Retrieved January 28, 2024, from https://cites.comercio.gob.es/es-es/convenio-cites/Paginas/default.aspx

Goettsch, B., Hilton-Taylor, C., Cruz-Piñón, G., Duffy, J. P., Frances, A., Hernández, H. M., & Gaston, K. J. (2015). High proportion of cactus species threatened with extinction. *Nature Plants, 1*(10), 1–7. https://doi.org/10.1038/nplants.2015.142

Goyes, D. R. (2019). *Southern green criminology: A science to end ecological discrimination*. Emerald Publishing.

Goyes, D. R. (2021). Environmental crime in Latin America and Southern green criminology. In H. Pontell (Ed.), *Oxford research encyclopedia in criminology and criminal justice*. https://doi.org/10.1093/acrefore/9780190264079.013.588

Goyes, D. R. (2023). First, they took the land: decolonizing nature to decolonize society. In C. Cunneen, A. Deckert, A. Porter, J. Tauri, & R. Webb (Eds.), *The Routledge international handbook on decolonizing justice*. Routledge.

Guardia Civil. (2017). *La Guardia Civil lleva a cabo una operación contra el comercio ilegal y el expolio de aves rapaces en Baleares*. Retrieved January 26, 2024, from https://www.guardiacivil.es/ga/prensa/historico_prensa/6416.html

Guardia Civil. (2018). *Desarticulada una red criminal dedicada a la exportación ilegal de angulas hacia mercados asiáticos*. Retrieved January 29, 2024, from https://www.guardiacivil.es/es/prensa/noticias/6551.html

Guardia Civil. (2019). *La Guardia Civil detiene e investiga a 16 personas dedicadas al tráfico ilegal de angula europea*. Retrieved January 26, 2024, from https://www.guardiacivil.es/es/prensa/noticias/7144.html

Hernández, O. (2019). *Intervenidas 76 crías de tortuga gigante en una maleta en El Prat*. Retrieved January 26, 2024, from https://www.elperiodico.com/es/barcelona/20190410/intervenidas-76-crias-de-tortuga-en-la-maleta-de-un-pasajero-en-el-prat-7400447

Hinsley, A., de Boer, H., Fay, M., Gale, S., Gardiner, L., Gunasekara, R., Kumar, P., Masters, S., Metusala, D., Roberts, D., Veldman, S., Wong, S., & Phelps, J. (2018). A review of the trade in orchids and its implications for conservation. *Botanical Journal of the Linnean Society, 186*(4), 435–455. https://doi.org/10.1093/botlinnean/box083

Hübschle, A. (2017). The social economy of rhino poaching: Of economic freedom fighters, professional hunters and marginalized local people. *Current Sociology, 65*(3), 427–447.

Hutchinson, A. (2023). *Wildlife we love to harm: How charisma impacts conservation responses to the illegal wildlife trade in Europe*. Retrieved January 26, 2024, from https://beastlybusiness.org/2023/02/06/launch-of-the-beastly-business-policy-briefs/

IPBES. (2019). *Global assessment report on biodiversity and ecosystem services of the Intergovernmental Science-Policy Platform on Biodiversity and Ecosystem*

Services. Retrieved January 26, 2024, from https://www.ipbes.net/global-assessment

La Vanguardia. (2023). *Penas de entre 6 meses y 3 años de cárcel por criar y vender halcones robados*. Retrieved January 26, 2024, from https://www.lavanguardia.com/local/lleida/20230117/8689776/cinco-acusados-criar-vender-falcones-robados-aceptan-penas-6-meses-3-anos-carcel.html

Lavorgna, A. (2015). The social organization of pet trafficking in cyberspace. *European Journal on Criminal Policy and Research, 21*(3), 353–370. https://doi.org/10.1007/s10610-015-9273-y

León Garrido, M. (2022). Una aproximación a la problemática del tráfico de especies protegidas en la UE y España. *Implicaciones para la seguridad. Revista de Pensamiento Estratégico y Seguridad CISDE, 7*(1), 41–66.

López-Tarabochia, M. (2017). *En peligro por su belleza: el tráfico de orquídeas en el Perú*. Retrieved January 26, 2024, from https://es.mongabay.com/2017/02/trafico-de-orquideas-en-el-peru/

Mackenzie, S. (2020). *Transnational criminology*. Bristol University Press.

Maher, J., & Sollund, R. (2016). Wildlife trafficking: Harms and victimization. In R. Sollund, C. Stefes, & A. Germani (Eds.), *Fighting environmental crime in Europe and beyond* (pp. 99–128). Palgrave.

Margulies, J. D. (2023). *The cactus hunters: Desire and extinction in the illicit succulent trade*. University of Minnesota Press.

Margulies, J. D., Bullough, L. A., Hinsley, A., Ingram, D., Cowell, C., Goettsch, B., Klitgård, B., Lavorgna, A., Sinovas, P., & Phelps, J. (2019a). Illegal wildlife trade and the persistence of 'plant blindness'. *Plants, People, Planet, 1*, 173–182. https://doi.org/10.1002/ppp3.10053

Margulies, J. D., Wong, R. W. Y., & Duffy, R. (2019b). The imaginary 'Asian Super Consumer': A critique of demand reduction campaigns for the illegal wildlife trade. *Geoforum, 107*, 216–219.

McMurray, C. A. (2009). Wildlife trafficking: U.S. efforts to tackle a global crisis. *Natural Resources & Environment, 23*(3), 16–18. http://www.jstor.org/stable/40925016

Moreto, W. (2015). Occupational stress among law enforcement rangers: Insights from Uganda. *Oryx, 50*(4), 646–654.

Nació Digital. (2023). *Cinc acusats de vendre cries de falcons robades al Solsonès accepten entre 6 mesos i 3 anys de presó*. Retrieved January 26, 2024, from https://www.naciodigital.cat/naciosolsona/noticia/40952/acusats-vendre-cries-falcons-robades-solsones-preso

Pacheco Calvache, M. A. (2022). *Tráfico ilegal de vida silvestre en Internet*. Centro Universitario de la Guardia Civil: IV Curso Universitario en Investigación de la Ciberdelincuencia.

Petrossian, G. A., Pires, S. F., & van Uhm, D. P. (2016). An overview of seized illegal wildlife entering the United States. *Global Crime, 17*(2), 181–201.

Pires, S. F., & Marteache, N. (2024). Robo ilícito de secuoyas en un parque estatal de California: un enfoque de análisis del guión del crimen. *Revista Española De Investigación Criminológica, 21*(2), e829.

Pons-Hernandez, M. (2023). Tráfico de Especies Silvestres en España: una exploración desde la criminología verde crítica no-especista. *Revista Española De Investigación Criminológica, 21*(2), e815.

Pons-Hernandez, M. (2024). 'Missing the Trees for the Forest?' An analysis of the harms to European Eels caused by their trafficking and trade. *Critical Criminology, 32*, 77–95.

Salinas, M., Altet, L., Clavel, C., et al. (2011). Genetic assessment, illegal trafficking and management of the Mediterranean spur-thighed tortoise in Southern Spain and Northern Africa. *Conservation Genetics, 12*, 1–13.

Segre. (2023). *Cinc acusats de vendre cries de falcons robades d'una empresa del Solsonès accepten entre 6 mesos i 3 anys de presó*. Retrieved January 26, 2024, from https://www.segre.com/noticies/comarques/2023/01/16/cinc_acusats_vendre_cries_falcons_robades_una_empresa_del_solsones_accepten_entre_mesos_anys_preso_196244_1091.html

Shelley, L. (2019). The globalization of crime. In M. Natarajan (Ed.), *International and transnational crime and justice*. Cambridge University Press.

Sollund, R. (2022). Wildlife trade and law enforcement: A proposal for a remodeling of CITES incorporating species justice, ecojustice, and environmental justice. *International Journal of Offender Therapy and Comparative Criminology, 66*(9), 1017–1035.

Sollund, R., & Runhovde, S. R. (2020). Responses to wildlife crime in post-colonial times. Who fares best? *British Journal of Criminology, 60*, 1014–1033.

Suazo Jiménez, I. (2023). CITES, mafias y tráfico de macacos en Tarragona. Retrieved February 2, 2024, from https://www.eldiario.es/caballodenietzsche/macacos-cites-camarles-tarragona-experimentacion-animal_132_10154827.html

TRAFFIC. (2019). *TRAFFIC warns of European Eel trafficking surge as fishing season gets underway*. Retrieved January 26, 2024, from https://www.traffic.org/news/traffic-warns-of-european-eel-trafficking-surge/

TRAFFIC. (2024). An overview of seizures of CITES-listed wildlife in the EU in 2022. Retrieved January 28, 2024, from https://www.traffic.org/publications/reports/an-overview-of-seizures-of-cites-listed-wildlife-in-the-eu-in-2022/

UNODC. (2020). *World Wildlife Crime Report*. Retrieved January 26, 2024, from https://www.unodc.org/unodc/en/data-and-analysis/wildlife.html

van Uhm, D. (2016a). *The illegal wildlife trade: Inside the world of poachers, smugglers and traders*. Springer.
van Uhm, D. (2016b). Monkey business: The illegal trade in Barbary macaques. *Journal of Trafficking, Organized Crime and Security, 2*(1), 36–49.
van Uhm, D. P. (2018). The social construction of the value of wildlife: A green cultural criminological perspective. *Theoretical Criminology, 22*(3), 384–401.
White, R. (2013). *Environmental harm: An eco-justice perspective*. Oxford University Press.
World Bank. (2019). *Illegal logging, fishing, and wildlife trade: The costs and how to combat it*. World Bank. Retrieved January 26, 2024, from https://openknowledge.worldbank.org/handle/10986/32806
WWF. (2018). *El negocio de la extinción en España*. Retrieved January 26, 2024, from https://d80g3k8vowjyp.cloudfront.net/downloads/negocioextincionespana_ok.pdf
Wyatt, T. (2021). *Is CITES protecting wildlife? Assessing implementation and compliance*. Routledge.
Wyatt, T. (2022). *Wildlife trafficking: A deconstruction of the crime, the victims and the offenders*. The Palgrave Macmillan.
Wyatt, T., & Arroyo Quiroz, I. (2018). Green harms and crimes in Mexico. In I. Arroyo Quiroz & T. Wyatt (Eds.), *Green crime in Mexico: A collection of case studies*. Palgrave Macmillan.
Wyatt, T., Johnson, K., George, R., & Gunter, R. (2018). Corruption and wildlife trafficking: Three case studies involving Asia. *Asian Journal of Criminology, 13*(1), 35–55. https://doi.org/10.1007/s11417-017-9255-8
Wyatt, T., Maher, J., Allen, D., Clarke, N., & Rook, D. (2021). The welfare of wildlife: An interdisciplinary analysis of harm in the legal and illegal trades and possible ways forward. *Crime, Law and Social Change, 77*(1), 69–89. https://doi.org/10.1007/s10611-021-09984-9

Not My Snake, Not My Circus: Customary Illegality and Moral Economies in the *Bitis* Trade

Janine Heim

Introduction

In this chapter, I present an exploration of the nuances of the trade in *Bitis*, a genus of venomous African vipers, using Nicolini's (2009) 'zooming-in, zooming-out' approach. This approach allows me to scrutinize the trade practices within their immediate context and consider their broader network and influences. The *Bitis* genus, encompassing 18 species according to the Integrated Taxonomic Information System (2017, p. 214), further categorized into clades of large-bodied *Bitis* and dwarf *Bitis* (Barlow et al., 2019), has been a consistent presence in the exotic pet market despite the inherent risks associated with these highly venomous snakes

J. Heim (✉)
Department of Criminal Law and Criminology, University of Maastricht, Maastricht, Netherlands

Global Risk Governance Programme, University of Cape Town, Cape Town, South Africa
e-mail: j.heim@maastrichtuniversity.nl

D. van Uhm, D. Siegel (eds.), *Global Green Crime and Ecojustice*, Palgrave Studies in Green Criminology, https://doi.org/10.1007/978-3-031-91557-4_8

(La Laina et al., 2021; Niiteta, 2022). My study aims to contribute to a more comprehensive understanding of the global exotic pet trade, a multifaceted issue that has significant implications for wildlife conservation and law. The ensuing descriptions of the *Bitis* trade draw on two critical theoretical perspectives—'customary illegality' and 'moral economy'. These perspectives not only provide a crucial framework for comprehending the intricacies of the *Bitis* trade, but also open avenues for examining other wildlife trades.

Customary illegality refers to the engagement in unlawful activities by typically law-abiding economic players—this phenomenon can be observed in various trades, including honey, olive oil, and meat (Gregson & Crang, 2017, pp. 213–216). The moral economy, on the other hand, is a concept comprising informal morals and attitudes held by participants in the trade that greatly impact its norms and operations (Gregson & Crang, 2017). One example of customary illegality is the wide-scale, inaccurate labelling in the virgin olive oil industry (Gregson & Crang, 2017, p. 214). Most of the tested extra virgin olive oil is revealed as a blend of various oils, reflecting the prevailing moral economy in which this practice is accepted in the trade, with Italy even having reclassified this activity as a misdemeanour (Gregson & Crang, 2017, pp. 213–214). In essence, customary illegality and moral economy consider practitioners' lived experiences and attitudes under a range of laws, informal norms and morals.

The *Bitis* trade often defies straightforward binary divisions into legal and illegal, necessitating a more nuanced understanding, partly due to the legal frameworks around the trade. Generally, the trade in both wild-caught and captive-bred *Bitis* is legal in Europe (Eurogroup for Animals, 2023; Gabehart et al., 2024; Gebhardt-Brinkhaus, 2023). More specifically, however, this legality is contingent on the specific species, its conservation status and the regulations of the country of origin. For instance, South Africa, where 12 of the 18 *Bitis* species are found, restricts exports (Wyk, 2023) with regulations, however, varying from state to state as KwaZulu-Natal can issue permits for the export of indigenous species (*Government Gazette Staatskoerant*, 2023; Marais, 2023). The European Union (EU), however, does not have a law equivalent to the Lacey Act of the United States, a 1900 act that restricts the import, transportation, and sale of illegally acquired wildlife from other nations (Amidzich, 2021). This discrepancy can lead to the legal import of illegally exported *Bitis*. Only the Kenyan *Bitis worthingtoni* is included in Appendix II of the Convention on International Trade in Endangered Species of Wild Fauna and Flora (CITES), an international agreement protecting endangered species, necessitating international export permits ("Wildlife Trade on the Internet | CITES," 2023). Regulations on keeping *Bitis* as pets differ greatly in Europe, both within and across countries (Eurogroup for Animals, 2020; Gebhardt-Brinkhaus, 2023). Germany, for example, has a federal system where some states forbid the keeping of venomous animals, others only require registration, and others require permits which can depend on regional governments (Gebhardt-Brinkhaus, 2023), while Switzerland has a national permit system in place (Schweizerische Bundesrat, 2008).

My research evaluates the 'zooming-in, zooming-out' method, typically used in organizational and management studies, as a tool to study wildlife trades such as the *Bitis* trade. This approach considers the context, relationships and socio-material activities that facilitate any given practice. It does this by observing the actions of individual practitioners involved in a specific practice, for instance, trading *Bitis*, and how these individual practices interlink to create a broader network (Nicolini, 2009). Through this method, I found that when instances of legal breaches occur in the *Bitis* trade, they are not mere individual transgressions of formal rules; they are influenced by prevailing customs and norms within the trade. This moral economy within the *Bitis* trade directly influences observable instances of customary illegality. The presence of this moral economy is particularly important considering the predominantly informal nature of the *Bitis* trade, coupled with the complicated legal landscape outlined above. While this chapter does not evaluate the impact of these practices on the conservation status of *Bitis* snakes, it does delve into attitudes surrounding this issue. This research thus adds to our comprehensive understanding of the intricacies inherent in the wildlife trade, thereby enriching scholarly dialogue on this multifaceted topic.

Methodology

The zooming-in, zooming-out methodology, as formulated by Nicolini (2009), underscores that practices are not standalone occurrences; rather, they are constructed by participants and interlinked, thus forming a complex network (Nicolini, 2009). I leveraged this methodology to explore practices in the *Bitis* trade within their immediate context (zooming-in), while simultaneously examining their broader network and the influences and impacts they produce (zooming-out). Nicolini (2009) suggests this continuous interplay between close examination of specific practices and their broader implications is crucial in understanding the interconnectedness between various practices. In the case of *Bitis*, this means observing the practitioners of the trade, observing their actions and following their interconnectedness to learn about the components that facilitate the *Bitis* trade.

For my research, I tailored the zooming-in, zooming-out methodology to focus on three areas: examining the links across virtual and physical spaces; investigating the trade as a holistic network without isolating its legal or illegal aspects; and applying a multitude of theoretical perspectives to understand the practice of *Bitis* trading. My observations are drawn from data collected from August 2021 until December 2023, including online observations through netnography in social media, attendance at international reptile trade fairs, and interviews with key stakeholders in the *Bitis* trade in Europe. Throughout the process, my identity as a researcher was fully disclosed to all participants and informed consent was obtained for all formal interviews. In the case of any form of a closed group on social media, I announced my presence to moderators or the group and made it clear in my profiles that I was an academic researcher, adding information on my research in posts and links.

I started my research by observing the *Bitis* community online, joining several specialized Facebook Groups and following key community members on various social media platforms which resulted in over 500 online observations. I conducted these observations across various platforms, including Facebook, Instagram, TikTok, and several forums. Although relevant content was found across all platforms, Facebook provided the most in-depth discussions around breeding *Bitis* in captivity with

Instagram featuring the second most content on *Bitis* breeding as well as the most content on *Bitis* imports and exports.

Via my online observations, I identified and attended international reptile trade fairs and smaller regional gatherings. These events were selected based on their prominence within the online reptile community and included Terraristika in Hamm, Germany, which is one of the largest international reptile trade fairs. This was followed by visits to Terraria and Snakeday in Houten, Netherlands, both of which are significant events in the herpetological calendar that attract reptile enthusiasts across Europe and the world. Lastly, the fieldwork includes a visit to Reptilienbörse Dreiländereck in Tettnang, Germany, another reptile trade fair that is, however, much smaller and more provincial, despite its location at the borders of Austria and Switzerland. A small amount of research was conducted at the archive of the Royal Zoological Society of London for historical information on the *Bitis* trade not found elsewhere.

Additionally, I conducted nine formal, semi-structured interviews with key stakeholders in the *Bitis* trade, including keepers, breeders, veterinarians and service providers, with most interviewees claiming many of these titles at once. Informal conversations were also held with a broader range of individuals involved in the trade, which included more keepers, breeders and veterinarians as well as workers at reptile sanctuaries, biologists, conservation experts, casual expo attendees and also other service providers. These interviews and conservations served as a valuable source of context for observations I had made online and provided different contexts, insights and opinions on the online and in-person observations I made.

To systematically code and organize the data I collected, a coding system was employed through the MAXQDA software. Additionally, traditional highlighters and sticky notes corresponding to the coding system in MAXQDA were used for annotating the material artefacts collected. After familiarizing myself with the data and going through multiple coding and re-coding stages, I grouped the codes into actions that I describe as the signs of customary illegality, as well as another group that I propose make up the implicit beliefs that form the moral economy of the *Bitis* trade.

Customary Illegality

Applying the concept of customary illegality was key to understanding the various practices I observed in the *Bitis* trade. The practices in the *Bitis* trade frequently challenge the clear binary categorizations of legality: while I would come across practices related to the *Bitis* trade that broke regulations or operated in a grey area, the majority of practices appeared to be legal. These tangible practices I observed, which sometimes clashed with regulations or laws, correspond to what I identified as customary illegality in the *Bitis* trade.

I found in archival records from the Zoological Society of London that the European *Bitis* trade is not a recent phenomenon, nor is its customary illegality. Memos and letters show that the London Zoo acquired *Bitis* species through a Southern African expedition in 1943—despite a war-related government ban on the keeping of venomous animals. However, most community members I talked to trace the origin of the current pet trade in Europe back to the 1980s, with some acknowledging that the import of some *Bitis*, such as Dwarf *Bitis*, was done without export permits. Publications outlining successful captive breeding across the world, including Europe, first emerged in the following decade (Egan & Grant, 1993; Fleck, 1999; Haagner, 1990; Rosi, n.d.).

Today, according to information available online, nearly all *Bitis* species, with the exception of *Bitis harenna*, *Bitis heraldica*, *Bitis inornata* and *Bitis albanica*, are now housed in zoos across the globe ("Zootierliste," 2024). I was able to confirm all *Bitis* to be present in the exotic pet market, with the apparent exceptions of *Bitis heraldica* and *Bitis harenna*. The breeders I spoke to emphasized the considerable time, money, and effort needed to rear young *Bitis* snakes. Hence it is important that breeding is fuelled not necessarily by the economic value of the *Bitis* species, but also by the challenge and achievement of successful breeding. This process demands significant effort, especially for novice breeders, suggesting that successful breeding is a skill earned and highly esteemed within the community. Indeed, the financial rewards of *Bitis* breeding in Europe can be unpredictable, dependent on the species bred and the breeder's experience and connections. First-time breeders often find it challenging to sell larger *Bitis* snakes at expos. In contrast, more seasoned breeders with a strong network tend to adapt their breeding strategies to meet market demands and rarely struggle to sell their snakes.

My research also revealed practices that carry the appearance of illegality but might not necessarily be illegal, showing the difficulty in accurately interpreting some observations of the practice. One such observation is the trading of animals in parking lots at large expos. While NGOs have raised concerns about the potential illegality of the animals traded in these transactions (Hruby, 2019), the individuals I interviewed stated that while this may be the case, it was also the case that some buyers preferred this method due to fears of disease transmission inside the expos. This illustrates that actions that may seem illegal are not necessarily so. Despite this, there was also an acknowledgement of the possibility of covert illegal trading at these large international events, highlighting an acceptance that some degree of illegality at such international events was inevitable.

Speculation was also rife within the community itself regarding the legality of some related practices, such as field herping activities. Herping, i.e. observing snakes in the wild, was considered vital to many keepers and breeders to understand the habitat the animals require. There was speculation that some used these expeditions to illegally relocate *Bitis* for photographic purposes or to enhance their tours by guaranteeing encounters which could also, however, be explained through notable experience with locating *Bitis*. On the other hand, some herpers raised concerns about tourists possibly collecting reptiles, with some herpers recounting instances of what at least appeared like an illegal collection of protected *Bitis* from South Africa and Namibia.

Another concern was the involvement in the *Bitis* trade of individuals who had previously been convicted of wildlife trafficking. The mention of these individuals elicited varying responses from members of the community, ranging from indifference to unease and disapproval. The *potential* for illicit activity appeared to contribute to an atmosphere of speculation *within* the community but also *about* the comparatively small *Bitis* community on the part of the larger reptile community.

This speculation does carry over to the snakes themselves, with people expressing different views on captive-bred or wild-caught *Bitis*. Based on the documentation of the breeding process, from mating to birth, by several prolific breeders on Facebook groups and Instagram, it appears that many *Bitis* specimens I saw later available online or at expos in Germany and the Netherlands were bred in captivity. This, however, does not conclusively eliminate the presence of wild-caught *Bitis* in the market. I observed numerous legally exported species of wild-caught *Bitis*, such as *Bitis rhinoceros*, *Bitis gabonica*, *Bitis nasicornis* and *Bitis arietans*, an observation that generally aligns with research on exporting countries (Jensen et al., 2019; Warner, 2009). The venomous reptile enthusiast community in general appears to be quite resourceful in obtaining wild-caught specimens, often sharing contacts for import and export from African countries, including those for *Bitis*.

However, besides these legally wild-caught *Bitis*, I also noted online discussions reporting signs of *Bitis caudalis* being collected from areas of South Africa and Namibia, despite export prohibitions in these regions (Ministry of Environment and Tourism; Wyk, 2023). There was also evidence of a *Bitis schneideri*, a species from the border region of Namibia and South Africa, carrying specific parasites indicative of their wild-caught status. Although there are legal restrictions in Ethiopia preventing the export (Legal Atlas, 2024), *Bitis parviocula* collection from the wild and subsequent export persists, inside and outside Europe, with some stating online they regularly import them. Since 2008, suggestions have emerged online asserting the successful private breeding of *Bitis parviocula* in the United States. However, these assertions have sparked considerable debate within the *Bitis* community regarding their authenticity. Several individuals have speculated that these snakes were illegally exported from Ethiopia rather than being legitimately bred. However, in 2013, captive reproduction of *Bitis parviocula* was documented by Maritz et al. (2013). It was also successfully bred in the London Zoo in 2022 (Kane et al., 2022), with the parent animals originating from a private seller. Unless the animals display clear signs of being wild-caught, such as the scars a new owner noted on his *Bitis Arietans*, their provenance is not apparent and sometimes a topic of speculation among the community.

Other observations were more clear-cut in terms of legality. Illegal aspects mostly relate to sales approaches and were the most visible to me on platforms such as Facebook, where advertising *Bitis* or any wildlife for sale violates the platform's policies ("Facebook terms and policies," 2024).

Sellers typically adopted a covert sales approach, using photos of *Bitis* and coded emojis to signal availability and/or inviting potential buyers to engage in private communications. Despite these practices, regulatory measures are in place. Moderators frequently reminded members of the rules, removing non-compliant posts and cautioning against open solicitation of prices with bans from the group as punishment. In informal comments, members were also cautioned against sourcing wild-caught snakes from countries where those permits would not be granted. Discussions highlighted the challenges or illegality of securing permits for certain *Bitis*, with less experienced members and newcomers often enquiring the most about the export possibilities of specific *Bitis*, such as various Dwarf *Bitis*, most of which cannot be legally exported from South Africa if they are wild-caught. Another clear breach of law occurred when a *Bitis* was sent via post in Germany instead of through an appropriate live-animal courier. These breaches, predominantly technical in nature, are generally part of an otherwise lawful transaction.

The Moral Economy

What shapes these visible actions that make up my observations of customary illegality is based on implicit, shared understandings and beliefs that make up the moral economy of the trade. This moral economy fundamentally informs the decision-making processes of practitioners of the *Bitis* trade, particularly concerning navigating the complexities of both existing and absent legislative frameworks.

In the *Bitis* and larger venomous reptile keeper community, the concept of personal responsibility is a cornerstone of the trade's moral economy. This is apparent in their reactions to breaches of rules and laws. While these violations are discussed and often called out, there is a reluctance to express judgement, suggesting a surrender to an in-group culture and morality. For instance, individuals often appended defusing statements such as '*Just saying, do what you want*' to their criticism of rule violations. In instances where veterinarians part of this community had to report clear violations, they stated they were cognizant of the potential negative impact on their standing within the group. Evidence of the importance of personal responsibility was also found in material artefacts related to the *Bitis* trade, such as a snake hunting guidebook I bought at one of the expos which states: "Keep in mind that collecting a snake removes the animal from the population much the same manner as running over it with your vehicle does. This does not mean that you should not collect the snake, but it is best to think the process through prior to collection" (Bird & Peak, 2009, p. 5). Overall, members of the *Bitis* community were hesitant to reprimand people for collecting snakes, whether legal or illegal, and expressed this was something each individual had to figure out for themselves and take responsibility for.

Despite the hesitance to reprimand individuals for collecting snakes, there is a strong ethic of environmental protection and conservation within the *Bitis* community. News on the conservation of *Bitis* and other reptiles is often shared within the community, sometimes leading to expressions of outrage at environmental damage. Occasionally, calls for donations to conservation efforts also appeared within the broader reptile keeper communities. This focus on conservation may seem counterintuitive given that it involves thinking beyond individual desires (Nageotte et al., 2022), such as acquiring wild-caught snakes or the offspring of wild-caught snakes which are ultimately all *Bitis*. However, the general consensus within the community is that *Bitis* enthusiasts are not significant contributors to illegal collection activities, although they acknowledge this as a problem in other wildlife trades. The community also tends to view small-scale poaching as negligible, provided the snakes are given superior enclosures that are deemed safer and more favourable than their natural habitats. This belief is linked to the community's strong stance on the welfare of *Bitis*. While some publications highlight the collective harm caused by wild-caught animals to both animal welfare and conservation efforts (Eurogroup for Animals, 2023), these two concepts are not directly related. Conservation generally emphasizes protecting species in their native habitats, whereas animal welfare prioritizes the well-being of individual animals (Greggor et al., 2018). In a legal context, animal welfare laws sanction the ownership of animals but also assert obligations on the owners to ensure their animals' welfare (Boniface, 2016). While many jurisdictions have recognized animals as sentient beings that can have emotions and feel pain, the prevailing legal view largely treats them as property and does not bar all types of harm (Boniface, 2016). The conservation and ecological approaches to animal welfare examine the effects of wildlife trade on species populations, genetic diversity, and overall ecosystem health (Paquet & Darimont, 2010; Sekar & Shiller, 2020). Some conservation scientists defend causing harm to individual animals if it serves research or species preservation (Sekar & Shiller, 2020). Researchers like Sollund (2019) argue that animals are the principal victims of the wildlife trade, asserting that there is always harm when they are traded as commodities. This criticizes prevailing laws and regulations, asserting that they might inadvertently allow or even cause more harm than protection to animals (Sollund, 2019).

In the *Bitis* community, there's a strong emphasis on prioritizing individual animal welfare, possibly because this is an immediate and tangible concern for breeders, the most prolific of which are often based in Europe away from *Bitis'* natural habitat. Despite the majority opinion not considering the trade as inherently harmful, there are varying views on wild-caught species. The community believes that if correct transportation methods are used, it's possible to prevent animal harm. The concern with individual animal welfare was echoed in the community's strong disapproval of keeping animals in substandard conditions or delaying veterinary visits. Interviewees also often suggested high prices result in greater value placed on the reptiles and hence their welfare, while mass-produced, inexpensive animals are seen as more prone to mistreatment due to their easy replaceability. Notably, seasoned breeders asserted they wouldn't give snakes to people unable to guarantee their welfare, which often necessitates costly equipment. Many European keepers and breeders have additional concerns and standards for the care of *Bitis* snakes that surpass the law's requirements. These extra measures include advanced UV lamps, better quality substrates as well as different types of heaters.

The community's attitude is also shaped by the breeding success of *Bitis* species in Europe. The successful breeding has been documented online, theoretically demonstrating both the feasibility of captive breeding and its potential to mitigate the demand for wild-caught specimens. My observations, although not exhaustive as they are restricted to posts during the observation period, include the successful breeding from mating to the birth of *Bitis armata*, *Bitis arietans*, *Bitis atropos*, *Bitis caudalis*, *Bitis cornuta*, *Bitis nasicornis*, *Bitis rhinoceros*, *Bitis rubida*, *Bitis schneideri*, *Bitis worthingtoni*, and *Bitis xeropaga*. As mentioned, *Bitis parviocula* was also successfully bred in the London Zoo during this time (Kane et al., 2022). However, it is important to note that reducing the demand for wild-caught specimens was rarely the explicit motivation for breeding efforts. I observed both instances that affirmed and challenged the proposition that captive breeding can serve as a protective measure for wild species. Specifically, I observed instances of wild-caught *Bitis gabonica* and *Bitis rhinoceros*, commonly known as Gaboon viper and West African Gaboon viper, being imported into Europe, a practice that is generally considered lawful given the proper paperwork is completed in the countries of origin. Despite the reputation of *Bitis gabonica* and *Bitis rhinoceros* as impressive 'showpiece' snakes, I was informed that they are not bred as frequently in Europe today. I indeed did not spot them as frequently as Dwarf *Bitis* at European expos, potentially lending credence to the theory that a decline in breeding might be a contributing factor to an increase in wild-caught snakes. Among the keepers and importers, I noticed a particular interest in improving the breeding success of *Bitis paviocula*. This interest was driven by the fear that the species might be phased out from the pet trade if breeding outcomes do not improve.

In contrast, *Bitis schneideri*, a dwarf adder species, displayed evidence of wild-caught origin through parasites. This was although well-known breeders have documented its successful breeding online, leading to the implication that wild collection of the species is ongoing. Similarly, I found multiple online accounts of herpers observing the collection of other Dwarf *Bitis* in the wild, even though captive-bred specimens were available. I also observed signs of wild-caught *Bitis arietans*, even though this species is bred successfully across Europe and the United States. Given the large number of offspring a single female *Bitis arietans* can produce, breeders have been known to sell them at discounted prices. These findings highlight the complex and seemingly contradictory dynamics at play within the *Bitis* trade practice. Overall, I observed a strong degree of acceptance towards practices in the grey area, alongside a hesitancy to criticize the actions of others.

Implications of the Findings from *Bitis* Trade

I believe that the application of the zooming-in, zooming-out method and analytical frameworks like customary illegality and moral economy provide a nuanced understanding of the *Bitis* trade and hold the potential to be applied to other wildlife trades. Despite the wildlife trade being a sizeable economy, it is rarely compared with other sectors, such as service economies with elements of illegality due to informal employment practices (Lloyd, 2020), trades such as honey and olive where origins are often disguised (Gregson & Crang, 2017), or gold mining, which sometimes operates outside the law (High, 2012). Emphasizing the need for more nuanced research becomes critical when the illegal facets of the trade have drawn significant academic interest (Anagnostou & Doberstein, 2022). The illegal wildlife trade has been likened to drug and weapons trades and even implicated in terrorist activity, the latter of which was later shown to be anecdotal (Anagnostou & Doberstein, 2022; Bolton, 2015; Felbab-Brown, 2018; Haenlein et al., 2016; South & Wyatt, 2011; Zucca et al., 2020). Across a range of animals from reptiles, amphibians and birds to mammals, the focus tends to lie on examining illicit activities, primarily concerning animal welfare and conservation concerns (Borgerson et al., 2019; Carpenter et al., 2005; Dobbins et al., 2020; Harrington et al., 2021; Nguyen & Jones, 2022; Peng & Broom, 2021; Siregar et al., 2023; Tudge et al., 2022). This focus on the illegal trade holds despite, as Duffy (2022, pp. 31–42) and Wyatt (2021, pp. 159–163) have noted, it being rare for entire wildlife or wildlife product trades to be illegal, even in the face of seemingly comprehensive bans. This holds particularly true in the reptile trade, as the most frequently owned pet reptiles in Europe are usually traded legally (Valdez, 2021).

I posit that examining the trade by trying to divide legal and illegal activities leads to fragmented pictures of the trade that leave large gaps in terms of how the trade really functions. An in-depth understanding of customary illegality and moral economy, as observed in the *Bitis* trade, can provide valuable insights into the dynamics of the trade and the experiences of practitioners. This research not only enriches our understanding of the *Bitis* trade but also enhances the wider wildlife trade research by shedding light on the intricacies and realities of practitioners' experiences.

Moreover, while a binary categorization of legal and illegal might on the surface provide a practical framework for analysis, the reality of such research is more complicated. Nijman et al.'s (2018) study on the Indonesian bird trade identified a degree of regulatory infringements in most transactions, while Wyatt's (2009) research on the Russian fur trade showed how illegally poached pelts infiltrated the legal market. Similarly, a study on the illegal trade in European eel highlighted the difficulty of addressing threats within the binary division of legality (Gutierrez & Duffy, 2023). Further complexities arise when wildlife products are deliberately mislabelled as illegally sourced to inflate their value, as demonstrated in the caviar trade (Van Uhm & Siegel, 2016). Adding another layer of intricacy, conflicting international laws make distinguishing between legal and illegal wildlife trades even more challenging (Broussard, 2017).

These complexities are particularly apparent in the context of the *Bitis* trade, especially in federal states like Germany, the largest buyer of live reptiles in the EU (Valdez, 2021), where wildlife laws and enforcement are fragmented (Gabehart et al., 2024). I found in my research that a separation between legal and illegal aspects is unviable, as practitioners in the trade do not operate within these binary categories. In the *Bitis* trade, for instance, the dynamics are not divided into legal and illegal trades. Instead, a network of practices exists, some of which contravene regulations or laws, but nonetheless integrate into the larger legal trade. The same individuals often adhere to the law in some areas while violating it in others for complex reasons. Even those practitioners who actively try to abide by the law must still manoeuvre the convoluted legal terrain and moral values of the broader trade and their community. Attempting to separate these elements results in a distorted view of the trade, detached from the realities experienced by practitioners.

My research on *Bitis* also underscores the necessity for more extensive research specifically into the exotic pet trade within the broader context of the wildlife trade. The ethical consumption of exotic pets, something I observed practitioners trying to grapple with in the *Bitis* trade, is a debated subject, intersecting with conflicting views on the definition of ethical practices and the morality of owning an exotic pet at all (Alexander et al., 2023; Altherr & Lameter, 2020; Eurogroup for

Animals, 2023; Noll & Davis, 2020). The ethics persist throughout the trade: unlike with other wildlife products, the appeal of exotic pets hinges on their being alive for human experience and interaction (Carr, 2017; Collard, 2020; Collard & Dempsey, 2013). The concept of animal fetishism, introduced by Collard (Collard, 2020, pp. 16–26), is particularly pertinent in this context, as these animals are commodified to satisfy the demand for exotic pets. The *Bitis* trade community, as per my observations, attempts to strike a balance between their desire for exotic pets and their preoccupation with animal welfare. An interesting paradox emerges here, where illegal activity is justified by a moral economy: The community's staunch commitment to animal welfare is exemplified in their acceptance of small-scale poaching, provided the animals are guaranteed seemingly better living conditions. This mirrors findings from studies on the illegal collection of rare orchids and antiquities, where collectors justify their actions by appealing to higher moral codes, despite legal prohibitions (Mackenzie & Yates, 2016). It further aligns with research that suggests that motivated buyers often harbour ethical self-perceptions, underpinned by the belief that pet shops are stringently regulated and that they can discern between animals bred in captivity and those captured in the wild (Moorhouse et al., 2021). The determination of what is considered acceptable or not has significant implications for policy initiatives in this field, particularly when NGOs and governments attempt to work with private keepers.

In the broader context of wildlife trade research, the *Bitis* trade offers a unique perspective on the role of personal responsibility in conservation efforts. Studies by Naito et al. (2022) and Nageotte et al. (2022) have connected a sense of personal responsibility to pro-conservation attitudes and wildlife conservation actions respectively. However, within the *Bitis* trade, the concept of personal responsibility is shaped differently. As it is interpreted through the lens of the trade's moral economy it does not exclude collection from the wild. This reflects the findings of Cetas and Yasué (2017), which asserts that internal motivations are strong drivers for wildlife conservation. My research suggests that these internal motivations are strongly influenced by the moral economy of the *Bitis* trade. This resonates with Sollund's (2019) argument for a personal and ethical responsibility towards animals, which many *Bitis* breeders and collectors would ostensibly agree with, but in the *Bitis* context, this doesn't necessarily preclude collection from the wild or trading them. This is again due to the belief that *Bitis* poaching is not widespread as the overall success of captive breeding is regarded as negating the need for large numbers of wild-caught specimens. This is very much a source of pride, with successful breeding seen as having excelled at providing the snakes with optimal conditions. Tyng this expertise back to conservation efforts is not unheard of, as the knowledge of private breeders has proven beneficial for the conservation of some species (Leiss et al., 2022; Mattioli et al., 2006; Owen et al., 2014). Concerning the success of private breeders, wider academic literature, including works by Drury (2011), Gao and Clark (2014), Geldenhuys (2022) and Nijman et al. (2023), has explored the role of prestige in wildlife consumption and the use of exotic pets as status symbols, but the prestige associated with captive wildlife breeding and how this ties back to trade dynamics, as I observed in the *Bitis* community, remains relatively unexplored.

Paradoxically, the demand for more captive-bred snakes seems to motivate the continued introduction of at least some wild-caught *Bitis* species, as it was in the case of *Bitis parviocula*, which is feared to disappear from the trade if its private captive breeding is not improved. To acquire this expertise, more wild-caught specimens might be required. This is in line with the wider academic research that provides mixed evidence on the impact of captive breeding on the trade of wild-caught animals (Fukushima et al., 2021; Nijman et al., 2018; Rizzolo, 2021). Regarding snakes, most research, however, focuses on farming for snake derivatives such as skin and venom, rather than the sale of snakes as pets. D'Cruze et al.'s (2020) study on West African python farms and Nijman's (2022) study on North Sumatran blood pythons, both of which are often kept as pets, indicate misuse of harvesting quotas. Research on venomous snake farming is also scant; however, Hossain et al.'s (2022) Bangladesh study pinpointed issues like poor record-keeping, improper enclosures and care as well as the dominant use of wild-caught snakes. The author recommends conducting surveys to ascertain the purpose of these farms and what they breed, with one potential category being pets (Hossain et al., 2022, p. 129). Despite the high rates of unintentional human contact resulting in bites from *Bitis arietans* (Farooq et al., 2022; Shah et al., 2022), and recurring antivenom shortages across Africa (Dalhat et al., 2023; Shah et al., 2022), there is little research into farming and breeding *Bitis* for antivenom production, despite the existence and necessity of these farms.

While focused primarily on Europe, my research findings on customary illegality also connect with the theme of contested illegality observed in South Africa, home to many *Bitis* species. Legal definitions, within the ambit of economic transactions that cross political borders, are often debated, especially as those involved may not agree with official state decisions on what's legal and what's not (Van Schendel & Abraham, 2005). In such contexts, legal definitions and their interpretation can be disputed, as actors may disagree with the state-sanctioned labels of legality and illegality (Van Schendel & Abraham, 2005). Such divergence can delegitimize state legislation, with alternative interpretations of legality deriving from societal or cultural norms instead (Satz, 2010). This is illustrated by the regulations of CITES. African and Asian delegates often view CITES as a tool of continued neocolonial control, arguing that it predominantly espouses a Western conservation and animal rights ethos, which fails to take into account the realities and developmental challenges faced by rural African communities (Hübschle, 2017, pp. 185–188). An example is the international rhino horn trade, banned in 1977 under CITES, with countries like South Africa allowing limited legal trophy hunting, including the export of rhino horns under strict conditions ('T Sas-Rolfes et al., 2019, p. 209; Hübschle, 2017, p. 187). In practice, however, differentiating between legal trophy hunting and poaching remains a daunting task, leading to conflicts in the international regulatory framework. Another contentious point revolves around the privatization of rhinos in South Africa, where private individuals can legally own them—a strategy framed towards conservation. However, this ownership is often limited to the economic elite, potentially exacerbating social inequalities and jeopardizing rhino conservation efforts (Hübschle, 2017, p. 179). In my upcoming research, I am planning to delve deeper into the *Bitis* trade in South Africa on the *Bitis* trade within the South African context.

Conclusion

This chapter provides a comprehensive analysis of the *Bitis* trade in Europe, based on observations collected through the 'zooming-in, zooming-out' method. The findings reveal that the trade is influenced not solely by individual participants, but also by the intricate interactions between actors and the object they trade, as well as the broader socio-economic context. The study challenges the oversimplified dichotomy of 'legal' and 'illegal' practices, instead introducing the notions of customary illegality and moral economies borrowed from economic geography to elucidate the dynamics of the trade. I found the concept of customary illegality to be instrumental in understanding the practice of trading *Bitis*, as many actions within the *Bitis* trade defy straightforward categorization into legal or illegal. Moreover, the moral economy of the trade, shaped by shared beliefs and understandings, was found to significantly inform the decision-making processes of the actors involved in shaping the trade.

The application of the constructs of customary illegality and moral economies has unveiled a nuanced perspective on the *Bitis* trade, facilitating a more holistic and empathetic understanding of the practices and conventions that shape it. These constructs allow for a deeper understanding of the motivations and behaviours of the trade participants and the influence of the broader network of the trade. Applying the concepts of customary illegality and moral economy also offers important contributions to our understanding of wildlife trade practices and their implications for conservation efforts. The study emphasizes the need for future research to continue exploring these dynamics to develop more effective conservation strategies and regulatory frameworks.

The findings of this study also underscore the necessity for a more contextual understanding of the practices involved in the *Bitis* trade. Concepts such as personal responsibility, animal welfare and conservation issues, as well as sustainability, were found to be significantly shaped by the perspectives of individual practitioners. Understanding these nuances that form the practice of trading *Bitis* is vital not just for understanding what drives the trade, but also for developing better communication with key stakeholders and more informed policy responses.

It is crucial to recognize the limitations of this study and the need for further research. While this research has provided valuable insights into the *Bitis* trade, it is not exhaustive. Future research should aim to explore other aspects of the trade and examine the transferability of these findings to other wildlife trades. The inherent complexities and nuances of the trade necessitate a continual refinement of our understanding, informed by ongoing observation and analysis.

References

T Sas-Rolfes, M., Challender, D. W. S., Hinsley, A., Veríssimo, D., & Milner-Gulland, E. J. (2019). Illegal wildlife trade: Scale, processes, and governance. *Annual Review of Environment and Resources, 44*(1), 201–228. https://doi.org/10.1146/annurev-environ-101718-033253

Alexander, S. D., Waters, S., Aldrich, B. C., Shanee, S., Clarke, T. A., Radford, L., Hansen, M. F., Gnanaolivu, S. D., & Dempsey, A. (2023). The past, present, and future of the primate pet trade. In *Primates in anthropogenic landscapes: Exploring primate behavioural flexibility across human contexts* (pp. 247–266). Springer.

Altherr, S., & Lameter, K. (2020). *Stolen Wildlife III: The EU is a main hub and destination for illegally caught exotic pets.* Report by Pro Wildlife (ed.), Munich, Germany, 40pp.

Amidzich, M. R. (2021). Stop pur-petuating the norm: Amending the Lacey Act to include a 'dangerous or potentially dangerous wildlife' definition for exotic pet protection. *Vermont Journal of Environmental Law, 24*(4), 23, 56.

Anagnostou, M., & Doberstein, B. (2022). Illegal wildlife trade and other organised crime: A scoping review. *Ambio, 51*(7), 1615–1631. https://doi.org/10.1007/s13280-021-01675-y

Barlow, A., Wüster, W., Kelly, C. M. R., Branch, W. R., Phelps, T., & Tolley, K. A. (2019). Ancient habitat shifts and organismal diversification are decoupled in the African viper genus *Bitis* (Serpentes: Viperidae). *Journal of Biogeography, 46*(6), 1234–1248.

Bird, W., & Peak, P. (2009). *Snake hunting guide II; Methods, tools, and techniques for finding snakes* (Updated Edition ed.). ECO / Serpents Tale NHBD. https://www.amazon.com/Hunting-Methods-Techniques-Finding-Snakes/dp/097889796X

Bolton, M. (2015). Using the arms trade treaty to address wildlife poaching in East Africa: A human security approach. *Contrôlez les armes*. http://controlarms.org/en/wp-content/uploads/sites/2/2015/12/BoltonPoaching-1.pdf.

Boniface, A. E. (2016). Animals: 'Objects' or 'sentient beings'? A comparative perspective of the South African law. *Journal of Advances in Humanities and Social Sciences, 2*(3), 143–155.

Borgerson, C., Razafindrapaoly, B., Rajaona, D., Rasolofoniaina, B. J. R., & Golden, C. D. (2019). Food insecurity and the unsustainable hunting of wildlife in a UNESCO world heritage site. *Frontiers in Sustainable Food Systems, 3*, 99.

Broussard, G. (2017). Building an effective criminal justice response to wildlife trafficking: Experiences from the ASEAN region. *Review of European, Comparative & International Environmental Law, 26*(2), 118–127. https://doi.org/10.1111/reel.12203

Carpenter, A. I., Robson, O., Rowcliffe, J. M., & Watkinson, A. R. (2005). The impacts of international and national governance changes on a traded resource: A case study of Madagascar and its chameleon trade. *Biological Conservation, 123*(3), 279–287. https://doi.org/10.1016/j.biocon.2004.11.015

Carr, N. (2017). Recognising the position of the pet dog in tourism. *Annals of Tourism Research, 62*, 112–113.

Cetas, E. R., & Yasué, M. (2017). A systematic review of motivational values and conservation success in and around protected areas. *Conservation Biology, 31*(1), 203–212.

Collard, R.-C. (2020). *Animal traffic: Lively capital in the global exotic pet trade*. Duke University Press.

Collard, R.-C., & Dempsey, J. (2013). Life for sale? The politics of lively commodities. *Environment and Planning A, 45*(11), 2682–2699.

D'Cruze, N., Harrington, L. A., Assou, D., Green, J., Macdonald, D. W., Ronfot, D., Segniagbeto, G. H., & Auliya, M. (2020). Betting the farm: A review of ball python and other reptile trade from Togo, West Africa. *Nature Conservation, 40*. https://ora.ox.ac.uk/objects/uuid:4eda3cbb-e875-47f2-bdb1-383ad5789219

Dalhat, M. M., Potet, J., Mohammed, A., Chotun, N., Tesfahunei, H. A., & Habib, A. G. (2023). Availability, accessibility and use of antivenom for snakebite envenomation in Africa with proposed strategies to overcome the limitations. *Toxicon X, 18*, 100152. https://doi.org/10.1016/j.toxcx.2023.100152

Dobbins, M., Sollmann, R., Menke, S., Almeyda Zambrano, A., & Broadbent, E. (2020). An integrated approach to measure hunting intensity and assess its impacts on mammal populations. *Journal of Applied Ecology, 57*(11), 2100–2111.

Drury, R. (2011). Hungry for success: Urban consumer demand for wild animal products in Vietnam. *Conservation and Society, 9*(3), 247–257.

Duffy, R. (2022). Crime, security, and illegal wildlife trade: Political ecologies of international conservation. *Global Environmental Politics, 22*(2), 23–44. https://doi.org/10.1162/glep_a_00645

Egan, V. T., & Grant, W. S. (1993). Breeding the striped puff adder *Bitis* arietans: Inbreeding avoidance. *South African Journal of Wildlife Research-24-Month Delayed Open Access, 23*(3), 78–81.

Eurogroup for Animals. (2020). *Analysis of national legislation related to the keeping and sale of exotic pets in Europe.*

Eurogroup for Animals. (2023). The current pet trade in the EU and its variation between member states | Eurogroup for Animals. https://www.eurogroupforanimals.org/library/current-pet-trade-eu-and-its-variation-between-member-states.

Facebook terms and policies. (2024). https://www.facebook.com/policies_center/commerce/animalsfiles/742/animals.html

Farooq, H., Bero, C., Guilengue, Y., Elias, C., Massingue, Y., Mucopote, I., Nanvonamuquitxo, C., Marais, J., Faurby, S., & Antonelli, A. (2022). Snakebite incidence in rural sub-Saharan Africa might be severely underestimated. *Toxicon, 219*, 106932.

Felbab-Brown, V. (2018). Wildlife and drug trafficking, terrorism, and human security: Realities, myths, and complexities beyond Africa. *PRism, 7*(4), 124–137. https://www.jstor.org/stable/26542711

Fleck, J. (1999). Bemerkungen zur Wüstenbergotter *Bitis* xeropaga HAACKE 1975 sowie Beobachtungen bei der Haltung und Zucht. *Herpetofauna, 21*(122), 5–9. https://reptile-database.reptarium.cz/species.php?genus=*Bitis* &species=xeropagafiles/732/species.html

Fukushima, C. S., Tricorache, P., Toomes, A., Stringham, O. C., Rivera-Téllez, E., Ripple, W. J., Peters, G., Orenstein, R. I., Morcatty, T. Q., Longhorn, S. J., Lee, C., Kumschick, S., de Freitas, M. A., Duffy, R. V., Davies, A., Cheung, H., Cheyne, S. M., Bouhuys, J., Barreiros, J. P., et al. (2021). Challenges and perspectives on tackling illegal or unsustainable wildlife trade. *Biological Conservation, 263*, 109342. https://doi.org/10.1016/j.biocon.2021.109342

Gabehart, K. M., Fullerton, A. H., & Stefes, C. H. (2024). Policy feedback and the enforcement of international wildlife treaties in Germany. *European Policy Analysis, 10*(1), 10–38. https://doi.org/10.1002/epa2.1192

Gao, Y., & Clark, S. G. (2014). Elephant ivory trade in China: Trends and drivers. *Biological Conservation, 180*, 23–30.

Gebhardt-Brinkhaus, R. (2023). Überblick über die rechtlichen Regelungen zur Gift- und Gefahrtierhaltung in den Ländern der Bundesrepublik. https://aspe-institut.de/pdf/RechtlicheRegelungenzuGiftuGefahrtieren_2014.pdf

Geldenhuys, K. (2022). Cheetahs as pets to the rich & famous. *Servamus Community-based Safety and Security Magazine, 115*(11), 22–26.

Government Gazette Staatskoerant. (2023). Issue 49470. https://www.dffe.gov.za/sites/default/files/legislation/2023-10/nemba_tops.listrepublication_freshwater_g49470gon3963.pdf

Greggor, A. L., Vicino, G. A., Swaisgood, R. R., Fidgett, A., Brenner, D., Kinney, M. E., Farabaugh, S., Masuda, B., & Lamberski, N. (2018). Animal welfare in conservation breeding: Applications and challenges. *Frontiers in Veterinary Science, 5*, 323.

Gregson, N., & Crang, M. (2017). Illicit economies: Customary illegality, moral economies and circulation. *Transactions of the Institute of British Geographers, 42*(2), 206–219.

Gutierrez, L., & Duffy, R. (2023). Harms and the illegal wildlife trade: Political ecology, Green criminology and the European eel. *Critical Criminology, 32*(1), 61–76.

Haagner, G. V. (1990). The maintenance of puffadders (*Bitis* a. arietans) in captivity, with notes on captive breeding. *The Journal of the Herpetological Association of Africa, 38*(1), 53–55.

Haenlein, C., Maguire, T., & Somerville, K. (2016). III. Poaching, wildlife trafficking and terrorism. *Whitehall Papers, 86*(1), 58–76.

Harrington, L. A., Auliya, M., Eckman, H., Harrington, A. P., Macdonald, D. W., & D'Cruze, N. (2021). Live wild animal exports to supply the exotic pet trade: A case study from Togo using publicly available social media data. *Conservation Science and Practice, 3*(7), e430.

High, M. (2012). The cultural logics of illegality: Living outside the law in the Mongolian gold mines. In J. Dierkes (Ed.), *Change in democratic Mongolia: Social relations, health, mobile pastoralism, and mining* (pp. 249–270). (Brill Inner Asian Library). Brill Academic Publishers.

Hossain, M. S., Saeed, M. A., Ahsan, M. F., Jaman, M. F., Chayan, H. A.-R., Hasan, S., Biswas, S., & Asaduzzaman, M. (2022). Present status, challenges and prospects of snake farming in Bangladesh. *Bangladesh Journal of Zoology, 50*(1), 121–133.

Hruby, D. (2019). The world's biggest reptile fair is also a hub for traffickers | *Earth Journalism Network*. https://earthjournalism.net/stories/the-worlds-biggest-reptile-fair-is-also-a-hub-for-traffickers.

Hübschle, A. (2017). Contested illegality. In J. Beckert & M. Dewey (Eds.), *The architecture of illegal markets: Towards an economic sociology of illegality in the economy* (pp. 177–197). (Oxford, 2017; online edition, Oxford Academic, 24 Aug. 2017), https://doi.org/10.1093/oso/9780198794974.003.0010

ITIS. (2017). https://www.itis.gov/servlet/SingleRpt/SingleRpt?search_topic=TSN&search_value=634420#nullfiles/773/SingleRpt.html

Jensen, T. J., Auliya, M., Burgess, N. D., Aust, P. W., Pertoldi, C., & Strand, J. (2019). Exploring the international trade in African snakes not listed on CITES: Highlighting the role of the internet and social media. *Biodiversity and Conservation, 28*, 1–19.

Kane, D., Tapley, B., Carter, K., & Michaels, C. (2022). Reproduction of the Ethiopian mountain adder, *Bitis* parviocula Böhme, 1976 (Reptilia: Viperidae), at ZSL London zoo. *Herpetology Notes, 15*, 423–429.

La Laina, D. Z., Nekaris, K. A. I., Nijman, V., & Morcatty, T. Q. (2021). Illegal online pet trade in venomous snakes and the occurrence of snakebites in Brazil. *Toxicon, 193*, 48–54.

Legal Atlas. (2024). Wildlife trade legal framework in Ethiopia. https://www.legal-atlas.net/Wildlife_Trade/Legal_Framework/Ethiopia?frn=1#files/789/Ethiopia.html.

Leiss, L., Rauhaus, A., Rakotoarison, A., Fusari, C., Vences, M., & Ziegler, T. (2022). Review of threatened Malagasy freshwater fishes in zoos and aquaria: The necessity of an ex situ conservation network—A call for action. *Zoo Biology, 41*(3), 244–262.

Lloyd, A. (2020). Working for free illegal employment practices, 'off the books' work and the continuum of legality within the service economy. *Trends in Organized Crime, 23*(1), 77–93. https://doi.org/10.1007/s12117-018-9351-x

Mackenzie, S., & Yates, D. (2016). Collectors on illicit collecting: Higher loyalties and other techniques of neutralization in the unlawful collecting of rare and precious orchids and antiquities. *Theoretical Criminology, 20*(3), 340–357. https://doi.org/10.1177/1362480615607625

Marais, J. (2023). *ASI Newsletter—May 2023*. Retrieved July 16, from https://www.africansnakebiteinstitute.com/news/newsletters/snake-conservation/

Maritz, B., Wallner, A., & Deans, R. (2013). *Bitis* parviocula (Böhme, 1977) reproduction. *African Herp News, 59*, 28–29.

Mattioli, F., Gili, C., & Andreone, F. (2006). Economics of captive breeding applied to the conservation of selected amphibian and reptile species from Madagascar. *Natura (Milan), 95*, 67–80.

Moorhouse, T. P., D'Cruze, N. C., & Macdonald, D. W. (2021). Information about zoonotic disease risks reduces desire to own exotic pets among global consumers. *Frontiers in Ecology and Evolution, 9*, 609547.

Nageotte, N. L., Eagle-Malone, R., & Searles, V. (2022). Uncovering zoo audience values necessary to create effective messaging about illegal wildlife trade. *Journal of Interpretation Research, 27*(2), 93–112. https://doi.org/10.1177/10925872221130319

Naito, R., Zhao, J., & Chan, K. M. A. (2022). An integrative framework for transformative social change: A case in global wildlife trade. *Sustainability Science, 17*(1), 171–189. https://doi.org/10.1007/s11625-021-01081-z

Nguyen, M. H., & Jones, T. E. (2022). Predictors of support for biodiversity loss countermeasure and bushmeat consumption among Vietnamese urban residents. *Conservation Science and Practice, 4*(12), e12822.

Nicolini, D. (2009). Zooming in and out: Studying practices by switching theoretical lenses and trailing connections. *Organization Studies, 30*(12), 1391–1418.

Niiteta, K. (2022). Clinicopathological study of snakebite injury resulting from Zebra spitting Cobra (*naja nigricincta nigricincta*) and Puff adder (*bitis arietans*) (Doctoral dissertation, University of Namibia).

Nijman, V. (2022). Harvest quotas, free markets and the sustainable trade in pythons. *Nature Conservation, 48*, 99–121.

Nijman, V., Langgeng, A., Birot, H., Imron, M. A., & Nekaris, K. A. I. (2018). Wildlife trade, captive breeding and the imminent extinction of a songbird. *Global Ecology and Conservation, 15*, e00425. https://doi.org/10.1016/j.gecco.2018.e00425

Nijman, V., Morcatty, T. Q., El Bizri, H. R., Al-Razi, H., Ang, A., Ardiansyah, A., Atoussi, S., Bergin, D., Bell, S., & Braga-Pereira, F. (2023). Global online trade in primates for pets. *Environmental Development, 48*, 100925.

Noll, S., & Davis, B. (2020). The invasive species diet: The ethics of eating lionfish as a wildlife management strategy. *Ethics, Policy & Environment, 23*(3), 320–335.

Owen, A., Wilkinson, R., & Sözer, R. (2014). In situ conservation breeding and the role of zoological institutions and private breeders in the recovery of highly endangered Indonesian passerine birds. *International Zoo Yearbook, 48*(1), 199–211.

Paquet, P., & Darimont, C. (2010). Wildlife conservation and animal welfare: Two sides of the same coin? *Animal Welfare, 19*. https://doi.org/10.1017/S0962728600001433

Peng, S., & Broom, D. M. (2021). The sustainability of keeping birds as pets: Should any be kept? *Animals, 11*(2), 582.

Rizzolo, J. B. (2021). Effects of legalization and wildlife farming on conservation. *Global Ecology and Conservation, 25*, e01390.

Rosi, R. (n.d.). Captive breeding of *Bitis* Gabonica Rhinoceros. https://snakesociety.nl/jaargangen/1988e/Litteratura%20Serpentium%208-4%20182-188%20Rosi,%20Captive%20breeding%20of%20Bitis%20gabonica%20rhinoceros.pdf

Satz, D. (2010). *Why some things should not be for sale: The moral limits of markets*. Oxford University Press.

Schweizerische Bundesrat. (2008). Tierschutzverordnung SR 455.1. https://www.fedlex.admin.ch/eli/cc/2008/416/defiles/787/de.html

Sekar, N., & Shiller, D. (2020). Engage with animal welfare in conservation. *Science, 369*(6504), 629–630.

Shah, M. M., Imam, T. S., Bala, A., & Tukur, Z. (2022). Survey of snakes bites among Snake endemic communities in north eastern Nigeria. *Snake Venom and Ecology, 47*. https://www.intechopen.com/chapters/82113

Siregar, B., Hasudungan, F., Widyasari, V., Lopes, J. P., Trainor, C. R., Jain, A., & Yong, D. L. (2023). Unsustainable exploitation of wild birds in Wallacea—An uncertain road for conservation emerging from the pandemic. *Emu-Austral Ornithology, 123*(1), 85–89.

Sollund, R. A. (2019). *The crimes of wildlife trafficking: Issues of justice, legality and morality*. Routledge.

South, N., & Wyatt, T. (2011). Comparing illicit trades in wildlife and drugs: An exploratory study. *Deviant Behavior, 32*(6), 538–561.

Tudge, S. J., Brittain, S., Kentatchime, F., Tagne, C. T. K., & Rowcliffe, J. M. (2022). The impacts of human activity on mammals in a community forest near the Dja Biosphere Reserve in Cameroon. *Oryx, 56*(6), 947–955.

Valdez, J. W. (2021). Using Google trends to determine current, past, and future trends in the reptile pet trade. *Animals, 11*(3), 676.

Van Schendel, W., & Abraham, I. (2005). *Illicit flows and criminal things: States, borders, and the other side of globalization*. Indiana University Press.

Van Uhm, D., & Siegel, D. (2016). The illegal trade in black caviar. *Trends in Organized Crime, 19*, 67–87. https://doi.org/10.1007/s12117-016-9264-5

Warner, J. K. (2009). *Conservation biology of the gaboon adder (Bitis gabonica) in South Africa*. Doctoral dissertation, University of the Witwatersrand.

Wildlife Trade on the Internet | CITES. (2023). https://cites.org/eng/news/world/19/6.phpfiles/136/6.html

Wyatt, T. (2009). Exploring the organization of Russia Far East's illegal wildlife trade: Two case studies of the illegal fur and illegal falcon trades. *Global Crime, 10*(1–2), 144–154.

Wyatt, T. (2021). *Wildlife trafficking: A deconstruction of the crime, victims and offenders*. Springer.

Wyk, S. V. (2023). *Snake conservation*. African Snakebite Institute. https://www.africansnakebiteinstitute.com/articles/snake-conservation-2/files/777/snake-conservation-2.html.

Zootierliste. (2024). https://www.zootierliste.de/en/?klasse=3&ordnung=305files/771/en.html

Zucca, P., Rossmann, M.-C., Osorio, J. E., Karem, K., De Benedictis, P., Haißl, J., De Franceschi, P., Calligaris, E., Kohlweiß, M., Meddi, G., Gabrutsch, W., Mairitsch, H., Greco, O., Furlani, R., Maggio, M., Tolomei, M., Bremini, A., Fischinger, I., Zambotto, P., et al. (2020). The 'bio-crime model' of cross-border cooperation among veterinary public health, justice, law enforcements, and customs to tackle the illegal animal trade/bio-terrorism and to prevent the spread of zoonotic diseases among human population. *Frontiers in Veterinary Science, 7*, 593683. https://doi.org/10.3389/fvets.2020.593683

Wildlife Poaching in the COVID-19 World: A Retrospective Analysis of Poaching Trends Inside a Private Nature Reserve During the COVID-19 Lockdown

Nick van Doormaal, Paul Allin, and Craig Spencer

Introduction

The COVID-19 pandemic, first identified in China in December 2019, rapidly evolved into a global outbreak. On 11 March 2020, the World Health Organization declared it a global pandemic (Carvalho et al., 2021). Governments worldwide responded by implementing restrictions, commonly known as 'lockdown' measures, to control the spread of the virus (World Health Organization, 2023a). Between March 2020 and March 2022, successive waves of infection necessitated prolonged lockdowns, compelling both citizens and governments to adapt or strengthen existing measures (World Health Organization, 2023a).

During the lockdowns, people's mobility and activity were rapidly restricted. Data collected by Apple and Google revealed that more than

N. van Doormaal (✉) • P. Allin • C. Spencer
Transfrontier Africa, Hoedspruit, South Africa
e-mail: research@transfrontierafrica.org

D. van Uhm, D. Siegel (eds.), *Global Green Crime and Ecojustice*, Palgrave Studies in Green Criminology, https://doi.org/10.1007/978-3-031-91557-4_9

half of the world's population reduced travel by more than 50% in April 2020, and mobility declined by at least 10% in almost all of the 125 countries tracked, with some countries showing a decline of 80% or more (Forster et al., 2020). These abrupt changes in human activities on a global scale offered a unique opportunity to examine the potential effects on the environment (World Health Organization, 2023b) and to provide insights into how large-scale shifts impact the trends and patterns of wildlife crimes.

So far, few studies have focused on the repercussions of the lockdown restrictions on wildlife security and poaching, likely because of disrupted or fragmented data collection during the pandemic. Some studies were conducted during the pandemic and acknowledged the constraints in data availability on poaching during these times (Behera et al., 2022; Ehlers Smith et al., 2023). For example, Ehlers Smith et al. (2023) reported that some forms of poaching, such as illegal hunting with dogs and illegal harvesting, increased during the pandemic compared to pre-pandemic years, while other illicit practices showed no significant change. Other studies took a broader perspective, examining overall impacts at a national or global level. These studies suggested that the economic downturns associated with COVID-19 mitigation policies and the collapse of international tourism during lockdown periods led to a loss of funding for wildlife protection and enforcement (Buckley, 2020; Lindsey et al., 2020; Spenceley et al., 2021). No studies, however, explored these patterns and behaviours from an environmental criminology perspective. This literature gap emphasises the need for a better understanding of the pandemic's impact on poaching behaviour, which could ultimately inform more effective intervention strategies for wildlife security in protected areas.

Poaching has traditionally been studied within the biological and wildlife conservation sciences. These studies are typically not concerned with the human dimensions of poaching and therefore criminological theories are rarely considered (Boratto & Gibbs, 2021; Kurland et al., 2017). In the last two decades, criminologists have started to focus more on environmental harms, including wildlife poaching and trafficking. Much of the research has been inductive or descriptive in nature, using case studies to describe an overarching phenomenon or to characterise participants and their motivation (Boratto & Gibbs, 2021; Kurland et al., 2017). This work has described, for example, themes in transnational wildlife crime and trafficking (Schneider, 2008; Sollund, 2016), drivers and motivations (Eliason, 1999; Rizzolo et al., 2017) and ranger and local community attitudes (Anagnostou et al., 2020; Moreto & Lemieux, 2015). Other scholars use traditional criminological theories to study criminal patterns and analyse the impacts of external variables on the offender's decision-making (Boratto & Gibbs, 2021; Kurland et al., 2017). The bulk of this research uses approaches from environmental criminology. Environmental criminology aims to understand and explain crime patterns as a function of opportunity structures that vary across time and space. Understanding the role of environments and being aware of the way that crime is patterned can ultimately help to design crime prevention strategies (Cornish & Clarke, 1986). Poaching problems can also be viewed through this lens to better understand crime patterns and trends. For example, these approaches have been successfully applied to study online wildlife trade (Lavorgna & Sajeva, 2021; Pires et al., 2021), target selection of poachers (Kurland et al., 2018; Petrossian et al., 2015; Skidmore, 2021) and patrol strategies (Hill et al., 2014; Moreto et al., 2014). More recently, some have also looked at the utility of situational crime prevention, for example through the use of technology (Aransiola & Ceccato, 2020), law enforcement strategies (Delpech et al., 2021; Hill et al., 2014), and community-based interventions (Delpech et al., 2021; Viollaz et al., 2022).

Study Aim

In this study, we quantitatively analysed poaching patterns and trends using data collected by patrolling rangers in a private nature reserve in South Africa before, during, and after the COVID-lockdown restrictions. First, we outline the different lockdown restrictions implemented in South Africa. Next, drawing on insights from the rational choice perspective, we describe our hypotheses concerning the potential impact of these restrictions on poaching activities within the reserve. We focused on rhino poaching and bushmeat poaching because patrol efforts within the reserve were mainly directed towards these two types of poaching. Through a comparative analysis of poaching patterns before, during, and after the lockdown period, we interpret the results from an environmental criminological perspective, allowing us to make informed assumptions and inferences about the decision-making processes of offenders.

Lockdown Restrictions in South Africa

In response to the global COVID-19 pandemic, the government of South Africa implemented a series of lockdown restrictions to mitigate the spread of the virus and protect public health. These measures evolved through distinct stages, each marked by specific regulations and restrictions. In this section, we provide a brief overview of the different lockdown stages, highlighting key restrictions and when they were implemented.

A level-5 lockdown (full lockdown) was officially implemented by the South African government from midnight on 26 March 2020. Level 5 was the most restrictive of all the alert levels, with only essential services permitted to operate. During this phase, both domestic and international travel for individuals was prohibited, while the transportation of goods continued, albeit at a reduced capacity. The government gradually eased the level-5 lockdown regulations, transitioning through levels 4, 3, 2, and finally reaching level 1 on 21 September 2020 (Fig. 1). Towards the end of 2020, a surge in infections prompted the reimplementation of tighter restrictions, returning to an adjusted level-3 lockdown starting on 29 December 2020. Throughout 2021, the country predominantly operated under lockdown levels 2 or 3. On 1 October 2021, a further easing of restrictions led to a move to lockdown level 1. The National State of Disaster was officially terminated on 4 April 2022.

Different industries in South Africa were subject to specific work and operational regulations corresponding to each lockdown level (for more details, see: https://www.gov.za/covid-19/about/about-alert-system). Notably, law enforcement patrols were designated as essential services

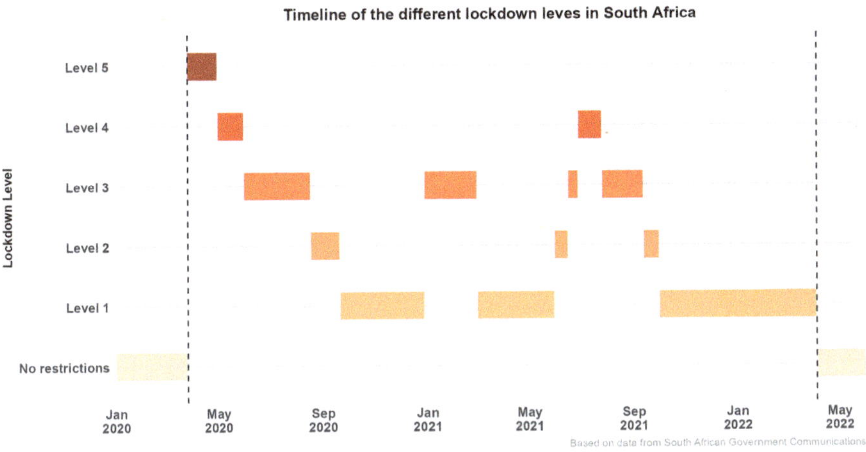

Fig. 1 An overview of when the different lockdown levels were active during the COVID-19 pandemic in South Africa

throughout the lockdown period (SA Corona Virus Online Portal, 2020). Consequently, patrol teams continued to monitor protected areas, although many law enforcement teams likely operated under reduced capacity. The lockdown regulations significantly affected the South African economy, employment levels, and individuals' ability to earn income (Van Heerden & Roos, 2021).

In response to the detrimental impact of the COVID-19 pandemic, the South African government took steps to mitigate socio-economic challenges by expanding its social assistance system. This included increasing the amounts of existing social grants and introducing a new temporary Social Relief of Distress (SRD)grant. The SRD grants were a provisional measure provided in the form of a food parcel, food voucher, or cash, targeting individuals facing challenges in meeting their families' basic needs. Despite challenges reported in the disbursement of social grants, they remained relatively well-targeted (Bhorat & Köhler, 2020). The government extended these grants multiple times between 2021 and 2022, even beyond the period of official lockdown restrictions. In a recent development, the government announced the extension of SRD grants until March 2025 (SAnews, 2023).

Predictions from an Environmental Criminology Perspective

The foundation of our predictions rests on established criminological theories, including the rational choice perspective, routine activity theory, and crime pattern theory. These theories assume that offenders are goal-oriented individuals who make (bounded) rational decisions influenced by the social environment and situational circumstances (Brantingham & Brantingham, 2008; Cornish & Clarke, 1986). These environmental characteristics are important in shaping the perceived rewards, costs, and risks associated with criminal activities. In the context of wildlife poaching, we assumed that poachers aim to avoid detection by rangers while simultaneously maximising their chances of obtaining valuable animal products, such as rhino horns or bushmeat. The subsequent predictions explore how shifts in environmental conditions, prompted by the COVID-19 lockdown restrictions, may have influenced the decision-making processes of poachers.

Predicted Impact on Rhino Poaching

Reward

Rhinos are arguably the most valuable species to poach in South Africa and poachers specifically target these animals due to the lucrative nature of the illegal international trade. The illegal hunting of rhinos is motivated by the international demand for rhino horns for medicinal uses and perceived sociocultural symbolic values in Southeast Asia (Hübschle, 2016; Milliken & Shaw, 2012; Wildlife Justice Commission, 2022). Hübschle (2016) found that 1 kg of rhino horn sells for approximately US$25,000 to $65,000 on Vietnamese black markets, although poachers likely sell the horn below optimum prices to secure immediate income (Milliken & Shaw, 2012). For example, poachers in South Africa typically receive around US$1750 to US$6500 for 1 kg of rhino horn (Eloff & Lemieux, 2014; Haas & Ferreira, 2016). While media reports and some organisations suggested a potential surge in rhino horn value during the COVID-19 pandemic, attributing it to unsupported claims as a cure for the coronavirus, these assertions lack empirical support and remain anecdotal (Wildlife Justice Commission, 2022). Some suggested that stockpiling of rhino horn had increased as the usual smuggling routes were temporarily unavailable to traffickers (Hübschle, 2022; Wildlife Justice Commission, 2020). This potentially made rhino horn more valuable for consumers in Asia, but is unlikely to have had an impact on the decision-making of rhino poachers. Therefore, we assumed that rhino horn did not became more valuable during or after the pandemic from the perspective of the rhino poacher and would not result in an increase in rhino poaching activity.

Effort

To minimise the effort required for rhino poaching, collaboration with rangers or personnel from tourist operators within the protected area has been reported as a strategy to obtain insider information on patrol regimes or rhino populations (Warchol, 2022). The implementation of lockdown restrictions and the subsequent ban on international travel compelled many wildlife-based tourism operators to either lay off personnel or cease business entirely (Ranchhod & Daniels, 2020; Spenceley et al., 2021). Although rangers were technically permitted to continue their operations during the lockdown, the reduced presence of wildlife-based tourism personnel may hinder poachers' opportunities to collude with individuals possessing insider knowledge. The impact of lockdown restrictions on the required effort for poaching is not straightforward, but based on current insights, we assumed that more effort would be necessary to poach a rhino during the lockdown period.

Risk

Poachers strategically make decisions to minimise their risk of detection, especially by patrolling rangers. Law enforcement patrols were deemed as essential services during the lockdown period and thus were allowed to continue operations (SA Corona Virus Online Portal, 2020).

The mobility constraints imposed by the lockdown, coupled with the decrease in overall human activity, has increased the scrutiny on any detected activities inside or at the borders of protected areas, making them more likely to be interpreted as suspicious. In theory, this reduced ambiguity provides law enforcement teams with a clearer signal to react faster to signs or activities associated with poaching. Furthermore, existing poaching networks were likely disrupted because of the sudden increase in security and controls at both the provincial and international borders. This would have also made rhino poaching more risky. Therefore, we assumed that the increased vigilance, coupled with law enforcement's essential status, would have deterred potential poachers and led to a decrease in rhino poaching activity during the lockdown period.

Predicted Impact on Bushmeat Poaching

Reward

The economic fallout from the lockdown restrictions has been particularly harsh on individuals dependent on the tourism industry, with many experiencing significant income losses (Ranchhod & Daniels, 2020; Spenceley et al., 2021). As previous studies have established a correlation between bushmeat poaching and poverty in South African communities (Grey-Ross et al., 2010; Nieman et al., 2019), concerns have been raised regarding a potential surge in bushmeat poaching since the lockdown was announced (Buckley, 2020; Lindsey et al., 2020). Despite social grants offering some short-term relief to those most affected by the lockdown restrictions, many rural communities may find themselves increasingly dependent on bushmeat to sustain their own livelihoods or those of fellow community members (Martins & Shackleton, 2019). Therefore, we assumed that the desirability of bushmeat has increased since the pandemic, potentially leading to an increase in snaring activity within the reserve. In practical terms, this shift should manifest as a rise in the number of fresh snares deployed within the reserve.

Effort

Bushmeat poaching stands out as one of the most accessible methods, requiring minimal resources that are both cheap and readily available, such as wires from perimeter fences or electric cables (Kendon et al., 2022; Lindsey et al., 2013). These materials were likely accessible even during the lockdown restrictions. Additionally, meat processing often takes place at the kill site or within the community itself (Martins & Shackleton, 2019), and these locations were likely still reachable during the lockdown period. Contrary to rhino poaching, bushmeat poaching typically does not require insider knowledge, since hunters are not targeting specific species. Furthermore, bushmeat poaching can be done alone and does not require a small group or interactions with a middleman before or after the hunt. Hence, we assumed that the effort required for bushmeat poachers remained relatively constant, with neither a significant increase nor decrease in effort needed.

Risk

Similar to rhino poachers, bushmeat poachers make decisions to minimise the risk of detection by patrolling rangers. However, in contrast to rhino poachers, bushmeat poachers may employ a stealthier approach by using wire snares, while rhino poachers typically use firearms. This distinction in methods contributes to a lower risk of detection for bushmeat poachers compared to rhino poachers. Despite the lowered risk associated with their covert methods, bushmeat poachers still need to navigate the protected area, entering and exiting undetected. During the lockdown period, any activity detected inside or at the borders of the protected area can be interpreted as suspicious by patrolling rangers. The impact of lockdown restrictions on the overall risk of bushmeat poaching is complex, but based on current insights we assumed that poaching for bushmeat became more risky during the lockdown period.

Materials and Methods

Study Area

Our study was conducted in Olifants West Nature Reserve, a private nature reserve in South Africa, part of the Greater Kruger Park. The reserve is fenced, except where it shares a boundary with adjacent reserves. The local communities surrounding the reserve, in Ba-Phalaborwa and Maruleng, consist mainly of people between 15 and 64 years old (approximately 62% in 2011) (Department of Statistics, 2011). Unemployment is high, especially among the youth population (50% in 2011), and a substantial portion of individuals over 20 years of age lack formal education (approximately 18% in 2011). Typically, the local population finds employment in sectors such as wildlife tourism, transportation, or mining. A majority of the anti-poaching staff was recruited from these local communities.

During the lockdown period, only essential services, critical maintenance staff, and private landowners were allowed access to the reserve. Lodges and other business operations were allowed to operate again from April 2022, when the last lockdown restrictions were officially terminated.

Data Collection

Poaching Data

The data on illegal activities in and around the reserve were reported and documented by rangers from Olifants West Nature Reserve and Transfrontier Africa NPC. All observations were digitally recorded through a law-enforcement monitoring system. While on patrol, teams logged any indicators of poaching-related activities, as well as signs or sightings of endangered species, including rhinos. In instances where patrol teams encountered evidence of poaching activity, such as human tracks entering the reserve, breaches in the fence, or gunshots, reactive teams were dispatched to provide assistance and conduct further investigations.

Furthermore, rangers conducted regular searches for snares and poacher camps, typically once or twice a week. These searches were strategically scheduled in the early morning to avoid the heat of the day and lasted approximately 2.5 hours, potentially extending if indications of illegal activity were discovered. When a snare was located, rangers recorded its position, took photographs, and attempted to estimate its age (fresh or old) before removing it from the field.

Only after confirming that poaching signs were indeed linked to a rhino or bushmeat poaching event were they included in our analyses. Additionally, historical data on rhino poaching events were digitised from written records maintained by reserve staff. Due to the sensitive nature of wildlife security information (Chapman, 2020), our data presentation deliberately avoids disclosing precise locations of poaching activities. This precautionary measure was taken to avoid disrupting or undermining any ongoing law enforcement operations.

Patrol Effort

Patrols within Olifants West Nature Reserve included both day and night operations. Day patrols were predominantly carried out on foot, whereas night patrols were typically conducted using vehicles and equipped with large spotlights. These spotlights not only facilitated the search for poaching activities but also served as a visual deterrent. Patrol teams searching specifically for snares were referred to as conducting a 'snare sweep'. Due to resource limitations and the multifaceted responsibilities of rangers, snare sweeps were conducted opportunistically when adequate time and resources were available, usually occurring one to two times per week. Rangers received a small financial reward for each snare they reported, in addition to their base salary.

Patrol effort was measured as the number of patrols conducted, the distance covered, and the duration of each patrol. During our study period, multiple law enforcement monitoring systems were utilised, with transitions between systems occasionally resulting in the loss of information on patrolling efforts. To overcome this challenge, we relied on aggregated data derived from quarterly reports provided by Transfrontier

Africa from 2017 Q1 to 2023 Q2. While more recent quarterly reports standardised the estimation of patrol effort, this was not consistently the case for older reports. In instances where the exact number of patrols was not specified, the wildlife security managers of Olifants West Nature Reserve provided informed estimates.

Data Analysis

All data cleaning, analyses, and visualisations were done using the R programming language (R Core Team, 2023) with the packages 'tidyverse' (Wickham et al., 2019), 'sf' (Pebesma, 2018), and 'glmmTMB' (Brooks et al., 2017).

To assess the impact of the COVID-19 lockdown on wildlife poaching in the study area, we analysed reported incursions, rhino poaching events, and snaring data. We categorised three distinct periods for comparison:

1. **Pre-Lockdown** (1 January 2017 to 25 March 2020): This period spans approximately 38 months, leading up to the announcement of the full lockdown in March 2020.
2. **Lockdown** (26 March 2020 to 3 April 2022): Encompassing just over two years, this phase corresponds to the period starting from when the lockdown was announced up until early April 2022.
3. **Post-Lockdown** (4 April 2022 onwards): This period captures the time after the lockdown was officially terminated.

Rhino Poaching

Within the dataset, rangers documented various indicators of rhino poaching, including footprints, gunshots, and breaches in the fence. To ensure a comprehensive assessment, all recorded rhino poaching activities within a 24-hour period were considered part of the same poaching event. A manual review of each data entry was conducted to validate that the observation was linked to an incursion by rhino poachers or an actual rhino poaching event. Instances in which poachers successfully shot a rhino were specifically categorised as a "rhino poaching event".

Given the relatively small sample size and localised scope of our study, Poisson z-scores were calculated following the methodology outlined by Wheeler (2015). This approach aimed to identify significant changes in crime patterns over time. The Poisson z-score calculation, based on the assumption that the crime series, is a Poisson-distributed random variable, was expressed as:

$$2 \times \left(\sqrt{\text{Current}} - \sqrt{\text{Historical}} \right)$$

The Poisson z-scores were rounded to the nearest integer, as the error introduced by using z-scores likely does not justify accuracy to tenths (Wheeler, 2015). We used a value of 3 as the threshold for a significant change, because this theoretically corresponds to a false positive rate of approximately 3 in 1000. To account for potential variations in patrolling efforts, Poisson z-scores were also computed for changes in the total number of patrols per year. This additional analysis aimed to discern whether alterations in rhino poaching activity could be attributed to shifts in patrolling intensity.

Bushmeat Poaching

The age of snares was estimated by rangers and classified as either 'fresh' or 'old'. When the age of a snare was not explicitly recorded, it was assumed to be old for the analyses. To ensure data quality, snares without recorded dates or coordinates were excluded from further analyses. Additionally, snares with invalid coordinates or those falling outside the study area boundary were excluded. Duplicate entries were removed by retaining only unique combinations of date and report description. A manual review of each reported snare description was conducted to ensure that only relevant snare data were included in the analyses.

The analysis focused on summarising the total number of reported snares categorised by age, year, and quarter. We modelled the number of fresh snares using a general linear mixed model with a generalised Poisson distribution. This model is suitable for count data analysis and addresses over-dispersion (when the variance exceeds the mean) (Consul & Famoye, 1992). We built a simple model with the number of fresh snares as the response variable and lockdown, a categorical variable with three levels (pre-lockdown, lockdown, and post-lockdown), as a predictor variable. To account for variation between quarters, the quarter of the year was included as a random effect. Furthermore, the number of patrols was included as an offset term to adjust for periods with varying patrol frequencies. Estimated coefficients for the lockdown predictor were transformed into incidence rate ratios of reported fresh snares using an exponential function, facilitating a more accessible interpretation of the results.

Results

Observed Rhino Poaching Trends

We examined rhino poaching trends by analysing the recorded number of incursions and rhino poaching events. A total of 93 incursions were documented between January 2017 and June 2023, with 29 events (31%) resulting in the loss of one or more rhinos. The highest frequency of incursions and poaching events occurred in 2017, and the numbers steadily declined in each subsequent year (Fig. 2). One rhino poaching event was recorded in September 2020, while Level-2 restrictions were in place. Following the easing of lockdown restrictions, another rhino poaching event was recorded in April 2023.

The Poisson z-scores indicate a significant decrease in incursions from 2018 to 2019 and again from 2019 to 2020 (Table 1). Although a general decline in rhino poaching was observed from 2018 onwards, no statistically significant changes were identified. The number of patrols conducted throughout the study period were relatively stable and were not statistically different between the years.

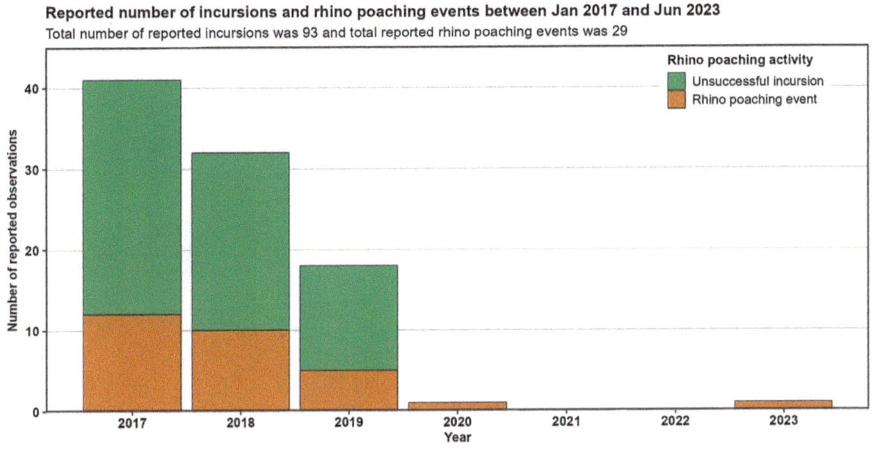

Fig. 2 Total reported number of incursions and rhino poaching events by year in the study area between January 2017 and June 2023

Observed Bushmeat Poaching Trends

We examined bushmeat poaching trends by analysing the number of fresh and old snares recorded by rangers during their patrols. Between January 2017 and June 2023, rangers reported 855 snares, with 143 snares (17%) estimated to be fresh. Before the implementation of lockdown restrictions, 49 fresh snares were reported (13% of all snares). During the lockdown, this figure dropped to 27 (13% of all snares) but increased to 67 (26% of all snares) after the restrictions were lifted.

A examination by year-quarter showed a shift in the proportion of fresh snares from the 2nd quarter of 2021 onwards, mid-lockdown (Fig. 3). This trend persisted post-lockdown, peaking in the 1st quarter of 2023, where an estimated 35% of reported snares were estimated to be fresh.

The results from the general linear mixed model showed a statistically significant difference in the rate of fresh snares discovered during different lockdown levels (Table 2). Specifically, the rate of fresh snares increased significantly during the post-lockdown period compared to the

Table 1 Comparison of year-on-year changes in incursions, rhino poaching and patrol effort by using Poisson z-scores

	2017–2018	2018–2019	2019–2020	2020–2021	2021–2022	2022–2023
All incursions	−1	**−3**	**−6**	−2	0	2
Rhino poaching events	−1	−2	−2	−2	0	2
Number of patrols	−1	−2	2	−2	1	−1

Values were rounded to the nearest integer. Significant values are shown in bold

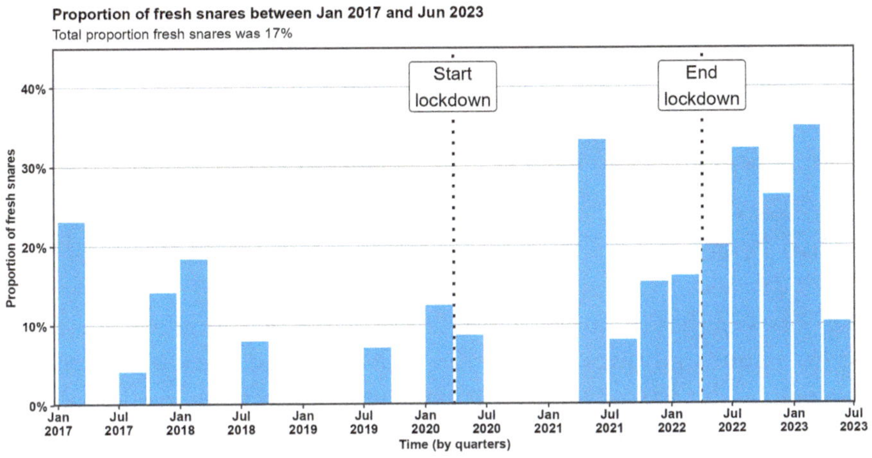

Fig. 3 Proportion of fresh snares by quarter-year in the study area between January 2017 and June 2023. Data was based on reports by patrolling rangers

Table 2 Result of general Poisson GLMM for with the number of fresh snares per quarter as response variable

	Incidence rate ratio	95% Conf.interval	
Pre-lockdown	0.68	0.23	1.99
Post-lockdown*	3.94	1.20	13.00

The estimated coefficients were transformed to a incidence rate ratio. Lockdown was used as the reference value

*$P < 0.05$

lockdown phase, with an incidence rate ratio of 3.94 (95% CI: 1.2–13.0). While the rate of fresh snares was lower in the pre-lockdown period compared to the lockdown, this difference did not reach statistical significance.

Discussion

This study focused on the trends in rhino poaching and bushmeat poaching during the COVID-19 lockdown period. Many studies and reports expressed concerns about an increase in poaching in South Africa, but few have examined the repercussions of the lockdown restrictions on wildlife poaching. We present our findings with caution and emphasise that the results must be interpreted in context. It is essential to recognise that similar restrictions may yield different outcomes in different protected areas in South Africa, as well as in other countries with different political-economic and socio-cultural characteristics.

Rhino Poaching

We observed that rhino poaching did not increase during the lockdown period in our study area. In fact rhino poaching had already been declining before the lockdown restrictions were announced. Anti-poaching units were declared as an essential service and continued operating consistently throughout the lockdown. The absence of recorded rhino poaching activities suggests that the lockdown restrictions did not lead to an increase in rhino poaching within the study area.

From an environmental criminology perspective, the lockdown restrictions did not make rhino poaching more rewarding, easier, or less risky. On the contrary, rhino poaching likely became riskier or required greater effort during the lockdown. A significant decrease in poaching activity was also reported for the Kruger National Park during the early months of the lockdown (Ferreira et al., 2021). Interestingly, rhino poaching in the study area remained at low levels even after the lockdown restrictions were lifted. In contrast, other protected areas reported an increase in incursions once the restrictions on inter-provincial and international travel were lifted (Department of Forestry, Fisheries and the Environment, 2024; Ferreira et al., 2021). While no such increase was observed in our study, the rise in rhino poaching activity could be related to the increased mobility and movements across the country. Alternatively, poaching networks that used to operate around the study area may not been able to recover after the lockdown restrictions were lifted. This could be especially true if those networks relied on corrupt individuals within protected areas who provided sensitive information or facilitated smuggling at international airports (Wildlife Justice Commission, 2020). It is also possible that poachers and traffickers operating in South Africa shifted their focus on other, more profitable illegal markets like the illegal trade of cigarettes and alcohol during the lockdown bans (Ross & Joossens, 2021). Although this study cannot provide evidence of crime type displacement or quantify the role of corruption, other research has shown that corruption is a major facilitator of the illegal wildlife trade (Hauenstein et al., 2019; Hübschle, 2016).

The observed decline in rhino poaching activities prior to the lockdown aligns with broader trends seen in other protected areas throughout South Africa. However, it's important to acknowledge the wider context—despite the drop in poaching incidents, the white rhino population has declined by approximately 64% since 2017 (SANParks, 2023). One potential factor contributing to the decrease in poaching may be the reduced availability of rhinos. Unfortunately, rhino population data could not be included in the analysis due to inconsistent monitoring efforts across the different regions of the study area during the lockdown. Including population data into future studies could improve our understanding of the relationship between population size and poaching dynamics.

Additionally, the study did not account for dehorning activities, which have been ongoing since 2019. Given the temporal overlap between dehorning and lockdown restrictions, it is difficult to separate their individual effects on rhino poaching. A more comprehensive investigation, possibly at a national scale, would be required to do so. This could involve comparing rhino poaching trends across multiple protected areas in South Africa through collaboration among researchers, law enforcement, and wildlife practitioners, while ensuring ongoing operations and investigations remain secure. Such studies can contribute significantly to the development and evaluation of specific conservation strategies and improve the protection of rhino populations.

Bushmeat Poaching

This study showed that significantly more fresh snares were found in the post-lockdown period, suggesting that bushmeat poaching increased in the reserve. From an environmental criminology perspective, bushmeat poaching in the study area became more rewarding, easier and/or less risky after the lockdown period.

Our hypotheses suggested that the economic hardships imposed by the pandemic would increase bushmeat poaching due to increased poverty levels in the local communities. However, the results indicate that the number of fresh snares during the lockdown was similar to numbers before the lockdown was implemented. The number of fresh snares was significantly higher in the post-lockdown period compared to the lockdown period, suggesting an increase in bushmeat poaching in the reserve. This is contrary to recent studies on the impacts of COVID-19 on bushmeat consumption (Enns et al., 2023; Kamogne Tagne et al., 2022). These studies found an increased dependence on subsistence hunting in rural areas across several African countries, especially during the early months of the pandemic. For our study area, it was possible that the government support funds may have prevented the impoverished communities from turning to bushmeat poaching at the early stages of the lockdown. In addition, staff from Olifants West Nature Reserve also identified a number of impoverished households in the local communities and provided food parcels to them between June 2020 and June 2021. The support from the local and national organisations may have helped local communities to provide for their immediate family members, although the (perceived) impact of these two forms of support has not been evaluated.

Another explanation could be related to the different drivers and motivations for bushmeat poaching. This can be explored, in part, through spatial analyses of the reported snares. While patrols and snare sweeps have been fairly consistent throughout the study period, a change in the spatial distribution of snares could suggest that different people with different motivations started hunting for bushmeat or that different areas with higher rewards, easier access, or lower risk were targeted. It is also important to point out that, while our study suggested that bushmeat poaching increased in the study area, this does not necessarily mean that it is unsustainable. The capacity of a given species to withstand hunting pressure depends greatly on its reproduction characteristics, habitat preferences, and geographical distribution (Enns et al., 2023; Grey-Ross et al., 2010; Nieman et al., 2019). To determine this, a specific study would need to research this more carefully, taking into account, among other things, estimates of the illegally hunted species, population dynamics, predator pressures, and protected area management. For example, based on anecdotal evidence and reports from patrol officers, it seems that new snaring hotspots were associated with activities of workers around building sites inside the reserve during the post-lockdown period. The assumption is that the workers are trying to supplement their income by illegally harvest bushmeat or other animal parts through snaring. The increase in snaring by workers on building sites could have been exacerbated due to the cumulative economic pressure of the lockdown restrictions. This study did not investigate this claim, but a follow-up study could explore the potential correlation between the locations of fresh snares and the spatial distribution of building sites.

While bushmeat poaching tends to correlate with poverty, it is unlikely that it is driven by poverty alone. Other studies within criminology found that locations with high levels of income inequality and a breakdown in social norms and values typically have high levels of crime (Harris & Vermaak, 2015; Nivette, 2011). Many of these factors were already present in the local communities bordering the study area before the lockdown restrictions were in place. The lockdown restrictions likely had a much more severe impact on these communities, which may have eventually led to an increase in bushmeat poaching despite the involved risks. However, the bushmeat trade and markets in South Africa is poorly studied and not much was known about both the short-term and longer-term drivers and effects (Martins & Shackleton, 2019). A better understanding of the drivers can be help to better design appropriate interventions. Law enforcement responses alone will be likely be insufficient to reduce wildlife crime, including bushmeat poaching (Delpech et al., 2021; Lemieux, 2014). Law enforcement efforts may be needed in the short term, but they are insufficient as a stand-alone response. In the long term, a wide range of evidence-based interventions are required to address the social, economic, and societal drivers of wildlife crime, such as community crime prevention, socio-economic improvement, conflict resolution, and behaviour change (Delpech et al., 2021; Sosnowski et al., 2021).

Conclusions

The consequences of the COVID-19 lockdown seem to impact the conservation arena in various ways. Many studies and reports expressed concerns about an increase in poaching in South Africa, but few studies have focused on the repercussions of the lockdown restrictions on wildlife security and poaching. Through a comparative analysis of poaching patterns before, during, and after the lockdown period, this study contributes towards a better understanding of the impact of the lockdown restrictions on wildlife poaching.

We observed that rhino poaching activity has been steadily declining since 2017. Rhino poaching did not increase during the lockdown or after the restrictions were lifted. From an environmental criminology perspective, our results suggest that the lockdown restrictions increased the risk of detection for rhino poaching. Anti-poaching units were declared as an essential service and have been operating consistently throughout the lockdown. The continued patrol efforts, in combination with the sudden increase in security and controls during the lockdown period, likely increased the risk for rhino poaching. Levels of rhino poaching remained low even after the lockdown restrictions were lifted. Rhino poaching events have been reported in other protected areas in South Africa with the easing of the lockdown and this could be related to increased mobility and movements across the country when some of the travel restrictions were lifted. It is unclear, however, why this increase did not happen in our study area and more research is needed to better understand the role of effort in the decision-making of rhino poachers.

The number of fresh snares significantly increased after the lockdown restrictions were lifted, suggesting an increase in bushmeat poaching in the reserve. The number of fresh snares found during the lockdown was similar to the pre-lockdown period, suggesting that the restrictions did not made bushmeat poaching more rewarding, riskier, or easier. However, bushmeat poaching did increase towards the end of the lockdown period, suggesting that the restrictions impacted the decision-making of bushmeat poachers. The economic hardships imposed by the pandemic would make bushmeat poaching more rewarding, especially as other sources of income became unavailable. More research is needed to better understand how these local communities were surviving at the early stages of the lockdown and what eventually drove the increase in bushmeat poaching. Above all, the local context and the behaviour of the actors involved in poaching must be understood properly to design effective interventions for reducing or mitigation poaching and other human–wildlife conflicts.

Acknowledgments We would like to thank the anti-poaching units and conservation staff of Olifants West Nature Reserve for doing all the hard work protecting the reserve, especially during the challenging times of the lockdown.

Author Contributions Conceptualisation: all authors; Data collection: PA and CS; data analyses: ND; manuscript draft: ND and PA. All authors: review and editing. All authors have read and agreed to the published version of the manuscript.

Conflicts of Interest The authors declare no conflict of interest.

Funding This research received no external funding.

Data Availability Given the sensitive nature of wildlife security information, the authors cannot make the data publicly available. The quarterly reports and data analysis scripts are available upon request.

References

Anagnostou, M., Mwedde, G., Roe, D., Smith, R. J., Travers, H., & Baker, J. (2020). Ranger perceptions of the role of local communities in providing actionable information on wildlife crime. *Conservation Science and Practice, 2*(6), e202. https://doi.org/10.1111/csp2.202

Aransiola, T. J., & Ceccato, V. (2020). The role of modern technology in rural situational crime prevention: A review of the literature. In *Rural Crime Prevention*. Routledge.

Behera, A. K., Kumar, P. R., Priya, M. M., Ramesh, T., & Kalle, R. (2022). The impacts of COVID-19 lockdown on wildlife in Deccan plateau, India. *Science of the Total Environment, 822*, 153268. https://doi.org/10.1016/j.scitotenv.2022.153268

Bhorat, H., & Köhler, T. (2020). *Social assistance during South Africa's national lockdown: Examining the COVID-19 grant, changes to the Child Support Grant, and post-October policy options*. Working Paper. https://ideas.repec.org//p/ctw/wpaper/202009.html

Boratto, R., & Gibbs, C. (2021). Advancing interdisciplinary research on illegal wildlife trade using a conservation criminology framework. *European Journal ofCriminology, 18*(6),777–798. https://doi.org/10.1177/1477370819887512

Brantingham, P. J., & Brantingham, P. L. (2008). Crime pattern theory. In R. Wortley & L. Mazerolle (Eds.), *Environmental criminology and crime analysis* (1st ed., pp. 78–93). Taylor & Francis.

Brooks, M. E., Kristensen, K., van Benthem, K. J., Magnusson, A., Berg, C. W., Nielsen, A., Skaug, H. J., Mächler, M., & Bolker, B. M. (2017). glmmTMB balances speed and flexibility among packages for zero-inflated generalized linear mixed modeling. *The R Journal, 9*(2), 378–400.

Buckley, R. (2020). Conservation implications of COVID19: Effects via tourism and extractive industries. *Biological Conservation, 247*, 108640. https://doi.org/10.1016/j.biocon.2020.108640

Carvalho, T., Krammer, F., & Iwasaki, A. (2021). The first 12 months of COVID-19: A timeline of immunological insights. *Nature Reviews Immunology, 21*(4), Article 4. https://doi.org/10.1038/s41577-021-00522-1

Chapman, A. D. (2020). *Current best practices for generalizing sensitive species occurrence data*. GBIF Secretariat. https://doi.org/10.15468/doc-5jp4-5g10

Consul, P. C., & Famoye, F. (1992). Generalized Poisson regression model. *Communications in Statistics - Theory and Methods, 21*(1), 89–109. https://doi.org/10.1080/03610929208830766

Cornish, D., & Clarke, R. (1986). Situational prevention, displacement of crime and rational choice theory. In K. Heal & G. Laycock (Eds.), *Situational crime prevention: From theory into practice* (pp. 1–16). Her Majesty's Stationary Office.

Delpech, D., Borrion, H., & Johnson, S. (2021). Systematic review of situational prevention methods for crime against species. *Crime Science, 10*(1), 1. https://doi.org/10.1186/s40163-020-00138-1

Department of Forestry, Fisheries and the Environment. (2024, February 27). *KwaZulu-Natal carried the brunt of rhino poaching in 2023, says Creecy*. Press release. https://www.dffe.gov.za/mediareleases/creecy_relasesrhinopoachingstats2024feb27

Department of Statistics. (2011). *Local municipality | statistics South Africa*. Statistics South Africa. https://www.statssa.gov.za/?page_id=964

Ehlers Smith, D. A., Ehlers Smith, Y. C., Davies-Mostert, H. T., Thompson, L. J., Parker, D. M., de Villiers, D., Ricketts, D., Coverdale, B., Roberts, P. J., Kelly, C., Macfadyen, D. N., Manqele, N. S., Power, R. J., & Downs, C. T. (2023). The impacts of a global pandemic on the efficacy and stability of contemporary wildlife conservation: South Africa as a case study. *Ambio, 52*(3), 598–615. https://doi.org/10.1007/s13280-022-01814-z

Eliason, S. L. (1999). The illegal taking of wildlife: Toward a theoretical understanding of poaching. *Human Dimensions of Wildlife, 4*(2), 27–39. https://doi.org/10.1080/10871209909359149

Eloff, C., & Lemieux, A. (2014). Rhino poaching in Kruger National Park, South Africa: Aligning analysis, technology and prevention. In *Situational prevention of poaching* (pp. 18–43). Routledge.

Enns, C., van Vliet, N., Mbane, J., Muhindo, J., Nyumu, J., Bersaglio, B., Massé, F., Cerutti, P. O., & Nasi, R. (2023). Vulnerability and coping strategies within wild meat trade networks during the COVID-19 pandemic. *World Development, 170*, 106310. https://doi.org/10.1016/j.worlddev.2023.106310

Ferreira, S. M., Greaver, C., Simms, C., & Dziba, L. (2021). The impact of COVID-19 government responses on rhinoceroses in Kruger National Park. *African Journal of Wildlife Research, 51*(1). https://doi.org/10.3957/056.051.0100

Forster, P. M., Forster, H. I., Evans, M. J., Gidden, M. J., Jones, C. D., Keller, C. A., Lamboll, R. D., Quéré, C. L., Rogelj, J., Rosen, D., Schleussner, C.-F., Richardson, T. B., Smith, C. J., & Turnock, S. T. (2020). Current and future global climate impacts resulting from COVID-19. *Nature Climate Change, 10*(10), Article 10. https://doi.org/10.1038/s41558-020-0883-0.

Grey-Ross, R., Downs, C. T., & Kirkman, K. (2010). An assessment of illegal hunting on farmland in KwaZulu-Natal, South Africa: Implications for Oribi (*Ourebia ourebi*) conservation. *South African Journal of Wildlife Research, 40*(1), 43–52. https://doi.org/10.3957/056.040.0104

Haas, T. C., & Ferreira, S. M. (2016). Conservation risks: When will rhinos be extinct? *IEEE Transactions on Cybernetics, 46*(8), 1721–1734. https://doi.org/10.1109/TCYB.2015.2470520

Harris, G., & Vermaak, C. (2015). Economic inequality as a source of interpersonal violence: Evidence from sub-Saharan Africa and South Africa. *South African Journal of Economic and Management Sciences, 18*(1), 45–57. https://doi.org/10.17159/2222-3436/2015/v18n1a4

Hauenstein, S., Kshatriya, M., Blanc, J., Dormann, C. F., & Beale, C. M. (2019). African elephant poaching rates correlate with local poverty, national corruption and global ivory price. *Nature Communications, 10*(1), Article 1. https://doi.org/10.1038/s41467-019-09993-2

Hill, J. F., Johnson, S. D., & Borrion, H. (2014). Potential uses of computer agent-based simulation modelling in the evaluation of wildlife poaching. In *Situational prevention of poaching*. Routledge.

Hübschle, A. M. (2016). *A game of horns: Transnational flows of rhino horn*. PhD, IMPRS-SPCE. https://doi.org/10.17617/2.2218357.

Hübschle, A. M. (2022). *The impact of the COVID-19 pandemic and lockdown regulations on wildlife poaching hotspots and trafficking routes in southern Africa*. United States Agency for International Development. https://law.uct.ac.za/media/265385

Kamogne Tagne, C. T., Brittain, S., Booker, F., Challender, D., Maddison, N., Milner-Gulland, E. J., Mouamfon, M., Roe, D., & Coad, L. (2022). Impacts of the COVID-19 pandemic on livelihoods and wild meat use in communities surrounding the Dja faunal reserve, south-East Cameroon. *African Journal of Ecology, 60*(2), 135–145. https://doi.org/10.1111/aje.12995

Kendon, T. A., Comley, J., Wilkinson, A., Grobler, M. J., Nieman, W. A., Leslie, A. J., Justin O'Riain, M., & Naude, V. N. (2022). Evaluating determinants of wire-snare poaching risk in the Boland Mountain complex of South Africa. *Journal for Nature Conservation, 70*, 126295. https://doi.org/10.1016/j.jnc.2022.126295

Kurland, J., Pires, S. F., & Marteache, N. (2018). The spatial pattern of redwood burl poaching and implications for prevention. *Forest Policy and Economics, 94*, 46–54. https://doi.org/10.1016/j.forpol.2018.06.009

Kurland, J., Pires, S. F., McFann, S. C., & Moreto, W. D. (2017). Wildlife crime: A conceptual integration, literature review, and methodological critique. *Crime Science, 6*(1), 4. https://doi.org/10.1186/s40163-017-0066-0

Lavorgna, A., & Sajeva, M. (2021). Studying illegal online trades in plants: Market characteristics, Organisational and Behavioural aspects, and policing challenges. *European Journal on Criminal Policy and Research, 27*(4), 451–470. https://doi.org/10.1007/s10610-020-09447-2

Lemieux, A. (2014). Introduction. In *Situational prevention of poaching* (pp. 1–17). Routledge.

Lindsey, P., Allan, J., Brehony, P., Dickman, A., Robson, A., Begg, C., Bhammar, H., Blanken, L., Breuer, T., Fitzgerald, K., Flyman, M., Gandiwa, P., Giva, N., Kaelo, D., Nampindo, S., Nyambe, N., Steiner, K., Parker, A., Roe, D., … Tyrrell, P. (2020). Conserving Africa's wildlife and wildlands through the COVID-19 crisis and beyond. *Nature Ecology & Evolution, 4*(10), Article 10. https://doi.org/10.1038/s41559-020-1275-6.

Lindsey, P. A., Balme, G., Becker, M., Begg, C., Bento, C., Bocchino, C., Dickman, A., Diggle, R. W., Eves, H., Henschel, P., Lewis, D., Marnewick, K., Mattheus, J., Weldon McNutt, J., McRobb, R., Midlane, N., Milanzi, J., Morley, R., Murphree, M., et al. (2013). The bushmeat trade in African savannas: Impacts, drivers, and possible solutions. *Biological Conservation, 160*, 80–96. https://doi.org/10.1016/j.biocon.2012.12.020

Martins, V., & Shackleton, C. M. (2019). Bushmeat use is widespread but under-researched in rural communities of South Africa. *Global Ecology and Conservation, 17*, e00583. https://doi.org/10.1016/j.gecco.2019.e00583

Milliken, T., & Shaw, J. (2012). *The South Africa—Viet Nam Rhino Horn Trade Nexus: A deadly combination of institutional lapses, corrupt wildlife industry professionals and Asian crime syndicates*. TRAFFIC. https://www.traffic.org/publications/reports/the-south-africa-viet-nam-rhino-horn-trade-nexus/

Moreto, W. D., & Lemieux, A. M. (2015). Poaching in Uganda: Perspectives of law enforcement rangers. *Deviant Behavior, 36*(11), 853–873. https://doi.org/10.1080/01639625.2014.977184

Moreto, W. D., Lemieux, A. M., Rwetsiba, A., Guma, N., Driciru, M., & Kirya, H. K. (2014). Law enforcement monitoring in Uganda: The utility of official data and time/distance-based ranger efficiency measures. In *Situational prevention of poaching*. Routledge.

Nieman, W. A., Leslie, A. J., Wilkinson, A., & Wossler, T. C. (2019). Socioeconomic and biophysical determinants of wire-snare poaching incidence and behaviour in the Boland region of South Africa. *Journal for Nature Conservation, 52*, 125738. https://doi.org/10.1016/j.jnc.2019.125738

Nivette, A. E. (2011). Cross-national predictors of crime: A meta-analysis. *Homicide Studies, 15*(2), 103–131. https://doi.org/10.1177/1088767911406397

Pebesma, E. (2018). Simple features for R: Standardized support for spatial vector data. *The R Journal, 10*(1), 439–446. https://doi.org/10.32614/RJ-2018-009

Petrossian, G. A., Marteache, N., & Viollaz, J. (2015). Where do "undocumented" fish land? An empirical assessment of port characteristics for IUU fishing. *European Journal on Criminal Policy and Research, 21*(3), 337–351. https://doi.org/10.1007/s10610-014-9267-1

Pires, S. F., Olah, G., Nandika, D., Agustina, D., & Heinsohn, R. (2021). What drives the illegal parrot trade? Applying a criminological model to market and seizure data in Indonesia. *Biological Conservation, 257*, 109098. https://doi.org/10.1016/j.biocon.2021.109098

R Core Team. (2023). *R: A language and environment for statistical computing*. Computer software. R Foundation for Statistical Computing. https://www.R-project.org/2023

Ranchhod, V., & Daniels, R. C. (2020). *Labour market dynamics in South Africa in the time of COVID-19: Evidence from wave 1 of the NIDS-CRAM survey.* Working Paper. http://hdl.handle.net/11090/981

Rizzolo, J. B., Gore, M. L., Ratsimbazafy, J. H., & Rajaonson, A. (2017). Cultural influences on attitudes about the causes and consequences of wildlife poaching. *Crime, Law and Social Change, 67*(4), 415–437. https://doi.org/10.1007/s10611-016-9665-z

Ross, H., & Joossens, L. (2021). Tackling illicit tobacco during COVID-19 pandemic. *Tobacco Induced Diseases, 19*, 10.18332/tid/137086.

SA Corona Virus Online Portal. (2020, March 20). *Essential services during the lockdown period.* SA Corona Virus Online Portal. https://sacoronavirus.co.za/2020/03/20/essential-services-during-the-lockdown-period/

SAnews. (2023, November 1). *COVID-19 social grant extended until 2025.* https://www.sanews.gov.za/south-africa/covid-19-social-grant-extended-until-2025

SANParks. (2023). *South African national parks annual report 2022/23.* Annual report. South African National Parks. https://www.sanparks.org/wp-content/uploads/2023/09/SANParks-Annual-Report-2022-2023.pdf

Schneider, J. L. (2008). Reducing the illicit trade in endangered wildlife: The market reduction approach. *Journal of Contemporary Criminal Justice, 24*(3), 274–295. https://doi.org/10.1177/1043986208318226

Skidmore, A. (2021). Using crime script analysis to elucidate the details of Amur tiger poaching in the Russian Far East. *Crime Science, 10*(1), 16. https://doi.org/10.1186/s40163-021-00150-z

Sollund, R. (2016). Wildlife trafficking in a globalized world: An example of motivations and modus operandi from a Norwegian case study. In F. M. Angelici (Ed.), *Problematic wildlife: A cross-disciplinary approach* (pp. 553–570). Springer International Publishing. https://doi.org/10.1007/978-3-319-22246-2_25

Sosnowski, M., Everatt, K., Pickles, R., Whittington-Jones, G., & Lemieux, A. (2021). *Wilderness problem-specific guide no. 2: Illegal and unsustainable hunting of wildlife for Bushmeat in Sub-Saharan Africa.* Center for Problem-Oriented Policing, Arizona State University. https://popcenter.asu.edu/sites/default/files/illegal_bushmeat_hunting_2021.pdf

Spenceley, A., McCool, S., Newsome, D., Báez, A., Barborak, J. R., Blye, C.-J., Bricker, K., Sigit Cahyadi, H., Corrigan, K., Halpenny, E., Hvenegaard, G., Malleret King, D., Leung, Y.-F., Mandić, A., Naidoo, R., Rüede, D., Sano,

J., Sarhan, M., Santamaria, V., et al. (2021). Tourism in protected and conserved areas amid the COVID-19 pandemic. *PARKS, 27*, 103–118. https://doi.org/10.2305/IUCN.CH.2021.PARKS-27-SIAS.en

Van Heerden, J., & Roos, E. L. (2021). The possible effects of the extended lockdown period on the south African economy: A CGE analysis. *South African Journal of Economics, 89*(1), 95–111. https://doi.org/10.1111/saje.12273

Viollaz, J., Rizzolo, J. B., Long, B., Trung, C. T., Kempinski, J., Rawson, B. M., Reynald, D., Quang, H. X., Hien, N. N., Dung, C. T., Huyen, H. T., Dung, N. T. T., & Gore, M. L. (2022). Potential for informal guardianship in community-based wildlife crime prevention: Insights from Vietnam. *Nature Conservation, 48*, 123–147. https://doi.org/10.3897/natureconservation.48.81635

Warchol, G. (2022). Policing wildlife crimes. A historical analysis of the development and impact of wildlife ranger units in sub-Saharan Africa. In D. Watson, S. N. Amin, W. C. Wallace, O. (Michael) Akinlabi, & J. C. Ruiz-Vásquez (Eds.), *Policing the global south: Colonial legacies, pluralities, partnerships, and reform* (pp. 353–366). Taylor & Francis. https://doi.org/10.4324/9781003126409-30

Wheeler, A. (2015). *Tables and graphs for monitoring temporal crime trends: Translating theory into practical crime analysis advice*. SSRN Scholarly Paper 2551472. https://doi.org/10.2139/ssrn.2551472.

Wickham, H., Averick, M., Bryan, J., Chang, W., McGowan, L. D., François, R., Grolemund, G., Hayes, A., Henry, L., Hester, J., Kuhn, M., Pedersen, T. L., Miller, E., Bache, S. M., Müller, K., Ooms, J., Robinson, D., Seidel, D. P., Spinu, V., et al. (2019). Welcome to the Tidyverse. *Journal of Open Source Software, 4*(43), 1686. https://doi.org/10.21105/joss.01686

Wildlife Justice Commission. (2020). *Rapid assessment of the impact of COVID-19 on wildlife trafficking*. https://wildlifejustice.org/wp-content/uploads/2020/04/WJC_Impact-of-COVID19-on-wildlife-trafficking_April2020.pdf

Wildlife Justice Commission. (2022). Use and consumption of rhino horn. In *Rhino horn trafficking as a form of transnational organised crime 2012–2021* (pp. 149–164). https://wildlifejustice.org/rhino-horn-trafficking-report/

World Health Organization. (2023a). *Considerations for implementing and adjusting public health and social measures in the context of COVID-19* (Interim Guidance WHO/2019-nCoV/Adjusting_PH_measures/2023.1). World

Health Organization. https://iris.who.int/bitstream/handle/10665/366669/WHO-2019-nCoV-Adjusting-PH-measures-2023.1-eng.pdf?sequence=1
World Health Organization. (2023b). *How coronavirus disease has changed the environment and health landscape: Web annex: A rapid review of reviews.* World Health Organization. https://iris.who.int/bitstream/handle/10665/368165/9789289060066-eng.pdf?sequence=3

Understanding and Navigating International Wildlife Trade Data: Avoiding Pitfalls Caused by Systemic Detail Obfuscation

Orion L. B. Goodman, Michael F. Tlusty, and Andrew L. Rhyne

Introduction

International wildlife crime threatens affected species with potentially irreversible population decline (Zimmerman, 2003; Duffy, 2022). While quantification is difficult, much of this threat is presumed to stem from pressure posed by wildlife trade, especially wildlife trafficking. Efforts to regulate wildlife trade, and therefore support and promote sustainable use, are often challenged by illegal and unreported commercial

O. L. B. Goodman (✉)
Criminal Justice, New York, NY, USA

M. F. Tlusty
Sustainability and Food Solutions, Bristol, VA, USA

A. L. Rhyne
University of Massachusetts, Amherst, MA, USA

D. van Uhm, D. Siegel (eds.), *Global Green Crime and Ecojustice*, Palgrave Studies in Green Criminology, https://doi.org/10.1007/978-3-031-91557-4_10

movement of wildlife commodities. At the international scale, this commerce can be worth several billion USD (van Uhm, 2016a; Tow et al., 2021).

This chapter reviews how relevant datasets are derived and examines their limitations and characteristics in order to provide readers with an improved understanding of the wildlife trade data environment, particularly regarding the Convention on the International Trade in Endangered Species of Wild Fauna and Flora (CITES) and the United States Fish and Wildlife Service's Law Enforcement Management System (LEMIS) databases. This chapter includes limited summary analysis of LEMIS and CITES data spanning January 1, 2000 to December 31, 2020. The CITES Trade database contains records representing international wildlife trade of CITES-listed species as reported annually by the treaty's party members. It is maintained by UNEP-WCMC and is available via download through the CITES website (Full CITES Trade Database Download, 2023). The LEMIS data represents international wildlife trade wherein either the importing country or the exporting country is the United States. The LEMIS data used in this analysis consists of two parts: data cleaned and organized by Eskew et al. (2020) and data resulting from a Freedom of Information Act (FOIA) request submitted by the authors on November 16, 2021 (DOI-FWS-2022-000708).

One of the major challenges to assessing the impacts of wildlife trade, let alone illegal wildlife trade (IWT), is the availability of reliable data. Many of the available datasets are complicated and require a high degree of familiarity for interrogators to avoid misinterpreting data. For example, CITES maintains a record of all trade events in CITES-listed species as reported by its subscribing parties (CITES Secretariat and UNEP-WCMC, 2022). Because this trade occurs as a subset of all potential wildlife trade, it cannot be used to evaluate biodiversity in trade, although it could be used to complement analyses of other data. Similarly, CITES parties report illegal activity unreliably, thus limiting access to IWT data through the publicly accessible CITES database (D'Cruze & Macdonald, 2016). When they are available, CITES database entries for illegal activity obfuscate the original source information for the associated record. This means that assertions of illegal trade being implicitly unsustainable require additional evidence.

The goal of this chapter is to present limited summary evidence of data obfuscation present in the LEMIS and CITES databases and to provide wildlife trade researchers with contextual information that will enable them to better navigate these issues when conducting their own investigations. These goals are met through assessing how data are created across different datasets, briefly describing how CITES and LEMIS data are related to international commercial trade, and by investigating patterns and trends across wildlife trade data that may provide useful metrics for further analyses.

Wildlife Trade and Research

Many researchers rely on the CITES Trade Database because it is available and is longitudinally robust. Hughes et al. (2021) used the CITES Trade Database, alongside reviews of other sources such as online marketplaces and LEMIS, to draw conclusions about the comparative biodiversity of amphibian species in trade. Tittensor et al. (2020) similarly used the CITES Trade Database, alongside US import data from LEMIS, to make inferences about illegal wildlife trade. Due to the nature of these databases, both studies struggled to infer meaningful conclusions that were free from confounding factors. Hughes et al. (2021) identified six times more amphibians present in trade events not regulated by CITES. Without the administrative costs necessitated by CITES, traders may be incentivized to engage in trade in unprotected species, but this is not necessarily unsustainable. Additionally, Hughes et al. (2021) fails to explore the present husbandry infrastructure for amphibians present in the international wildlife trade, which may further affect volume, frequency, and other metrics for trade intensity in these species. In the case of Tittensor et al. (2020), the authors found interesting correlations between annual increases in reported trade and reported law enforcement activity. While they speculate on the causes, they fail to realize that LEMIS is an evolving system attached to an evolving law enforcement program, particularly over the study period (Weissgold, 2024; Kolby & Goodman, 2025). Changes in port coverage, staffing, political interests, trader compliance, and standardization requirements can inject noise into the analyzed trends. Despite these issues, both studies lay important groundwork for future wildlife trade analyses—and they are not alone in this importance. Petrossian et al. (2016), Berec et al. (2018), Stringham et al. (2021), and a host of other studies identify important characteristics of international wildlife trade and IWT despite attempting to navigate datasets steeped in caveats and complications. Many of these common challenges are discussed in Robinson & Sinovas (2018), Challender et al. (2021), and Weissgold (2024).

While data on illegal wildlife trade exist, their availability is often limited to government stakeholders, non-governmental organizations (NGOs), and their partners (United Nations, 2022). When these data are available to researchers, they can be plagued by issues such as incomplete or incorrect taxonomy, missing values, a lack of standardization, arbitrary categorization, and other data quality challenges (Stringham et al., 2021). Due to the illicit nature of IWT, researchers often turn to datasets comprised of known illegal activity. These are often ad-hoc and their dependability can be reliant on access to funding (OG *pers. obs.*; WWF-UK, 2019). Furthermore, like LEMIS, these datasets are prone to evolution and often contain more information on species of major conservation or political interest, such as elephants, tigers, or rhinoceros, than on various reptile, insect, or plant species (Hinesley et al., 2024). Trends can be similarly difficult to capture because they are confounded by changes in taxonomic interest over time. For example, bird trafficking has recently piqued the interest of many European NGOs and government authorities, so seized bird records may increase in both number and detail as shipments are subject to heightened scrutiny (Ribeiro et al., 2019; Zsigmond, 2020). In order to successfully interrogate these data, it is crucial that analysts understand their origins.

Wildlife Trade and International Trade

The legal international wildlife trade exists as a subset of commercial international trade with some key differences (Hughes, 2021). Live and fresh wildlife generally have a heightened risk of perishability along their supply chains, and trade in wildlife and wildlife products may carry associated ecological, environmental, and epidemiological risks not posed by trade in other commodities (Keskin et al., 2023). Additionally, transporting live wild animals between countries necessarily includes transporting their waste and soiled shipping substrates which may act as fomites (Swift et al., 2007; CITES Secretariat, 2023). Escaped animals can also become invasive if not recaptured (Lockwood et al., 2019). Additionally, wildlife-based supply chains that depend on the wild harvesting of plants or animals can threaten wild populations, increasing the likelihood of extirpation, and potentially affecting ecosystem services and ecological resilience (Robinson et al., 2019).

The risks posed by wildlife trade are largely controlled through regulatory and monitoring protocols, many of which are parallel to those implemented by customs authorities monitoring and regulating other commercial trade. These mechanisms are supported by legal frameworks such as CITES (CITES, 2024a). However, CITES only applies to species listed in one of its three appendices (Challender et al., 2021; Easter & Carter, 2024). It is otherwise up to individual States or regions to implement regulatory processes for species not listed by CITES. Most publicly available data on international wildlife trade is generated through the implementation of these protocols by customs authorities or their partner agencies. For many countries, customs authorities are responsible for executing CITES obligations, but this frequently falls outside of their agency's primary function.

The primary goal of most customs authorities is to protect their State's economic interests by enforcing tariffs and regulating imports and exports (Widdowson, 2007). Often, they work alongside other agencies who are responsible for ensuring that incoming commodities pose an acceptable level of health or environmental risk (USCBP, 2024). One crucial aspect of customs regulation is the widespread adoption of the World Customs Organization's Harmonized System (HS), a means of categorizing all internationally traded commodities (WCO, n.d.). These categories help to simplify trade risks and determine the appropriate import or export process for the commodity in question. The HS uses numbers consisting of six to ten digits to categorize commodities based on their level of manufacture. Some countries have expanded upon the HS with their own national standards. The United States (US) and states in the European

Union (EU) developed the Harmonized Tariff System (HTS) and the Integrated Tariff of the European Communities (TARIC), respectively (USITC, 2024; European Commission, n.d.). These longer codes provide more detailed categories for commodities.

When wildlife products are imported into a country, they are assigned a harmonized code by the relevant customs authority. Though these codes describe commodity categories, they typically provide little to no information on species, origin, or other details potentially relevant to investigating sustainability or risk (Biondo & Burki, 2019; Tlusty et al., 2023). A shipment of ornamental fish imported to the US, whether it contains cardinal tetra (*Paracheirodon axelrodi*), goldfish (*Carassius auratus*), or Asian arowana (*Scleropages formosus*), will either be assigned HTS code 0301.11.0090, indicating the contents contain "freshwater fish, ornamental, live", or 0301.10.0000, indicating the contents contain "freshwater fish, unspecified". This designation is recorded by customs authorities and later enshrined in data accessible through third-party platforms, such as Comtrade or Panjiva (United Nations, 2024; Panjiva, n.d.). In the US, Customs and Border Protection (USCBP) employ a *de minimis* rule, whereby shipments below a value threshold are exempt from more stringent regulatory requirements (USCBP, 2024). To further compound the issue, HTS guidelines recommend describing certain commodities with standard units of measurement (USITC, 2024). For live fish, HTS guidelines recommend using kilograms. However, shipments of live fish contain varying proportions of water weight and packing materials (The Fish Site, 2024). Similar issues affect records documenting trade in mammals, reptiles, and amphibians. Leathers made from reptile skins fall under various headings in HTS Chap. 4, all of which serve to describe the product's economic value and not the biological characteristics of the wildlife involved (USITC, 2024). Dozens of reptile species are harvested from countries around the world to satiate the demand for luxury leather products and this type of reporting makes it challenging to estimate the number of animals impacted and also their true origins (Marshall et al., 2020; Sosnowski & Petrossian, 2020).

Some mechanisms exist to fill gaps in wildlife trade monitoring. CITES provides a legal framework for all countries party to the convention. Due to the temporal consistency of reporting obligations, CITES can provide historical data through the CITES Trade Database that may be useful for evaluating conservation issues posed by wildlife trade (Berec & Šetlíková, 2019). However, only ~10.5% of all wildlife trade is covered by CITES (Watters et al., 2022). Some individual countries and regions have developed wildlife trade regulation data management systems outside of those required by CITES. One of the most prominent wildlife trade regulatory authorities is the United States Fish and Wildlife Service (USFWS) Office of Law Enforcement (OLE), which employs wildlife inspectors (WIs) to support customs officers (USFWS OLE, n.d.-b). Imports and exports of wildlife products are required to follow specific additional processes beyond those required by USCBP, including additional declaration processes and required submission of permits and certificates (USC Title 50 C.F.R. § 14.61; USC Title 50 C.F.R. § 14.63; Rhyne et al., 2017). The USFWS ostensibly records all wildlife trade through LEMIS, a multi-purpose resource management tool that, among other things, records wildlife trade data for wildlife commodities imported to and exported from the US (Eskew et al., 2020; Weissgold, 2024). LEMIS data is generally considered to be one of the most robust wildlife trade data sources available (Smith et al., 2008; Eskew et al., 2020). However, despite being developed and applied to specifically address international wildlife trade, both LEMIS and CITES Trade data have severe limitations that must be understood to properly evaluate the data they contain (Weissgold, 2024).

CITES in Brief

CITES represents a cooperative agreement between signatories to control trade in certain wildlife species (CITES, 2024a). The CITES framework functions by providing a blueprint for monitoring and regulating international wildlife trade alongside customs protocols. While the precise monitoring mechanisms vary between countries, similarities in import procedures enable relatively simple ways to implement CITES requirements. When a commodity enters a country's borders, it likely does so under bond. The freight is removed from its conveyance and enters a physical location, such as a warehouse (Johnson & Bade, 2010). The commodity remains in bond until an agent representing the importer provides documentation to warehouse staff that proves all legal requirements for importing the commodity are satisfied. This usually includes fulfilling CITES requirements by furnishing the appropriate CITES export permit to the Division of Management Authority (DMA) representatives or their delegates, as well as satisfying any additional declaration obligations, such as those required by USFWS. Export processes are slightly different in that the CITES export permits must be presented to officials alongside the commodity prior to entering bond in the country of export. Once all administrative conditions are satisfied the item enters bond and may be exported.

One especially important facet of CITES compliance is that it only applies to species listed on one of the three Appendices (Tlusty et al., 2023; Easter & Carter, 2024).

Appendix I confers the highest level of protection, where wildlife or wildlife products require a CITES export permit, issued by the exporting country, *and* a CITES import permit, issued by the destination country. Generally, no commercialized trade in Appendix I specimens is allowed. Some exceptions exist, whereby Appendix I specimens may be commercialized if originating from a registered captive breeding institution. Examples of CITES I species that are produced from registered captive breeding operations include Asian Arowana (*Scleropages Formosus*) and Gyrfalcons (*Falco rusticolus*). Appendix II consists of species that may be threatened by international wildlife trade, but regulated trade is allowed for any purpose. CITES Appendix II also includes "look-alike" species that are not themselves threatened by trade, but closely resemble those that are and would provide enforcement challenges if not also listed (Kolby, 2019). Consistent monitoring of trade in these species is relevant to ensuring long-term conservation and economic interests. A CITES export permit is required for these species.

Species listed in Appendix III do not typically meet the same "proof-of-threat" requirements as those in Appendices I and II. Appendix III consists of species that individual countries list unilaterally in order to protect native species that might only be threatened by trade through part of their range. Trade in Appendix III species must be accompanied by a Certificate of Origin document rather than a CITES export permit.

Like a CITES export permit, the Certificate of Origin facilitates monitoring and regulation of species that are traded internationally. Party countries are expected to enforce Certificate of Origin requirements for Appendix III species coming from the specified populations.

Trade in CITES-listed wildlife specimens typically occurs as follows: anticipating an export, an exporter or broker applies for a CITES export permit. Once obtained, the freight is prepared for export and both the CITES permit and the freight are made available to a relevant authority acting on behalf of the exporting country's DMA (USFWS, 2008; CITES, 2024a). This authority confirms the veracity and accuracy of the permit information and stamps the permit, thus validating it. The validation process may include a physical inspection of the freight at the validating officer's discretion. Once validated, the permit accompanies the freight along its transit route, although the reality of international business processes results in exceptions. Once the freight reaches its import destination, the CITES permit and the freight are made available to the importing country's DMA or delegated authority, where the accuracy and the veracity of the permit are once again compared to the contents of the imported freight (AHPA, 2023). Like before, this may include physical inspection at the discretion of authorities. The original CITES permit is retained by the authority, a copy is issued to the importer or their agent, and the remaining importation process is allowed to proceed (CITES, 2024a).

CITES Trade data is primarily collected at two points; export data is collected by the issuing or validating authority for the country of export and import data is derived from the documentation process employed by the importing country's inspection authority. Both records are sent to the CITES secretariat as part of each country's annual reporting requirements. Within the CITES Trade Database, the import and the export may be displayed as two separate trade events denoted by the "ReporterType variable" (Full CITES Trade Database Download, 2023). They are linked in the CITES Trade Database by a randomized CITES permit number stand-in. Discrepancies between linked trade events can occur for several reasons (Berec et al., 2018).

The volume of internationally traded wildlife commodities is impossible to physically inspect in full with the current distribution of resources reserved for wildlife trade control (Tlusty et al., 2023). When a physical inspection fails to occur during the export, import, or both phases of the trade process, the possibility for a data discrepancy increases. A validated CITES export permit includes two quantity fields (CITES, 2016). One field indicates the maximum quantity of a specified wildlife product that is allowed to be contained in the relevant shipment. It also denotes the appropriate units of measurement for describing said quantity. This field is typically pre-printed when the permit is created (OG *pers. obs.*). The second quantity field is completed by the validating official and represents the actual contents of the shipment. The validating official can determine the actual quantity by inspecting shipment documentation such as invoices and packing lists or by completing a full physical inspection.

Half-measures exist for both processes. Quantity verification is often completed on the basis of minimal information. Overwhelmed inspectors lack the time to complete exhaustive reviews of all documentation, let alone full physical inspections, and so must judge shipments based on the amount of risk they identify within a timeframe deemed appropriate by their agency standards (Tlusty et al., 2023). Additionally, it is not uncommon for the exporter to provide one set of supporting documents to the export-side inspecting officer while the import-side inspecting officer receives a different set of documents (Rhyne et al., 2012; Slábová et al., 2021). Such an event is not always nefarious, and may result from miscommunication between brokers and traders or even last-minute modifications to drafted invoices. The high volume of trade and the necessary expediency of international business interfere with inspectors' abilities to fully verify the accuracy of every CITES permit, and so the second quantity field may be completed inaccurately. When an inspector encounters a CITES permit that is completed incorrectly, they conduct a physical inspection of the freight and document the actual contents they find. *If the actual contents of the freight are less than the allowed quantity, this will not invalidate the legality of the permit or the transaction.* This usually occurs because exporters submit stock applications with prefilled whole number values or there are last-minute changes to the shipment due to the exporter's available inventory (*OG pers. obs.*). If the actual contents of the freight exceed the permitted amount, the entire shipment may be subject to seizure, although actual enforcement actions may vary between parties. Each country submits an annual report to the CITES Secretariat based on their records, resulting in mismatching records between the exporter and importer reported quantities in the CITES Trade Database (Slábová et al., 2021). Missing records on either side can occur when either or both responsible authorities are denied the opportunity to process the freight accordingly. This is typically a violation of CITES guidelines, and thus illegal, but processes exist to allow for the correction of honest mistakes or events that fall outside the control of the importer (CITES, 2018).

Interpreting CITES Data

The practical implementation of the CITES framework means unique characteristics of the CITES Trade Database must be considered when assessing wildlife trade. For example, reported illegal activity only represents events that were identified and acted upon by empowered authorities. Even then not all illegal activity that is identified is reported to the CITES Secretariat when parties submit their annual reports ('tSas-Rolfes et al., 2019). This may be due to concern over criticism or funding. To overcome this challenge, the CITES Secretariat recently implemented the CITES Seizure Database, which, unlike the CITES Trade Database, is not available for public interrogation (United Nations, 2022). Another challenge is the completeness, or lack thereof, of annual report submissions. Parties are required to submit annual reports every year and may be subject to sanctions if delinquencies carry on too long (CITES, 2019). Nevertheless, delinquencies inevitably occur. Missing annual reports from one or two countries can have severe impacts on the completeness of available data, particularly when they are major importers or exporters of wildlife (CITES, 2024b). CITES Trade data is also used to measure biodiversity present in international wildlife trade (Sheffers et al., 2023). This type of analysis conflicts with the design of the CITES framework.

While there are over 40,000 species of plants and animals listed under CITES across various taxonomic levels, biodiversity is not evenly represented in the CITES Appendices (Hughes et al. 2021; CITES, 2023; Donald et al., 2024). Approximately 6600 of CITES listed species are animals. This is important because the CITES Trade Database is not capable of reflecting trade in any wildlife species that are not listed in its appendices. Species will also not appear in the CITES Trade Database unless they appear in international trade after they are listed in the appendices. Similarly, species will cease to appear in the database once they are un-listed. Not all species listed in the CITES appendices are subject to regular trade, or even any trade at all (Harfoot et al., 2018; Morton et al., 2022). Figures 10.1 and 10.2 suggest that ~12–16 of CITES-listed species are traded annually. A CITES listing is meant to confer legal protections and required monitoring to species who, being subject to any number of human, environmental, and geographic threats, may be pushed further towards extinction by trade-based pressures. Furthermore, CITES listings may be extended towards "look-alike" species, which can be perceived to be economically substitutable

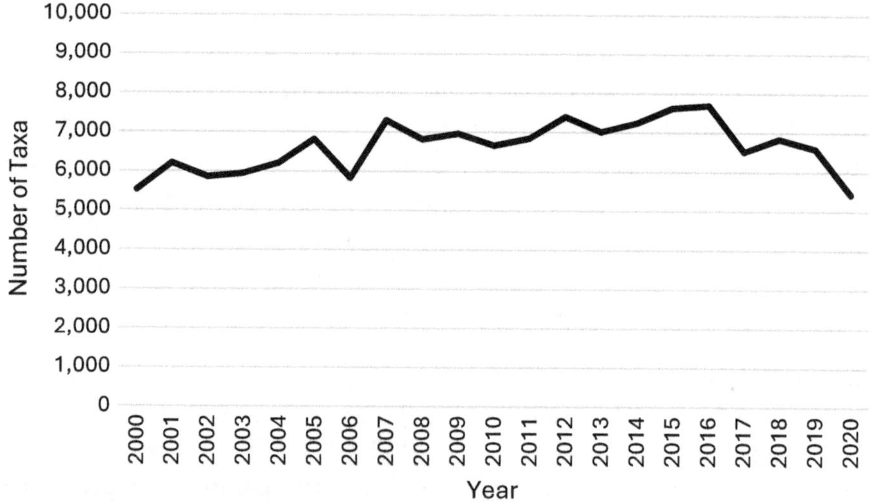

Fig. 1 Annual reported CITES export permits as reported by exporting country, inclusive of all-purpose types. Only exporter-reported data were included as known failure to document imports has been observed by Robinson & Sinovas (2018)

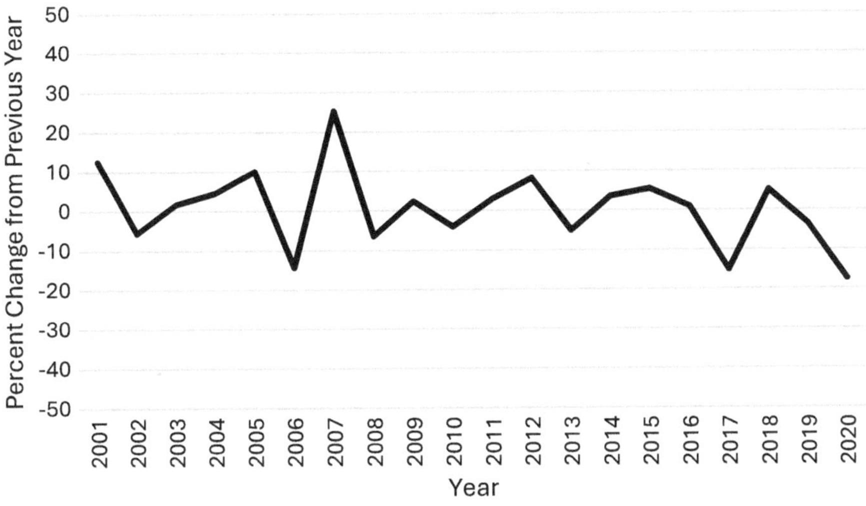

Fig. 2 Average rows of data per CITES permit as reported by exporting countries to the CITES Secretariat. "Rows", or the number of different commodities listed on a permit, provide an imperfect indicator of shipment complexity

for—or difficult to visually distinguish from—threatened species and therefore pose challenges for enforcement (Alfino & Roberts, 2019). Due to these factors, analyses of the biodiversity present in international trade using the CITES Trade Database will likely indicate, as observed by Hughes et al. (2021), that the number of CITES-listed species in trade is largely static, with major fluctuations apparently correlating with listing/de-listing decisions made at Conference of the Parties (CoP) events. This phenomenon occurs even as volumetric indicators of trade suggest overall increases in the frequency and size of individual shipments (see Figs. 1, 2, 3, and 4). Lastly, the CITES database provides limited means to measure trade volume by shipment. Rows of data are generally linked to permits, which are randomized to preserve anonymity. However, a single shipment can contain wildlife linked to several separate CITES permits. In the last 20+ years, the number of items per CITES Export permit has decreased substantially while the number of Export permits used to conduct trade has risen (see Figs. 3 and 4).

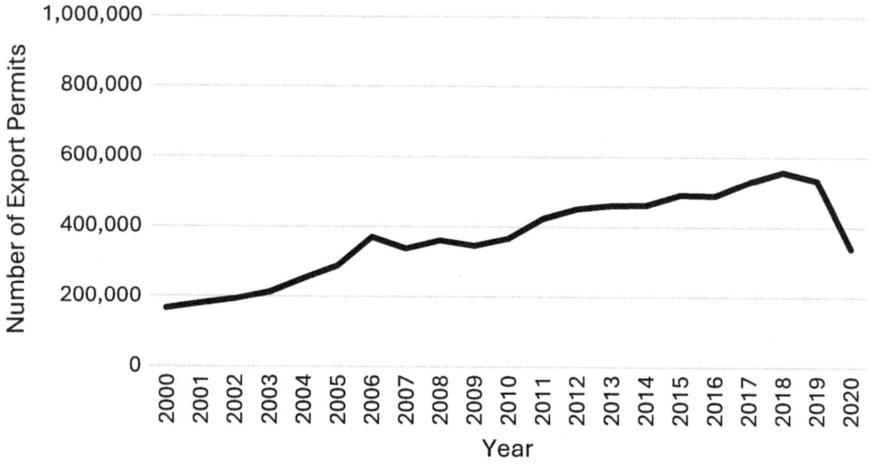

Fig. 3 Annual reported unique CITES-listed taxa traded as reported by both importers and exporters. Both reporter types are used to more effectively represent the biodiversity recorded through CITES Trade monitoring

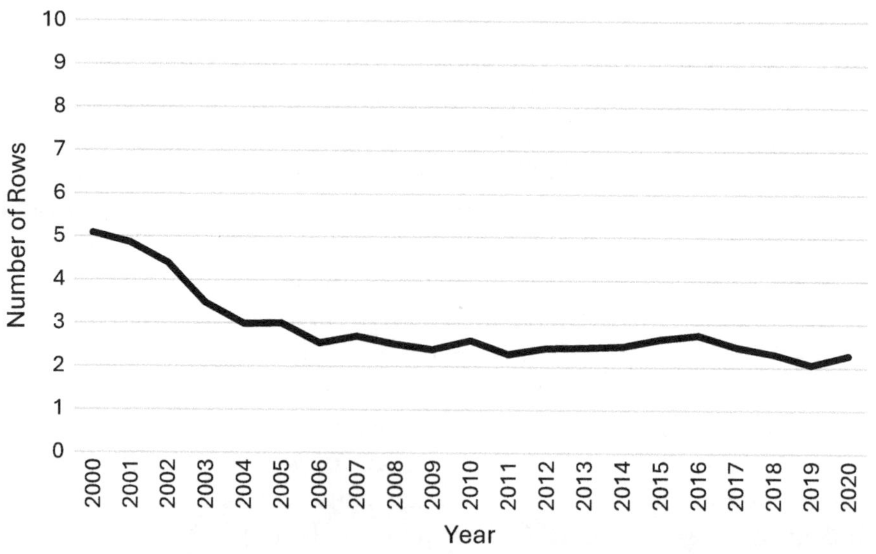

Fig. 4 Year-on-year percent change in unique CITES-listed taxa records. Major changes correspond to CoP events

LEMIS

LEMIS is another source of wildlife trade data commonly used to evaluate international wildlife trade. LEMIS is a multi-purpose resource management platform used by USFWS OLE to keep track of investigations, evidence, case subjects, violations, and a range of other information (Tlusty et al., 2023: Weissgold, 2024). LEMIS allows WIs to interface with the online declaration system used by importers and exporters. It enables the USFWS to process and store declaration information and it houses a repository of pre-built codes that provide crude, but standardized species categorization without overburdening the system with typos, synonyms, and errata. Additionally, it necessarily only reflects wildlife trade data where a US actor is either the importer or the exporter (Eskew et al., 2020). LEMIS data may still be reflective of global trends as the US is one of the largest single-country importers of wildlife products in the world (Eskew et al., 2020; Center for Biological Diversity, 2023).

USFWS and US Wildlife Trade

In order to better understand LEMIS trade data, a fundamental understanding of how the US manages wildlife trade is necessary. Like any other commodity, when a wildlife part or product is imported to the US, it must remain in bond until the legal requirements for release are met (USCBP, 2023b). USCBP makes the final determination to release any trade goods in bond. USCBP is assumed to verify that a declaration has been provided to the USFWS, and, if so, to further verify that the importer has proof that the USFWS reviewed and cleared (approved) the declaration (USCBP, 2023a; USC Title 19 C.F.R. § 113).

When wildlife or wildlife products are imported to the US, the importer or their agent must complete a declaration form known as a USFWS Form 3–177 (USC Title 50 C.F.R. § 14.61). This form documents characteristics of the proposed shipment, including taxonomic classifications and commodity types (USFWS, 2017). A Form 3–177 must be accompanied by supporting documentation, including relevant CITES permits, invoices, bills of lading, waybills, and customs entry forms. A USFWS WI will review the declaration form and the document package and assign it one of several statuses: "cleared", "refused", or "rejected" (Eskew et al., 2020; Easter & Carter, 2024). A cleared declaration has been approved and can continue through the remainder of the import process. A refused declaration has likely been found to be in violation of US law, subjecting the property to enforcement action. A rejected declaration must either be corrected or presented to the USFWS for physical inspection so the contents can be verified against the declaration information. Once a declaration is processed, that record is stored in LEMIS where it can be accessed by the public in accordance with the US Freedom of Information Act (FOIA) 1967 (Eskew et al., 2020).

Interpreting LEMIS Data

Despite the relative robustness of LEMIS data compared to other wildlife trade data sources, effective analyses require careful consideration of a number of caveats. When an importer completes a declaration form, they are expected to describe the species in box 16a of the Form 3–177 (USFWS, 2017). A WI uses this information, in conjunction with supporting documentation and a physical inspection, to verify the species in the shipment and assign an appropriate Species Code. A Species Code is a two- to four-digit alphanumeric sequence assigned by USFWS to represent a species, genus, and/or family name (USFWS, 2023a). These codes enable USFWS to standardize taxonomic information in declarations and can include detailed information differentiating sub-species populations.

The breadth of biodiversity recordable in LEMIS is superficially limited by Species Codes. When LEMIS data is obtained through a FOIA request, the Genus and Species columns (and any other taxonomic data columns) are populated based on the corresponding Species Code, not the actual information provided by the importer, exporter, or WI (OG *pers. obs.*). For some taxonomic records, it can appear that wildlife trade is not being recorded at a species level. This does not necessarily mean WIs are approving declarations that lack species information. If a Species Code for a species does not exist, a WI may have to use a code corresponding to the genus, family, or class in order to process the declaration record. In such an instance, the WI can write the precise species name in box 16a of the Form 3–177. However, correct species names written in this way have not typically been made available when LEMIS datasets are requested through FOIA (OG *pers. obs.*). Instead, the taxonomic fields are filled according to the standardized information associated with the Species Code. This creates a disparity between the quality of wildlife trade data that is recorded and the quality of wildlife trade data that is shared for research and policy development. Furthermore, sometimes a code of "****" is assigned to a declaration record. This code indicates that no Species Code exists for the recorded taxonomic information and enables the information entered into box 16a to be returned in a FOIA query exactly as written.

In other instances, the overreliance on Species Codes causes data-loss. Until 2005, both freshwater and marine ornamental fish were recorded using "TROP" (all tropical fish including goldfish) (Smith et al., 2008). After 2005, new codes were created to provide alternatives to TROP: "FWTF" (freshwater tropical aquarium fish), "MATF" (marine aquarium tropical fish), and a diversity of codes for *Carassius* species (USFWS 2023a). Thousands of species are imported annually under MATF alone, but the use of these bundled codes allows importers to obfuscate data (Rhyne et al., 2012, 2017).

Species Codes limit the extent to which biodiversity in wildlife trade can be measured in LEMIS. However, the difference between CITES and LEMIS recording parameters allows unique insights into trends affecting non-CITES specimens. For one, biodiversity in the international wildlife trade appears more varied than that recorded by the CITES Trade Database (see Fig. 5). Additionally, biodiversity appears to increase annually, although this pattern is confounded by the limitations of Species Codes. New Species Codes are created over time, and limited information is available to indicate when a Species Code was first created, so actual changes in biodiversity may vary from what can be measured currently (Smith et al., 2008; Rhyne et al., 2012; Tlusty et al., 2017).

Procedural issues also affect LEMIS data. While most declaration forms are submitted electronically, all ports provide paper declaration forms upon request. Some ports receive more paper declarations than

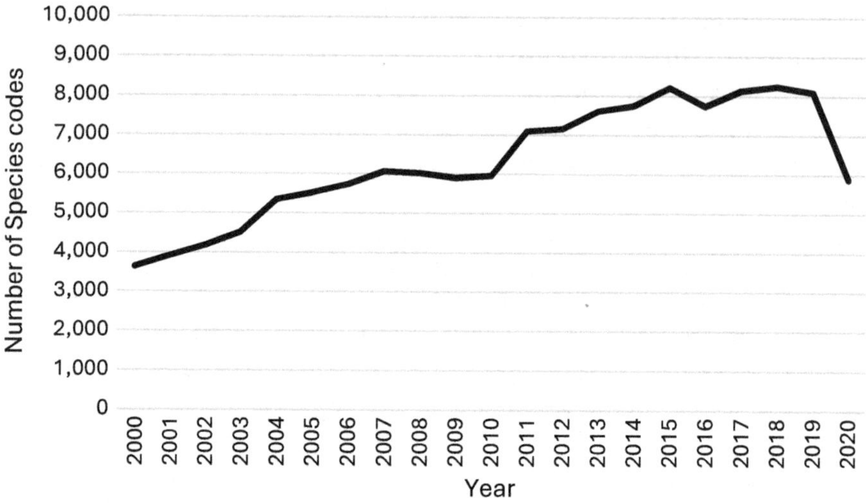

Fig. 5 Annual reported unique taxa, as indicated by LEMIS Species Codes. Unique species codes may be duplicative, whereby some codes record taxa at the genus or family level when LEMIS has the capacity to record them at the species or sub-species level. Additionally, some Species Codes are duplicates of each other

others. Summary analyses of LEMIS data reveal time lags between three events: shipment, disposition, and electronic record creation. Processed LEMIS declarations are assigned a ten-digit code that begins with a four-digit assignation indicating the year in which it was created. LEMIS data suggests some shipments have not been processed for up to 13 years after the documented shipment date. Comparison between LEMIS and CITES trade data suggests time lags can distort annual trade reports. For example, the authors identified 130 *Chlorocebus saebus,* imported into the US in 2015 per LEMIS, that appear to be documented as traded in 2017 in the CITES Trade Database, the same year the electronic LEMIS record was created (UNEP-WCMC, 2023). These desynchronizations largely affect ports which rely more heavily on paper declarations.

LEMIS presents a robust record of reported wildlife trade with US partners, but it passes through too many opaque filters to be consistently reliable for driving quantifications of actual wildlife trade. Approximately 100 WIs are tasked with regulating international wildlife trade in the US, and it would be unreasonable to assume that these officers are experts in identifying all taxa they intercept (USFWS, 2020). While these officers are experts in the international wildlife trade process, the expectation that such a small number of officers can detect, inspect, and identify hundreds of thousands, if not millions of specimens annually without interfering with the necessary expediency of international commercial trade is optimistic (Tlusty et al., 2023). Furthermore, except when published through FOIA requests or academic articles, these data are not subject to audits for quality or accuracy. This leads to the continued creation of records, such as those for American Black Bear trophies, *Ursus americanus*, that, although originally from Canada, are codified in both LEMIS and the CITES Trade Database as imported from Namibia (Berec et al., 2018; UNEP-WCMC, 2023). In this case, the issue results from erroneously entering permit values of 'N/A' as NA, the ISO-2 code for Namibia when the coder most likely meant to convey that the Country of Export was "Not Applicable" as the trophies were exported from their Country of Origin.

Seizure Data

Many published reports of trends in wildlife trade rely on records of seizures and other enforcement actions to form conclusions. Records of seized wildlife are available in both LEMIS and the CITES Trade Database, as well as a range of other sources. Generally, seizure records provide limited information regarding the purported violations (USFWS, 2022; TRAFFIC International, 2024; C4ADS, 2024b). Additionally, smuggling, trafficking, administrative errors, and exceeded permit allowances can all warrant similar enforcement action, despite having disparate criminality (Moreto & Pires, 2018; Margulies, 2022). Seizure databases largely fall into two categories: those developed for government/law enforcement use and those that are developed and maintained by NGOs. Examples of databases that fall into the latter category include the Center for American Defense's (C4ADS) Wildlife Defense Platform and TRAFFIC's Wildlife Trade Portal. Records added to these databases are largely informed by media articles and other publications documenting law enforcement responses to illegal wildlife trade (C4ADS, 2024a;

TRAFFIC International, 2024). As a non-primary source of data, these publications are strongly affected by the quality of reporting, which can frustrate the codification of accurate and reliable records. While these databases typically have processes in place to improve the overall truthfulness of added records, some inaccuracies inevitably fall outside of the control of the curating team (Esmail et al., 2020; C4ADS, 2024a). For example, customs authorities in India frequently intercept captive-bred reptiles, indicated by morphologies specific to captive-bred animals (Madhav, 2023; Venkadasan, 2023). Without sufficient contextual detail, such interceptions can be indistinguishable from interceptions of smuggled wild-caught animals, leading to potentially misinformed analyses of stress on native populations.

Examples of government initiatives to document wildlife seizures include the EU's TWIX, the United Nations Office of Drugs and Crime's World WISE database, and various national-level seizure record compilations. Many of these databases are unavailable for public inquiry or are only selectively available. When these databases *are* available, they can be rife with errors and incomplete documentation, such as those described in Petrossian et al. (2016) and van Uhm (2016b). Brazil's environmental authority (IBAMA) make their documentation of infractions publicly available; however, these infraction records are not always well-organized, evidenced by the categorization of *Sporophila, Podocnemis, Caiman,* and *Paroaria* species as 'anfíbio' (Amphibians) (IBAMA, 2023). The use of seizure data as a tool to evaluate wildlife trade and trafficking is also complicated by the scope of infractions addressed—wildlife trafficking is a specific illegal activity, and international wildlife trafficking necessarily differs from domestic wildlife trafficking. National laws that enable the enforcement of CITES requirements might not empower authorities to regulate domestic wildlife trade. When seizure records fail to differentiate between these types of infractions, they can easily be used to misrepresent wildlife trade (Kolby, 2019; Esmail et al., 2020; Challender et al., 2021).

Besides being subject to issues with detail and accuracy, sometimes these seizure databases conflate international wildlife trafficking with regional, or even domestic wildlife trafficking. Sometimes they are interrelated, such as jaguar poaching in Bolivia—but other times they are not;

the illegal trade in sea turtle eggs in Costa Rica apparently caters mostly to domestic demand, although such trends are subject to change (Goetschel, 2021; Pheasey et al., 2023). Seizure records often lack the kind of context that enables analysts to navigate these complexities effectively.

Specific Examples of Data Issues

Grouped Codes

As discussed previously, the USFWS uses some Species Codes that provide for the grouping of species with taxonomic or function-linked "species groups". Previously mentioned examples include FWTF and MATF, but others exist (see Table 1). The USFWS even allows for the use of family- or genus-level codes, some of which see use despite containing CITES- or ESA-listed species. In a previous analysis, Rhyne et al. (2012) obtained invoices and other documentation for marine ornamental fish shipments to compare against LEMIS records. Using optical character recognition, they identified several challenges affecting the accuracy of LEMIS records documenting trade in "species groups". Through their automated invoice scanning process, they identified CITES Appendix II Humphead wrasse, *Cheilinus undulatus,* present on invoices for shipments that were declared as only containing non-CITES marine ornamental fish. Additionally, they counted the number of individual fish described on the invoices and found an almost 50% difference between that and the invoice records (Rhyne et al., 2012; Rhyne et al., 2017). This difference is partially attributed to the use of MATF to obfuscate importations of live goldfish, *Cassius auratus,* which require commercial importation permissions from USDA that marine ornamental fish do not.

Misdeclaration

Analyses of wildlife trade and wildlife trafficking records are substantially hampered by their vulnerability to human error. In the legal wildlife trade, declarations for import, whether into the United States or another

Table 1 Some of the Species Codes used by USFWS LEMIS platform to record trade records at less detailed taxonomic levels and the number of times each was used between 2015 and 2020

Species Code	Class_ Code	"Genus" Description	"Species" Description	Common Name Information	Frequency
B000	B	UNKNOWN	CITES BIRD	BIRD	5
BI##	B	AVES		ALL BIRDS	5
BUTT	I	NONCITES	BUTTERFLIES		6693
FWTF	F	TROPICAL_ FISH	FRESHWATER		48,994
M000	M	UNKNOWN	CITES MAMMAL	MAMMAL	12
MA##	M	MAMMALIA		ALL MAMMALS	12
MATF	F	TROPICAL_ FISH	MARINE		55,173
NONA	A	NONCITES	AMPHIBIANS		1031
NONB	B	NONCITES	BIRDS		1263
NONC	Z	NONCITES ENTRY		NON CITES	3131
NOND	D	NONCITES	CORAL		1035
NONF	F	NONCITES	FISH		11,094
NONI	I	NONCITES	INSECTS		27,766
NONJ	J	NONCITES	ARACHNIDS		4008
NONL	C	NONCITES	CRUSTACEANS		18,080
NONM	M	NONCITES	MAMMALS		1719
NONR	R	NONCITES	REPTILES		1095
NONS	S	NONCITES	MOLLUSCS		1956
NONV	Z	NONCITES	INVERTEBRATES		9132
RE##	R	REPTILIA		ALL REPTILES	4
TROP	F	ALL LIVE TROP. FSH	(INCL. GOLDFISH)		8

Less descriptive codes such as these are used disproportionately. MATF, FWTF, NONI, and NONV are used frequently while M000 and MA## are primarily used to describe seizure records. These Species Codes account for 6.5% of all trade events documented in this period

country, are usually handled by customs brokers or freight forwarders (Reneau, 2011; USFWS OLE, n.d.-a). These are agents whose expertise lies in navigating the administrative process of international trade. Outside of a small number of specialist brokerages, they are not experts in wildlife, so when they receive shipping documents, they declare the species taxonomy as they are written in invoices or packing lists provided by manufacturers or other agents along the supply chain (OG *pers. obs.*).

Gerson and Remmal (2021) identified a case where an agent imported live flounder into Canada from Korea as *Paralichthys californicus*, California Flounder. *P. californicus* is native to the Pacific Coast of North America and does not extend to Korean waters. Further scrutiny of the shipping documents suggested the correct species was Olive Flounder, *Paralichthys olivaceous*, which cannot be imported live into Canada without an Aquatic Animal Health Import Permit issued by the regulatory authority of Canada. Such a misdeclaration could be intentional or a result of negligence. The author, in a previous career as a wildlife inspector, identified a case where an importer was bringing hundreds of pounds of dried sea cucumber into the US declared as a single species (OG *pers. obs.*). After genetic testing, it was apparent that more than five species of sea cucumber were in the shipment, four of which were not described in any of the shipping documentation, nor on the exporter's website.

The trade in seafood and other fisheries products exemplifies some of the most challenging aspects of wildlife trade regulation. An environment where high-value commodities that are shipped in high volumes bear little to no risk of punishment for violating laws invites a standard operating procedure that leaves little room for effective regulation. Large shipments of perishable or live goods must be processed immediately, or risk spoiling, or killing, the cargo (Tlusty et al., 2023). If a violation is discovered, USFWS looks to partners to provide facilities for perishable goods or care for live animals (USFWS, 2023b). If partners are unavailable, the USFWS must consider other options, such as re-export or civil penalties. Trafficking in species that are not considered conservation priorities but are nevertheless common in wildlife trade is typically met with the least severe response.

Conclusion

Quantitative analyses of international wildlife trade are critical to the sustainability of wildlife as a vital component of the environment as well as economic resources. These analyses depend on informed investigation of reliable data. Fortunately, there are several robust datasets available for public use, and still more available through cooperative relationships. The increasing abundance of data sources means that triangulation is becoming increasingly easy, although every dataset is bound to have its quirks. Additionally, the diversity and accessibility of available secondary data ensures that researchers are not limited by geography or funding as much as they might be capturing data in the field—although this point further stresses the need to promote cooperation as much as possible.

While researchers may find their ability to navigate wildlife trade data hampered by their lack of familiarity with the corresponding dataset, they can overcome these challenges by familiarizing themselves with the process through which these data are derived. For example, conservative analyses of LEMIS data can certainly provide insight into temporal and geospatial trends in international wildlife trade, and possibly even wildlife trafficking, but analysts are recommended to remember that LEMIS data, and CITES Trade data, can only ever describe recorded wildlife trade events. Recorded wildlife trade events may not always accurately reflect reality (as in the case of the Namibian Black Bears). As discussed previously, these data are prone to inaccuracies that can influence perceptions of trends. Researchers are encouraged to investigate outliers and other "weird" elements in order to better expand their comprehension of wildlife trade. In a recent study, Hughes et al. (2023) found millions of individuals imported into the US in single shipments. Further scrutiny revealed that these shipments were not typographical errors and consisted of agricultural mites meant to protect industrial crop fields from pests. Ultimately, the suitability of the data is most impacted by the appropriateness of the research question. Robust and appropriate wildlife trade analyses can provide valuable information crucial to research, conservation, and policy development.

Forcing wildlife trade data to conform to inappropriate research questions can negatively affect the ability of researchers and policy analysts to decipher trends. Scheffers et al. (2019) and Hughes et al. (2023) showcase robust analyses of phylogenetic and geographic characteristics of select species in trade but they do so by assuming that a species' presence in the CITES Appendices is proof of its presence in trade. This chapter provides clarification necessary to improve these analyses. Furthermore, analyses can have significant real-world policy implications. Despite a consistent failure to provide evidence of accurate record-keeping of illegal wildlife trade interventions, Homeland Security Investigations (HSI), a US agency under the Department of Homeland Security, continues to receive an increasing budgetary allowance for combatting wildlife trafficking (DHS, 2021; Fed Agent, 2023). This leads to jurisdictional overlap between HSI and USFW and results in, among other things, inconsistent documentation and disposal of biological materials that might pose zoonotic or conservation risk. This continued jurisdictional overlap can also further contribute to the obfuscation of IWT events, excluding them from relevant databases such as LEMIS and CITES.

These issues matter as security and conservation are becoming increasingly intertwined, and few issues exemplify this relationship quite as well as international wildlife trade (Massé et al., 2018; Duffy, 2022). This only makes the reliability and accessibility of wildlife trade data platforms that much more important. Without data-informed strategies, security-based interventions run the risk of introducing unwarranted harms, igniting tensions between law enforcement, NGOs, and local communities. Some organizations, such as LINI, have found success through working with local communities to enhance the benefits provided by trading wildlife resources, indicating that militarized strategies may not always be warranted (Watons & Roberts, 2015; OATA, n.d.).

Competent analyses of appropriate data support the development of effective interventions and help to ensure that policies designed to curb illegal, unsustainable, or risky wildlife trade are fair to the people affected, effective for the wildlife affected, and yield tangible results. This requires coordinated advocacy for improving monitoring capacity and transparency globally. Without transparent processes, wildlife trade regulatory processes will remain unreliable and ineffectual for successfully conserving at-risk species.

References

Alfino, S., & Roberts, D. L. (2019). Estimating identification uncertainties in CITES 'look-alike' species. *Global Ecology and Conservation., 18*, e00648. https://doi.org/10.1016/j.gecco.2019.e00648

American Herbal Products Association. (2023). *Primer on importing and exporting of CITES-listed species*. AHPA.

Berec, M., & Šetlíková, I. (2019). Important step to understanding the CITES trade database: A reply to Pavitt et al. *Biological Conservation, 230*, 197–198. https://doi.org/10.1016/j.biocon.2018.12.018

Berec, M., Vršecká, L., & Šetlíková, I. (2018). What is the reality of wildlife trade volume? CITES trade database limitations. *Biological Conservation, 224*, 111–116. https://doi.org/10.1016/j.biocon.2018.05.025

Biondo, M. V., & Burki, R. P. (2019). Monitoring the trade in marine ornamental fishes through the European trade control and expert system TRACES: Challenges and possibilities. *Marine Policy, 108*(8). https://doi.org/10.1016/j.marpol.2019.103620

Center for Advanced Defense Studies (C4ADS). (2024a). *About the dashboard.* Retrieved January, 2024, from https://wildlifedashboard.c4ads.org/home/about

Center for Advanced Defense Studies (C4ADS). (2024b). C4ADS *wildlife seizure dashboard.* Retrieved January, 2024, from https://wildlifedashboard.c4ads.org/home

Center for Biological Diversity. (2023). *Wildlife data: massive U.S. mammal imports threaten biodiversity*. Retrieved February, 2024, from https://biologicaldiversity.org/w/news/press-releases/wildlife-data-massive-us-mammal-imports-threaten-biodiversity-2023-05-03/

Challender, D. W. S., Brockington, D., Hinsley, A., Hoffman, M., Kolby, J. E., Massé, F., Natusch, D. J. D., Oldfield, T. E. E., Outhwaite, W., 't Sas-Rolfes, M., & Milner-Gulland, E. J. (2021). Mischaracterizing wildlife trade and its impacts may mislead policy processes. *Conservation Letters, 15*(1), e12832.

CITES. (2016). *Resolution conf. 12.3 (Rev. CoP16) Annex 2.* Retrieved March, 2024, from https://cites.org/sites/default/files/eng/res/12/E-Res-12-03R16-A2.pdf

CITES. (2019). *Resolution conf. 11.17 (Rev. CoP19).* Retrieved February, 2024, from https://cites.org/sites/default/files/documents/COP/19/resolution/E-Res-11-17-R19.pdf

CITES. (2023). *Appendices*. Retrieved January, 2024, from https://cites.org/eng/app/appendices.php

CITES. (2024a). *How CITES works*. Retrieved March, 2024, from https://cites.org/eng/disc/how.php

CITES. (2024b). *What is CITES?* Retrieved March, 2024, from https://cites.org/eng/disc/what.php

CITES Secretariat. (2023). *International trade in amphibians: a summary of the status, management, and trade in amphibians (amphibia spp.) to support the implementation of CITES decision 19.197*.

CITES Secretariat and UNEP-WCMC. (2022). *A guide to using the CITES trade database. Version 9*.

Convention on International Trade in Endangered Species (CITES). (2018). *Resolution conf. 12.3 (RevCoP18)*. Retrieved March, 2024, from https://cites.org/sites/default/files/document/E-Res-12-03-R18.pdf

D'Cruze, N., & Macdonald, D. W. (2016). A review of global trends in CITES live wildlife confiscations. *Nature Conservation, 15*, 47–63. https://doi.org/10.3897/natureconservation.15.10005

Donald, P. F., Fernando, E., Brown, L., Busana, M., Butchart, S. H. M., Chng, S., de la Colina, A., Ferreira, J. M., Jain, A., Jones, V. R., Lapido, R., Malsch, K., McDougall, A., Muccio, C., Nguyen, D., Outhwaite, W., Petrovan, S. O., Stafford, C., Sutherland, W. J., Tallowin, O., & Safford, R. (2024). Assessing the global prevalence of wild birds in trade. *Conservation Biology, e14350*. https://doi.org/10.1111/cobi.14350

Duffy, R. (2022). *Security and conservation: The politics of the illegal wildlife trade*. Yale University Press.

Easter, T., & Carter, N. (2024). Analysis of 20 years of turtle exports from the US reveals mixed effects of CITES and a need for better monitoring. *Conservation Science and Practice, e13092*. https://doi.org/10.1111/csp2.13092

Eskew, E. A., White, A. M., Ross, N., Smith, K. M., Smith, K. F., Rodríguez, J. P., Zambrana-Torrelio, C., Karesh, W. B., & Daszak, P. (2020). United States wildlife and wildlife products imports from 2000-2014. *Scientific Data, 7*(22). https://doi.org/10.1938/s41597-020-0354-5

Esmail, N., Wintle, B. C., t' Sas-Rolfes, M., Athanas, A., Beale, C. M., Bending, Z., Dai, R., Fabinyi, M., Gluszek, S., Haenlein, C., & Harrington, L. A. (2020). Emerging illegal wildlife trade issues: A global horizon scan. *Conservation Letters, 13*(4), e12715.

European Commission. (n.d.). *TARIC*. Retrieved January, 2024, from https://taxation-customs.ec.europa.eu/customs-4/calculation-customs-duties/customs-tariff/eu-customs-tariff-taric_en

Fed Agent. (2023). *New HSI unit takes aim at wildlife trafficking*. Retrieved February 10, 2024, from https://www.fedagent.com/news/new-hsi-unit-takes-aim-at-wildlife-trafficking

Full CITES Trade Database Download. Version. (2023). *Compiled by UNEP-WCMC*. Cambridge, UK for the CITES Secretariat. trade.cites.org

Gerson, H., & Remmal, Y. (2021). Report on use of Nature Intelligence System: automated screening of commercial import documentation – simulation. Canada Border Services Agency unpublished, 36 pgs.

Goetschel, A. M. M. (2021). *Illegal jaguar trade in Latin America: An evidence-based approach to support conservation actions*. PhD thesis, University of Oxford.

Harfoot, M., Glaser, S. A. M., Tittensor, D. P., Britten, G. L., McLardy, C., Malsch, K., & Burgess, N. D. (2018). Unveiling the patterns and trends in 40 years of global trade in CITES-listed wildlife. *Biological Conservation, 223*, 47–57.

Hinesley, A., Huges, A., & Margulies, J. (2024). Creating a more inclusive approach to wildlife trade. *Conservation Biology, e14360*. https://doi.org/10.1111/cobi.14360

Hughes, A. (2021). Primer: Wildlife trade. *Current Biology, 31*(19), R1218–R1224.

Hughes, A. C., Marshall, B. M., & Strine, C. T. (2021). Gaps in global wildlife trade monitoring leave amphibians vulnerable. *eLife, 10*. https://doi.org/10.7554/elife.70086

Hughes, L. J., Massam, M. R., Morton, O., Edwards, F. A., Scheffers, B. R., & Edwards, D. P. (2023). Global hotspots of traded phylogenetic and functional diversity. *Nature, 620*(7973), 351–357. https://doi.org/10.1038/s41586-023-06371-3

Hughes, L. J., Morton, O., Scheffers, B. R., & Edwards, D. P. (2023). The ecological drivers and consequences of wildlife trade. *Biological Reviews, 98*(3), 775–791. https://doi.org/10.1111/brv.12929

Instituto Brasileiro do Meia Ambiente e dos Recursos Naturais Renováveis (IBAMA). (2023). *Fiscalização—Auto de infração*. Retrieved January, 2024, from https://dadosabertos.ibama.gov.br/dataset/fiscalizacao-auto-de-infracao

Johnson, T. E., & Bade, D. (2010). *Export/import procedures and documentation*. AMACOM.

Keskin, B., Griffin, E. C., Prell, J. O., Dilkina, B., Ferber, A., Macdonald, J., Hilend, R., Griffis, S., & Gore, M. L. (2023). Quantitative investigation of wildlife trafficking supply chains: A review. *Omega, 115*(102780). https://doi.org/10.1016/j.omega.2022.102780

Kolby J. (2019). *Misuse of wildlife trade data jeopardizes efforts to protect species and combat trafficking (commentary)*. Retrieved January, 2024, from https://news.mongabay.com/2019/10/misuse-of-wildlife-trade-data-jeopardizes-efforts-to-protect-species-and-combat-trafficking-commentary/

Kolby, J. E., & Goodman, O. L. B. (2025). Implementation biases in wildlife trade regulation foster unscientific and inequitable intervention strategies. *Frontiers in Conservation Science, 6.* https://doi.org/10.3389/fcosc.2025.1488946

Lockwood, J. L., Welbourne, D. J., Romagosa, C. M., Cassey, P., Mandrak, N. E., Strecker, A., Leung, B., Stringham, O. C., Udell, B., Episcopio-Sturgeon, D. J., Tlusty, M. F., Sinclair, J., Springborn, M. R., Pienaar, E. F., Rhyne, A. L., & Keller, R. (2019). When pets become pests: The role of the exotic pet trade in producing invasive vertebrate animals. *Frontiers in Ecology and the Environment., 17*(6), 323–330. https://doi.org/10.1002/fee.2059

Madhav P. (2023). Passenger lands with 47 pythons, 2 lizards at Trichy airport, detained. *India Today*. https://www.indiatoday.in/india/story/passenger-lands-with-47-pythons-2-lizards-at-trichy-airport-detained-2413967-2023-07-31

Margulies, J. D. (2022). *The cactus hunters: Desire and extinction in the illicit succulent trade*. Minnesota University Press.

Marshall, B. M., Strine, C., & Hughes, A. C. (2020). Thousands of reptile species threatened by under-regulated global trade. *Nature Communications, 11*(4738). https://doi.org/10.1038/s41467-020-18523-4

Massé, F., Dickinson, H., Margulies, J., Joanny, L., Lappe-Osthege, T., & Duffy, R. (2018). Conservation and crime convergence? Situating the 2018 London illegal wildlife trade conference. *Journal of Political Ecology, 27*(1), 23–43. https://doi.org/10.2458/v27i1.23543

Moreto, W., & Pires, S. (2018). *Wildlife crime: An environmental criminology and crime science perspective*. Carolina Academy Press.

Morton, O., Scheffers, B. R., Haugaasen, T., & Edwards, D. P. (2022). Mixed protection of threatened species traded under CITES. *Current Biology, 32*, 999–1009. https://doi.org/10.1016/j.cub.2022.01.011

Panjiva. (n.d.). *Panjiva supply chain intelligence*. Retrieved March, 2024, from https://panjiva.com/

Petrossian, G., Pires, S. F., & van Uhm, D. P. (2016). An overview of seized illegal wildlife entering the United States. *Global Crime, 17*(2), 181–201.

Pheasey, H., Griffiths, R. A., Matechou, E., & Roberts, D. L. (2023). Motivations and sensitivities surrounding the illegal trade of sea turtles in Costa Rica. *Ecology and Society, 28*(4), 15. https://doi.org/10.5751/ES-14296-280415

Reneau, R. (2011). *Global trade compliance: Managing U.S. customs compliance*. Outskirts Press, Inc..

Rhyne, A. L., Tlusty, M. F., Schofield, P. J., Kaufman, L., Morris, J. A., Jr., & Bruckner, A. W. (2012). Revealing the appetite of the marine aquarium fish trade: The volume and biodiversity of fish imported into the United States. *PLoS One, 7*(5), e35808.

Rhyne, A. L., Tlusty, M. F., Szczebak, J. T., & Holmberg, R. J. (2017). Expanding our understanding of the trade in marine aquarium animals. *PeerJ, 5*, e2949.

Ribeiro, J., Reino, L., Schindler, S., Strubbe, D., Vall-llosera, M., Araújo, M. B., Capinha, C., Carrete, M., Mazzoni, S., Monteiro, M., Moreira, F., Rocha, R., Tella, J. L., Vaz, A. S., Vicente, J., & Nuno, A. (2019). Trends in legal and illegal trade of wild birds: A global assessment based on expert knowledge. *Biodiversity and Conservation, 28*(12), 3343–3369. https://doi.org/10.1007/s10531-019-01825-5

Robinson, J. E., & Sinovas, P. S. (2018). Challenges of analyzing the global trade in CITES-listed wildlife. *Conservation Biology, 32*(5), 1203–1206.

Robinson, J. P. W., McDevitt-Irwin, J. M., Dajka, J. C., Hadj-Hammou, J., Howlett, S., Graba-Landry, A., Hoey, A. S., Nash, K. L., Wilson, S. K., & Graham, N. A. J. (2019). Habitat and fishing control grazing potential on coral reefs. *Functional Ecology, 34*(1), 240–251.

Scheffers, B. R., Oliveira, B. F., Lamb, I., & Edwards, D. P. (2019). Global wildlife trade across the tree of life. *Science, 366*(6461), 71–76. https://doi.org/10.1126/science.aav5327

Slábová, M., Šetlíková, I., Vodrážková, M., Richtrová, Z., & Berec, M. (2021). The number of shipments in the CITES Trade Database does not accurately reflect the volume of traded wildlife. *Biological Conservation, 253*, 108917. https://doi.org/10.1016/j.biocon.2020.108917

Smith, K. F., Behrens, M. D., Max, L. M., & Daszak, P. (2008). U.S. drowning in unidentified fishes: Scope, implications, and regulation of live fish import. *Conservation Letters, 1*, 103–109.

Sosnowski, M., & Petrossian, G. A. (2020). Luxury fashion wildlife contraband in the USA. *EcoHealth, 17*(1), 94–110. https://doi.org/10.1007/s10393-020-01467-y

Stringham, O. C., Moncayo, S., Thomas, E., Heinrich, S., Toomes, A., Maher, J., Hill, K. G., Mitchell, L., Ross, J. V., Shepherd, C. R., & Cassey, P. (2021). Dataset of seized wildlife and their intended uses. *Data in Brief, 39*. https://doi.org/10.1016/j.dib.2021.107531

Swift, L., Hunter, P. R., Lees, A. C., & Bell, D. J. (2007). Wildlife trade and the emergence of infectious diseases. *EcoHealth, 4*(1), 25. https://doi.org/10.1007/s10393-006-0076-y

't Sas-Rolfes, M., Challender, D. W. S., Hinsley, A., Veríssimo, D., & Milner-Gulland, E. J. (2019). Illegal wildlife trade: Scale, processes, and governance. *Annual Review of Environment and Resources., 44*, 201–228.

The Fish Site. (2024). *Shipping fish in boxes*. Retrieved March, 2024, from https://thefishsite.com/articles/shipping-fish-in-boxes

The Ornamental Aquatic Trade Association (OATA). (n.d.). *Fishing for facts: An introduction to the UK ornamental fish trade*. Retrieved February, 2024, from https://ornamentalfish.org/wp-content/uploads/Fishing-for-facts-report-ONLINE-SPREADS.pdf

Tittensor, D. P., Harfoot, M., McLardy, C., Britten, G. L., Kecse-Nagy, K., Landry, B., Outhwaite, W., Price, B., Sinovas, P., Blanc, J., Burgess, N. D., & Malsch, K. (2020). Evaluating the relationships between the legal and illegal international wildlife trade. *Conservation Letters, e12724*. https://doi.org/10.1111/conl.12724

Tlusty, M. F., Bayina, N., Rhyne, A. L., Brown, C., & Smith, M. (2017). *Public aquaria: Marine ornamental species aquaculture* (1st edn.) (R. Calado, I. Olivotto, M. P. Oliver, & G. Joan Holt, Eds.). Published by John Wiley & Sons Ltd.

Tlusty, M. F., Cawthorn, D. M., Goodman, O. L. B., Rhyne, A. L., & Roberts, D. L. (2023). Real-time automated species level detection of trade documents systems to reduce illegal wildlife trade and improve data quality. *Biological Conservation, 283*(110022). https://doi.org/10.1016/j.biocon.2023.110022

Tow, J. H., Symes, W. S., & Carrosco, L. R. (2021). Economic value of illegal wildlife trade entering the USA. *PLoS One, 16*(10), e0258523.

TRAFFIC International. (2024). *Wildlife trade portal*. www.wildlifetradeportal.org. Retrieved February, 2024, from https://www.wildlifetradeportal.org

United Nations. (2022). *CITES illegal trade database*. Retrieved February, 2024, from https://citesdata.un.org/

United Nations. (2024). *UN comtrade database*. Retrieved February, 2024, from https://comtradeplus.un.org/

United Nations Environment Programme—World Conservation Monitoring Centre. (2023). *CITES trade database*. UNEP-WCMC. Retrieved February, 2024, from https://trade.cites.org/

US Customs and Border Protection. (2023a). *Bonds—Types of bonds (knowledge article 000001074)*. Retrieved February, 2024, from https://help.cbp.gov/s/article/Article-213?language=en_US

US Customs and Border Protection. (2023b). *What are the requirements for importing wildlife (fish, lizard, snakes, etc) and animal parts? (knowledge article 000001020)*. Retrieved February, 2024, from https://help.cbp.gov/s/article/Article-61?language=en_US

US Customs and Border Protection. (2024). *Basic importing and exporting*. Retrieved January, 2024, from https://www.cbp.gov/trade/basic-import-export

US Department of Homeland Security (DHS). (2021). *DHS needs to better demonstrate its efforts to combat illegal wildlife trafficking (OIG memorandum OIG-22-02)*. USDHSOIG.

US Fish and Wildlife Service (USFWS). (2008). 443 FW 1 wildlife inspection policy and procedures.

US International Trade Commission (ITC). (2024). *Harmonized tariff schedule*. Retrieved January, 2024, from https://hts.usitc.gov/

USC Title 19 C.F.R. § 113 Bond Application and Approval of Bond. (2024).

USC Title 50 C.F.R. § 14.61 Import declaration requirements. (2024).

USC Title 50 C.F.R. § 14.63 Export declaration requirements. (2024).

USFWS. (2017). *U.S. fish and wildlife service declaration for importation or exportation of fish or wildlife*. Retrieved February, 2024, from https://www.fws.gov/sites/default/files/documents/Wildlife%20Shipments%20-%203-177%202022%281%29.pdf

USFWS. (2022). *Law Enforcement Management Information System (LEMIS) recorded wildlife trade data from 2015–2020*. (Obtained under the Freedom of Information Act from USFWS; FOIA request 2022–000708; requested November 2021, received May 2022).

USFWS. (2023a). LEMIS wildlife trade species code database. (Obtained under the Freedom of Information Act from USFWS; FOIA request 2023–005337; requested July 2023, received August 2023).

USFWS. (2023b). *New network gives hope to animals trafficked through illegal wildlife trade*. Retrieved February, 2024, from https://www.fws.gov/press-release/2023-10/new-network-gives-hope-animals-trafficked-through-illegal-wildlife-trade

USFWS Office of Law Enforcement (OLE). (2020). *Annual report FY 2019*. Retrieved February, 2024, from https://www.fws.gov/sites/default/files/documents/2019-Office-of-Law-Enforcement-Annual-Report.pdf

USFWS Office of Law Enforcement (OLE). (n.d.-a). *About us*. Retrieved January, 2024, from https://www.fws.gov/program/office-of-law-enforcement/about-us

USFWS Office of Law Enforcement (OLE). (n.d.-b). *Information for importers & exporters*. Retrieved January, 2024, from https://www.fws.gov/program/office-of-law-enforcement/information-importers-exporters

van Uhm, D. (2016a). The illegal wildlife trade: Inside the world of poachers, smugglers and traders. *Springer*. https://doi.org/10.1186/s13550-015-0108-0

van Uhm, D. (2016b). Illegal wildlife trade to the EU and harms to the world. *Environmental Crime in Transnational Context: Routledge*, 59–82. https://doi.org/10.1186/s13550-015-0108-0

Venkadasan, S. (2023). 11 pythons seized from passenger at Chennai airport. *The Times of India*. https://timesofindia.indiatimes.com/city/chennai/11-pythons-seized-from-passenger-at-chennai-airport/articleshow/103402383.cms?from=mdr

Watons, I., & Roberts, D. (2015). *Literature review: The benefits of wild caught ornamental aquatic organisms*. Retrieved January, 2024, from https://ornamentalfish.org/wp-content/uploads/Literature-Review-The-Benefits-of-Wild-Caught-Ornamental-Aquatic-Organisms.pdf

Watters, F., Stringham, O., Shepherd, C. R., & Cassey, P. (2022). The U.S. market for imported wildlife not listed in the CITES multilateral treaty. *Conservation Biology, 36*(6), e13978. https://doi.org/10.1111/cobi.13978

Weissgold, B. J. (2024). US wildlife trade data lack quality control necessary for accurate scientific interpretation and policy application. *Conservation Letters, e13005*. https://doi.org/10.1111/conl.13005

Widdowson, D. (2007). The changing role of customs: Evolution or revolution? *World Customs Journal, 1*(1). https://doi.org/10.55596/001c.09038

World Customs Organization. (n.d.). *The HS: A multi-purpose tool. l*. Retrieved January, 2024, from https://www.wcoomd.org/en/topics/nomenclature/overview/hs-multi-purposes-tool.aspx

WWF-UK. (2019). More than words: are commitments to tackle illegal wildlife trade being met? a review of progress against commitments made at intergovernmental conferences. WWF-UK, Woking, UK. https://www.wwf.org.uk/sites/default/files/2019-08/WWF_IWT_Report_v8.pdf

Zimmerman, M. E. (2003). Black market for wildlife: Combating transnational organized crime in the illegal wildlife trade. *The Vanderbilt Journal of Transnational Law, 36*, 1657–1666.

Zsigmond, C. (2020). Trends in illegal wildlife trade across the EU between 2015 and 2020. *Sustainable Future, 1*(2), 161–170.

Culture, Politics and Social Constructions

Cockroach Races and Cricket Fights

Dina Siegel

Introduction

On one of my visits to Berlin at the end of the 1990s I was invited by my Russian acquaintances to attend a cockroach race at their apartment in the centre of the city. A small group of Russian-speaking immigrants were gathered around a two-metre-long wooden desk, and cockroaches were placed in four grooves, one end of which was illuminated (the start) and the other of which was kept dark (the finish). As explained by the participants, cockroaches usually try to escape from the light. Each cockroach had a name and number: (1) Elena, (2) Olga, (3) Piotr, and (4) Nikita. Spectators placed bets (30 marks per session), receiving a note with the name and number of 'his/her' cockroach. Music was turned on and cheerful shouts filled the space: '*Davai, druzhok*' ('Come on, little friend'), '*Dvigai*' ('Move!') etc. Whilst Nikita was the quickest, Piotr suddenly stopped in the middle and did not wish to continue the race, despite the

D. Siegel (✉)
Willem Pompe Institute for Criminal Law and Criminology,
Utrecht University, Utrecht, The Netherlands
e-mail: d.siegel@uu.nl

D. van Uhm, D. Siegel (eds.), *Global Green Crime and Ecojustice*, Palgrave Studies in Green Criminology, https://doi.org/10.1007/978-3-031-91557-4_11

cheerful cries of people around the table. One of the spectators even demanded to check the route as he suspected that something distracted Piotr during the race. He loudly accused the organizers of spreading jam on the desk to manipulate the results. The dispute almost ended in violence, had one of the organizers not swabbed the desk with a rag and showed it to the audience to prove the race was fair. The winner got 500 marks and opened a bottle of champagne to celebrate the victory with the other spectators.

Attending this event raised in me many questions. How can insects become the centre of such a furious quarrel? Why is there such a great fascination with insect races and fights? Why is this illegal in some countries? Should this event be considered a sport or entertainment—or perhaps even animal abuse and exploitation in the pursuit of gambling profits? And how ethical are these events?

Such questions encouraged me to dive into existing knowledge about insects, to try to understand their social, political, and cultural meaning for humans.

Green Cultural Criminology

Nigel South, one of the founders of green criminology, has observed that researchers must be sensitive to cultural beliefs, norms, values, customs, and the political and cultural symbols that can play a major and often decisive role in green criminology. New empirical research focusing on cultural aspects is proving important in the development of green criminology. Similarly, the cultural criminological 'godfather', Jeff Ferrell, also encouraged researchers to look for new ways to interpret the intersectionality between 'culture, crime and justice' and to develop new theoretical approaches, research methods, and empirical evidence to achieve these ends (Ferrell, 1999).

Despite these foundational remarks, 'green' studies remain a minority in the field of cultural criminology, and only a small number of green criminologists have tried to analyze the cultural meanings of green crimes (Wolfe et al., 2005; Brisman & South, 2013; Van Uhm & Siegel, 2016; Van Uhm, 2018; Siegel, 2024). In contrast, the complex interplay between nature and culture has become a prevalent theme in recent cultural anthropological research (e.g. Strang, 2006; Nance, 2013; Ingold & Palsson, 2013; Descola, 2013; Pauwelussen, 2017), and it is cultural anthropology specifically that has encouraged cultural criminologists to conduct empirical, primarily ethnographic studies related to the nexus of nature and culture.

For generations, different critical thinkers in criminology claimed we should re-consider the position of humans in regard to non-humans to avoid speciesism, respect animal rights and welfare, and change our proprietary attitude towards non-humans (Singer, 1975; Beirne, 2007; Sollund, 2011). Today a large amount of green criminological research is focused both on animals in general and on the relationship between animals and people specifically (Sollund, 2011; White, 2013; De Waal, 2016; Janssen & Schuilenburg, 2021). Wildlife trafficking is one of the most popular topics in contemporary green criminology and there exists much research on big animals, birds, and reptiles (Hubschle, 2017; Van Uhm, 2016; Van Uhm & Spapens, 2018; Wyatt et al., 2018; Sollund, 2019).

There are, however, almost no studies on insects, on their role in human lives, their influence on human behaviour, their representation in art, literature, and music and on their religious and symbolic meaning for millions of people around the world. The aim of this chapter, therefore, is to address this gap in green cultural criminology by analyzing two specific cases in which insects are the main actors, focusing on their historical legacy and traditional importance for the particular populations in question.

Insect Culture

In different cultures around the world, insects have typically been divided into two categories. They are considered either as positive, elegant, beautiful creatures, such as ladybirds, bees, butterflies, and singing crickets; or as negative, dangerous, and harmful, such as spiders, flies, fleas, cockroaches and worms, associated with slums, dirt, stink, and desperate wretchedness. Comparing someone with an insect from the second category is usually seen as pejorative, intended to offend and humiliate, and to associate the person in question with dirtiness, pauperism, immorality, beggary, and meanness, all of which cause disgust, fear and aversion to the public. In the classic works of Kafka (2000) and Camus (2012), insects represent a symbol, a metaphor, and a myth, of tragic lives and hostile invasions. In Russian literature, the world of humans is a reflection of the world of insects (Chechov, 1978; Krestovsky, 1990).

The attribution of this negative image to insects by humans has played a great role in social attitudes and even official policies. In Nazi propaganda, insect symbolism went so far as to validate the inhumane treatment of specific groups of people. Based on the theory of hierarchical race and social Darwinism, Jews were considered as inferior beings comparable to insects and pests, who should be exterminated to protect the 'healthy and pure' Aryan race against the 'infection' (Wiebe, 2008). Although these racial theories have been universally condemned and ridiculed after World War II, insect symbolism remains a part of our language, especially in propaganda and hate speech.

Conversely, humans have always maintained constant contact with insects—they have played a pivotal role in ritual acts, religious lore, traditional medicines, and entertainment. The ancient Greeks were especially interested in metamorphosis, with the butterfly becoming the symbol of the soul set free from the earthly caterpillar.[1] Insects are used as human food because of their highly nutritious quality, as 'tools' for cleaning bones in Australia, for producing flashlights in Venezuela, and as an effective fish bait. They have even been used as weapons of war. 'The oldest tactic in biological warfare was the heaving of beehives and wasp nests at an entrenched enemy' (Lockwood, 2008, p. 4). Hornets, bees, and wasps were catapulted in battles across Europe, and such entomological warfare also included the use of insects to spread diseases, cause agricultural harms, and mount physical attacks.

It therefore stands that there has always been a paradoxical attitude towards insects. In medicine, they are often identified as the source of contagion—many of the illnesses and tragedies afflicting mankind have been blamed on insects. Locusts, for example, have been frequently

---

[1] In Greek 'psyche' means both a butterfly and a soul.

associated with destruction and plagues. On the other hand, insects are considered a folk remedy for different diseases (Morris, 2020). Roasted cockroaches are used in food in Indonesia to cure asthma and in Thailand water beetles are eaten to stop diarrhea (Yi et al., 2010).

In some cultures, insects are viewed as forecasters of death, such as singing crickets found in a house. In other cultures, however, crickets are kept in special tiny bamboo cages precisely for their singing, which has been highly praised, especially by aristocrats (Morris, 2020). Negative attitudes towards insects in the West are rooted historically in the prejudices, whilst in Eastern Asia they are much more respected and appreciated.

Cultural entomology analyses the attribution of souls to insects. In several countries, they are used as totemic figures, symbolizing kinship with humans (Hogue, 2009). The insect totem is ascribed supernatural powers and the ability to contact gods, to request the protection of communities from famine, harm, epidemics, and other miseries. Various different insects are worshipped as totems. In his famous work on magical philosophy *The Golden Bough*, James Frazer uses insects as examples of imitative magic (Frazer, 1998). The most well-known mythological insect association is the sacred scarab beetle of ancient Egypt, which was always buried within every Egyptian tomb to guarantee rebirth of the dead. The scarab beetle represented the Egyptian god Khepera, the god of the morning sun, who was believed to move the sun across the sky and give birth to every new day at sunrise (Kritsky, 1991).

The use of insects for entertainment are also diverse. From the sixteenth century until the 1970s fleas 'performed' in circuses around Europe (Lerch & Murphy, 1997). Similarly, the fighting of crickets and rhinoceros beetles has also appeared historically as a gambling opportunity and way to manifest wealth. In China, more than one hundred traditional festivals are still related to insects, such as 'Silkworm Day' and 'Insects Delivery Day', at which cricket fights are a central feature (Yi et al., 2010, p. 148).

Besides crickets, other insects have been used to entertain people of all ages in various Asian countries. In South Korea, the water beetle is used in a roulette-like game, locally known as *mul bang gae nori* (Pemberton, 2003). Similarly, in parts of Thailand, fights between lucanid beetles are taking place on a competitive basis, with considerable stakes sometimes amounting to entire properties.

The behaviour of mantids has also inspired an entire style of kung fu, in which many of the postures and movements mimic the behaviour of the repetitive side-to-side gait of the 'praying' mantis (Berenbaum, 1996).

The legitimacy of these events has to do with ethical considerations, which are based on animal welfare ideas and/or religious perspectives. The most radical religious approach in this regard is Jainism, which demands from its followers extreme precaution in avoiding the harm of any living being, including insects. Jain monks carry a broom to move insects from their way without injuring them and also wear masks to avoid the unintentional inhalation of flying insects. Conversely, in animal welfare protocols there are almost no references to insects, and in agricultural ethics there is even a justification for their extermination due the harms they pose to harvest (Frolov, 2022). This ambivalent attitude towards insects remains one of the understudied issues in green and cultural criminology.

Cockroach Races

Cockroaches are amongst the oldest insects on the planet, with several studies having documented their initial evolution approximately 300 million years ago (Boudreaux, 1979). Of the 4400 species, only a few have adapted to indoor environments and human habitation (Brenner & Kramer, 2019, p. 61). Their presence is often viewed as a sign of poor hygiene and sanitation. Though they have not been proven to transmit diseases to humans cockroaches are stigmatized as dangerous and scary insects due to their dark varnish appearance, their behaviour as 'natural scavengers' and the presence of the *E. Coli* bacterium in their digestive system. This stigma is strengthened by horror movies, which often present (a giant black) cockroach as an enemy for humans (such as *The Nest*, 1987; *Mimic*, 1997; and *Terra Formars*, 2016). These negative images justify the killing of cockroaches and the perpetual fear of the potential harms they pose to humans.

However, some cockroaches, such as the Madagascan Hissing Cockroach, are also kept as pets. In some Slavic countries, such as Russia and Ukraine, there is a folk belief that keeping cockroaches at home will bring wealth. The so-called *prusaki* (big black cockroach) is said to also protect the home; when one changes residence, cockroaches should also move with the family. On the other hand, if cockroaches themselves are leaving the house, this is viewed as a sign of misfortune, meaning that one can expect a fire, or even death (Marjanić, 2019).

In addition to superstition and folk customs, cockroaches are also viewed as service providers for humans, like many other insects. They serve as food, recreation, tourism and, entertainment (Rai et al., 2023). The most popular of these leisure pursuits is cockroach racing.

In the mid- to late 1990s cockroach races were popular in Germany amongst immigrants from the former Soviet Union. Some even considered them to be a Russian legacy and part of an 'immigration culture', citing the description of cockroach races in the famous literary works of Aleksei Tolstoi ('The Adventures of Nevzorov, or Ibicus', 1924) and Mikhail Bulgakov ('Flight', 1927). After the 1917 October Revolution and the Russian Civil War which followed, hundreds of thousands of immigrants arrived in Europe, Turkey, the United States, Latin America and Australia, escaping from the revolution, economic reforms, and the Bolsheviks' repressive regime. The life of émigrés was not always easy, even to those who grabbed millions of rubles when leaving Russia. These immigrants found themselves in a country where their only contacts were other émigrés and had little to no knowledge of local language and a strong longing for home. For the poorer immigrants, such as dispossessed peasants, impoverished White Army officers, and other opponents of the new regime who had to flee abroad quickly to survive, life in emigration was even harder.

Cockroach races became one of the few entertainments of the émigrés, and for some their only income. For others, the races were a social occasion to meet compatriots, share memories, and justify their escape from the Motherland. These races became part of their cultural life and identity, a combination of old folk elements and imported Western European cultural trends (Raeff, 1990, p. 10). As Osis (2020, p. 11) notes, 'since the Russian word for "race" is the same as the word for "flight", these *bega* strongly suggest that the Russians émigrés themselves, and perhaps also those left behind in the midst of social upheavals—revolutions, civil war, etc.—are little better than cockroaches'.

Cockroach races have always been, and indeed remain, popular in Russia. In contrast to the immigration culture in the West, the races in Russia are linked to the nostalgic pre-Soviet times, being considered as a 'Tsarist amusement' and 'fun for aristocrats', and as a revival of a 'good tradition' for all Russians, who can participate through betting and enjoying, just as pre-Revolutionary aristocrats used to do.

Cockroaches are imported to Russia from Madagascar especially for these races. They are trained, follow a special diet of vegetables and fruits, and are starved several days before races to manipulate them into running faster in search of food. According to the organizers of these events, the training of cockroaches begins from their earliest days and is controlled by the Russian Cockroach Racing Federation. In Moscow, a Madagascan cockroach costs more than $50 and is usually considered a pet.

Most of the participants in Russian cockroach races are women. This fact challenges widespread stereotypes about female fear of cockroaches. The length of the treadmill is about 1.5 metres and the width 10 centimetres, which includes four running tracks. Each participant pays an entrance fee of 500 rubles (about 5 euros).[2]

Traditionally, cockroach races were considered to be a European entertainment; however, they are now popular all around the world. Although cockroaches are not migratory insects, they have still reached as far as Australia, perhaps by travelling in the luggage of immigrants. There is no record of who first organized cockroach races in Australia, which were primarily connected to gambling. Since the 1980s, however, cockroach racing has become the traditional way to celebrate national Australian Day. In Brisbane, Queensland, the World Championship Cockroach Races take place annually during the Australian Day festivities on the 26th of January. The championship includes several racing events, including the Story Bridge Hotel 'Gold Cup' in which everyone can participate, either by bringing their own cockroach from home or purchasing one at the contest. Each cockroach must be named, and each participant pays a small

[2] http://mkkuzbass.ru/2008/01/30/tarakan-razvlechenie-dlja.html.

entrance fee. There is a Hall of Fame which includes celebrities—the winners of the previous contests, which have names such as Soft Cocky, Cocky Dundee and Priscilla—the so-called 'Queen of the Drains' (Freeman et al., 1999). The names given to the cockroaches reflect Australian and world politics and popular culture. Other examples are Crawline Hanson, a cockroach named after right-wing politician Pauline Hanson, or Guns 'n' Roaches, after the American rock group Guns 'n' Roses.

Australians now consider the cockroach to be one of their national animals, like the koala or kangaroo, and have adopted it as a mascot. As Frew and White remark (2007, p. 11), 'this reflects the role that cockroaches have in Australian society, suggesting that every household experiences cockroaches irrespective of the owners' class and standing, so that they have become an integral part of a supposedly egalitarian Australian society'. Cockroach races have thus become a legitimate and popular way to celebrate their national day.

Gambling on cockroach races is also popular in the United States, especially at entomological exhibitions within university laboratories. The main purpose in this case is to promote the science of entomology to the public, especially to children (De Jong, 2004). The event 'Madagascar Madness' has taken place regularly at Loyola University in Maryland. The races are also popular prior to presidential elections, conducted every four years by the Pest Control Association in New Jersey. In 2008 two cockroaches, called Barack Obama and John McCain, went head-to-head in a cockroach race at Insectopolis bug museum, organized by the Rotary Club of Central Ocean Tom River's Bugfest in New Jersey.[3] The preparation for cockroach racing is taken very seriously as a sporting event. On occasion, cockroaches are fed sugar water (De Jong, 2004).

Giving human names to cockroaches, caring for their comfort, awarding them with special treats when they win, stroking their back, talking kindly to them and giving instructions and advice before and during the contest, and also taking care of their injuries or grieving if they die—these are examples of strong owner attachments to their racing cockroaches. It is a part of a humanization process, when non-humans are

[3] https://books.google.se/books?id=JTMEAAAAMBAJ&pg=PA14&redir_esc=y#v=onepage&q&f=false.

viewed as human and where the boundaries between the human and animal worlds become blurred.

Cricket Fighting

Cricket fighting is a popular form of entertainment in many Asian countries. It is a part of the so-called *hua niao yu chong* (flower, bird, fish, insect) approach, which typifies the Chinese tradition of cultivating and raising small plants and animals that are easy to care for (Suga, 2006). Much is written about the historical significance of this tradition, particularly the raising of crickets for competitive fights, but less is known about the daily practices of those who facilitate these events.

Cricket fighting dates back to the Tianbao era (742–756), when contestants would place two crickets in a basin and goad them to fight with each other until one insect became disabled or dropped dead. Bets were placed on the predicted winner. Under the Song dynasty (960–1279), cricket fights were popular among all social classes, from royal family members to poor peasants. The southern Song chancellor, 'Jia Sidao 賈 (1213–1275), whose obsession with concubines and fighting crickets famously led him to make light of an offensive by invading Mongol armies, is associated with a (now lost) Classic on Crickets (Cuzhi jing 促織經)' (Sterckx, 2012, p. 4).

The 'Cricket Emperor', Ming Xuanzong 宣宗 (1427–1464), was known for his passion for cricket fighting: thousands of crickets were sent to his castle annually and fighting was considered a serious event that justified numerous experts on insects being involved as advisers. Decorated ceramic cricket jars and cages remain a collector item to this day (Sterckx, 2012). Under the Ming (1368–1644) and Qing (1644–1911) dynasties, cricket fighting took place regularly in Beijing at special cricket markets located in Beixinqiao Residential Area, Dongsi Decorated Archway, Xisi Decorated Archway, and Tianqiao Market.

With the mass emigration of Chinese to the United States in the middle of the nineteenth century, cricket fighting became a popular form of entertainment in several American cities, including New York and Philadelphia (Costa Neto, 2003, p. 21).

Before the Chinese Revolution, cricket culture was flourishing. There were specialized shops selling crickets, decorated jars and cages and other related items. After the Revolution in 1949, cricket fights were banned as remnants of the capitalist past, specialized shops were closed, cricket cages were destroyed and 'cricket masters', namely experts in cricket culture, were persecuted (Pemberton, 2003). As a result of the ban, cricket fighting retreated to clandestine settings (Kevan & Hsiung, 1976).

Since the 1980s, however, cricket fighting has undergone a resurgence, especially in Shanghai, Beijing, Hong Kong and Guangzhou. There are various cricket fighting clubs and professional associations responsible for most of the events. There are two methods of gambling on cricket fights: firstly, the individual gambling, when cricket owners wager between $10 to $20 per fight against each other; secondly, there is the organized gambling which takes place at certain clubs, the so-called 'cricket hunting meets' with pin ('cake') bets equal to $5 and chu ('pig') bets at $200 (Costa Neto, 2003, p. 22). The Association for Cricket Fighting in Beijing is the main sponsor of national tournaments, which are televised using advanced video cameras to zoom in and project the fighting onto large screens (Berenbaum, 1996).

The preparations start long before a tournament. Farmers in remote rural areas hunt the crickets. These cricket hunters know the patterns of crickets' behaviour and develop the necessary skills required to catch them, preferably in late August–beginning of September, when they are fully grown and strong. They are then sold to local cricket merchants at the trade centres in villages, who further travel to sell them at urban markets. Some hunters visit these markets and sell crickets directly to customers (Suga, 2006, p. 80).

To keep them healthy and in good physical condition for fighting, there is a requirement for 'food', 'home' and 'bath'. Crickets are given the same food as their owners, fresh water, and are 'bathed' regularly to wash any dirt from their body and to make them feel cooler in hot days (Suga, 2006, p. 88). If any slight change is observed in their condition, Chinese traditional herbal medicine is used for treatment; it is also used for injuries during and after the fight. Preparations include regular health checks, training, and 'light trial fights' to observe their condition and determine their best time for being entered into fighting events.

Shortly before the fight, the crickets are either fed a tonic (e.g., honey) or starved to make them become more violent. There are different weight classes (heavy, middleweight, and lightweight) as fighters should be of the same size and weight. Small scales are used to weigh them publicly before each fight. The fight takes place in an oval or round ring and finishes when the winner spreads out his wings as a sign of victory, whilst his opponent is either dying or flying away from the fight. The 'losers' are usually set free and released into the wild.

The most legendary winning cricket was Genghis Khan, which earned the title of 'conquering cricket' (*shou lip*) following numerous victories. Such heroes are usually honoured and solemnly buried in small silver coffins when they die (Costa Neto, 2003, p. 23). Live champions can be sold for hundreds of euros.

Although China is considered the traditional centre for cricket fights, it is also a very popular and widespread form of entertainment in other Asian countries such as Indonesia and Vietnam. Cricket fighting is common in Indonesia, for example, despite the fact that it is illegal due to the gambling entailed. Professional gamblers and cricket fights take place daily, and Balinese fighting cricket culture includes collection, training, secret diets to enhance their fighting ability, and the construction of sophisticated cages (Pemberton, 2003, p. 153).

Chinese communities in the US, Canada, and Australia also consider cricket fights as sport and tournaments are organized on a regular basis. In contrast, there is a decrease in popularity of cricket fights in China. In Hong Kong, where cricket fighting was popular in the 1960s, it has recently become rarer. According to my respondents, one explanation can be the widespread use of pesticides, but another reason I heard during fieldwork was that young people consider cricket fights to be an entertainment of the older generations, which does not fit into their modern lifestyle.

Conclusion

This chapter offered a green cultural criminological analysis of the relationship between humans and insects, focusing on two specific practices—cockroach races and cricket fights. Although much is written about these specific insects in other disciplines, green criminologists have so far skipped or neglected the insects from their studies, perhaps due to their cultural stigma as a 'inferior animals'. Focusing on these two phenomena, on the one hand, opens up new perspectives for our understanding of the nexus between nature/culture, and, on the other, breaks the existing taboos and stereotypes in regard to these allegedly 'dirty and dangerous' small creatures.

This chapter also demonstrated the many paradoxes present in the relationship between humans and insects. Whilst there is a long and rich tradition of favourable human contact, where insects are praised and considered as pets, friends, and to a certain degree athletes, they are also treated as property, and, like many other animals, used for food, medicine, entertainment, and profit.

Gambling, despite being prohibited in many countries, remains an important element of both cockroach racing and cricket fighting. The legitimacy of these events has to do with ethical considerations, which are based on animal welfare ideas. There are, however, still no regulations, or even references to insects in this regard. In agricultural ethics, there remains a justification for their extermination due to the harms they pose to harvests and food, and their propensity for spreading disease.

Ultimately, the cultural significance of cockroach racing and cricket fighting cannot be ignored. These events are a continuation of tradition and identity, and an integral part of our global society. These aspects are especially relevant in the context of a general trend of extinction of species, including insects. Despite numerous warnings and increasing concern over the fate of insects, mainly as a result of pesticides, industrial pollution, and climate change, the importance of insects is still underestimated. Limited research and the low priority given to studies of insects remains problematic and cultural biases still prevail in our attitude towards insect populations. Many scholars from different disciplines are deeply worried by the current trend towards extinction (Cardoso et al., 2020). Green cultural criminological research must therefore play a pivotal role in improving knowledge and breaking taboos within this entomological context.

References

Beirne, P. (2007). Animal rights, animal abuse and green criminology. In P. Beirne & N. South (Eds.), *Issues in green criminology. Confronting harms against environment, humanity and other animals* (pp. 55–83). Willan Publishing.

Berenbaum, M. (1996). *Bugs in the system: Insects and their impact on human affairs*. Basic Books.

Boudreaux, H. B. (1979). *Arthropod phylogeny, with special reference to insects* (pp. viii–320). https://www.cabidigitallibrary.org/doi/full/10.5555/19790563244

Brenner, R. J., & Kramer, R. D. (2019). Cockroaches (blattaria). In *Medical and veterinary entomology* (pp. 61–77). Academic Press.

Brisman, A., & South, N. (2013). A green-cultural criminology: An exploratory outline. *Crime, Media, Culture, 9*(2), 115–135.

Camus, A. (2012). *The plague*. Vintage.

Cardoso, P., Barton, P. S., Birkhofer, K., Chichorro, F., Deacon, C., Fartmann, T., et al. (2020). Scientists' warning to humanity on insect extinctions. *Biological Conservation, 242*, 108426. https://www.sciencedirect.com/science/article/pii/S0006320719317823

Chechov, A. (1978). Sakhalin Island (Остров Сахалин). In *Полное собрание сочинений и писем в 30 томах*. T. 14, 15. Nauka (Наука).

Costa Neto, E. (2003). Entertainment with insects: Singing and fighting insects around the world. A brief review. *Etnobiología, 3*, 21–22.

De Jong, G. D. (2004). Betting on Blaberus: Beating the odds at cockroach races. *American Entomologist, 50*(3), 130–132.

De Waal, F. (2016). *Zijn we slim genoeg om tew eten hoe slim dieren zijn?* Uitgeverij Atlas Contact.

Descola, P. (2013). *Beyond nature and culture*. University of Chicago Press.

Ferrell, J. (1999). Cultural criminology. *Annual Review of Sociology, 25*(1), 395–418.

Frazer, J. (1998). *The golden bough*. Oxford University Press.

Freeman, D., Teplica, N., & Coonce, J. (1999). *100 things to do before you die: Travel events you just can't miss*. Rowman & Littlefield.

Frew, E. A., & White, L. (2007, July). Australia Day alternative events: An exploration of national identity and nationalism. *Proceedings of the Fourth International Event Research Conference*, Victoria University, Melbourne.

Frolov, A. N. (2022). Controlling the Behavior of Harmful Insects: Light and Chemical Signals and Their Combined Action. *Entomological Review, 102*(6), 782–819.

Hogue, J. N. (2009). Cultural entomology. In *Encyclopedia of insects* (pp. 239–245). Academic Press.

Hubschle, A. (2017). The social economy of rhino poaching: Of economic freedom fighters, professional hunters and marginalized local people. *Current Sociology, 65*, 427–447.

Ingold, T., & Palsson, G. (Eds.). (2013). *Biosocial becomings: Integrating social and biological anthropology*. Cambridge University Press.

Janssen, J., & Schuilenburg, M. (Eds.). (2021). Antropoceen. Special Issue. *Tijdschrift over cultuur en criminaliteit, 11*(1).

Kafka, F. (2000). *The metamorphosis* (p. 239). Fictionwise, Incorporated.

Kevan, D. K. M., & Hsiung, C. (1976). Cricket fighting in Hong Kong. *Bulletin Entomological Society of Canada, 8*(3), 11–12.

Krestovsky, V. (1990). *Peterburgs' slums (Петербургские трущобы)*. Художественная литература.

Kritsky, G. (1991). Beetle gods of ancient Egypt. *American Entomologist, 37*(2), 85–89.

Lerch, M. M., & Murphy, J. (1997). A circus flea. *The Lancet, 349*(9048).

Lockwood, J. A. (2008). *Six-legged soldiers: Using insects as weapons of war*. Oxford University Press.

Marjanić, S. (2019). Cockroaches: From belief narratives to the contemporary visual practice of Catherine Chalmers, or how cockroaches have survived on earth for more than 320 million years. *Folklore: Electronic Journal of Folklore, 77*, 139–158.

Morris, B. (2020). *Insects and human life*. Routledge.

Nance, S. (2013). *Entertaining elephants. Animal agency and the business of the American circus*. Johns Hopkins University Press.

Osis, N. (2020). The road and the dream: Mikhail Bulgakov's flight and contemporary Russian playwriting. *Irish Slavonic Studies, 27*, 1–21.

Pauwelussen, A. (2017). *Amphibious anthropology: Engaging with maritime worlds in Indonesia*. PhD dissertation, unpublished, Wageningen University, Wageningen.

Pemberton, R. (2003). Persistence and change in traditional uses of insects in contemporary East Asian cultures. In E. Motte-Florac & J. Thomas (Eds.), *Les 'insecets' dans la tradition orale ('Insects' in oral literature and tradition)*. Peeters.

Raeff, M. (1990). *Russia abroad: A cultural history of the Russian emigration, 1919-1939*. Oxford University Press.
Rai, S., Mishra, G., & Omkar. (2023). Insects in culture. In O. Omkar (Ed.), *Insects as service providers* (pp. 307–333). Springer Nature Singapore.
Siegel, D. (2024). Camel races—Tradition, sport and politics. In F. Massé, T. Wyatt, R. Wong, A. Hubschle, & L. Gutierrez (Eds.), *Handbook of wildlife crime*. Routledge.
Singer, P. (1975). *Animal liberation. A new ethics for our treatment of animals*. Harper Collins.
Sollund, R. (2011). Expressions of speciesism: The effects of keeping companion animals on animal abuse, animal trafficking and species decline. *Crime, Law and Social Change, 55*(5), 437–451.
Sollund, R. (2019). *The crimes of wildlife trafficking. Issues of justice, legality and morality*. Routledge.
Sterckx, R. (2012). Animals, gaming and entertainment in traditional China. In V. Lo (Ed.), *Perfect bodies: Sports, medicine and immortality* (pp. 31–38). British Museum Research Publications.
Strang, V. (2006). Substantial connections: Water and identity in an English cultural landscape. *Worldviews: Global Religions, Culture, and Ecology, 10*(2), 155–177.
Suga, Y. (2006). Independent paper. Chinese cricket-fighting. *International Journal of Asian Studies, 3*(1), 77–93.
Van Uhm, D. P. (2016). Monkey business—The illegal trade in Barbary macaques. *Journal of Trafficking, Organized Crime and Security, 2*(1), 36–49.
Van Uhm, D. P. (2018). The social construction of the value of wildlife: A green cultural criminological perspective. *Theoretical Criminology, 22*(3), 384–401.
Van Uhm, D., & Siegel, D. (2016). The illegal trade in black caviar. *Trends in Organized Crime, 19*(1), 67–87.
Van Uhm, D., & Spapens, T. (2018). Drie perspectieven op de illegale vogelhandel in Nederland. *Tijdschrift voor Criminologie, 60*(2), 231–245.
White, R. (2013). *Environmental harm. An eco-justice perspective*. Policy Press.
Wiebe, M. (2008). Extermination and euthanasia: Animal symbolism in Nazi Germany. *The Mirror—Undergraduate History Journal, 28*(1), 21–38.
Wolfe, N., Daszak, P., Kilpatrick, A., et al. (2005). Bushmeat hunting, deforestation, and prediction of zoonotic disease. *Emerging Infectious Diseases, 11*(12), 1822–1827.

Wyatt, T., Johnson, K., Hunter, L., George, R., & Gunter, R. (2018). Corruption and wildlife trafficking: Three case studies involving Asia. *Asian Journal of Criminology, 13*, 35–55.

Yi, C., He, Q., Wang, L., & Kuang, R. (2010). The utilization of insect-resources in Chinese rural area. *Journal of Agricultural Science, 2*(3), 146–152.

The Devastation of Sea Otters in the American Northwest. A Look From Eco Global History and Criminology (18th and 21st Centuries)

Matilde Souto Mantecón and Inés Arroyo Quiroz

This study* deals with a socio-environmental catastrophe: the devastation of sea otters in the North Pacific by the conversion of their skins into valuable commodities in international trade during the 18th and 19th centuries. This is just one case, but throughout history many ecological harms have been caused by human activities. It is true that they have not

* Original title in Spanish: Souto Mantecón, M. y Arroyo Quiroz, I. (2022). La devastación de las nutrias marinas en el noroeste americano: una mirada desde la historia y la criminología ecoglobales (siglos XVIII al XXI). En: Souto Mantecón, M. y Kent Carrasco, D. (Eds.). Miradas globales desde América Latina: estudio histórico más allá de lo nacional (pp. 89–114).

M. Souto Mantecón
Instituto de Investigaciones Dr. José María Luis Mora, Mexico City, Mexico
e-mail: msouto@institutomora.edu.mx

I. A. Quiroz (✉)
Centro Regional de Investigaciones Multidisciplinarias (CRIM, UNAM), Cuernavaca, Mexico
e-mail: inesaq@crim.unam.mx

D. van Uhm, D. Siegel (eds.), *Global Green Crime and Ecojustice*, Palgrave Studies in Green Criminology, https://doi.org/10.1007/978-3-031-91557-4_12

occurred deliberately and that, in many cases, they occurred without there being a clear awareness of what was happening; today, however, we do know that many human activities disrupt the environment, and we also know that continuing along this path, life, as we know it today, in all its biodiversity, will disappear. With this concern in mind, we have come together to study some of these processes from conservation biology, green criminology and history to make them known and, above all, contribute to forming a critical consciousness that, as far as possible, contributes to the avoidance of new ecological disasters.

The case of sea otters is emblematic for several reasons. It was a species used by humans since prehistory; in the early modern age, in a period of less than a hundred years, Europeans hunted it relentlessly to sell its skins in China and satisfy the vanity of the mandarins with them under the Qing dynasty. It is also an emblematic species because it was precisely the trade in otter skins that was one of the main economic activities in supporting the European advance and dominance over the North Pacific, one of the last maritime regions which was to come into dispute by the empires and that was key to establishing the global connections. In other words, the sale of these precious skins was an aspect of the expansion of capitalist world trade with which the process of globalization began.[1] It was a trade in luxury goods that linked three continents, created great fortunes, built empires, destroyed several indigenous peoples, and drove sea otters almost to extinction.[2]

Studied historically, this is a topic that must be considered from the perspective of global history and from oceanic history, understanding the Pacific as a space of articulation in which different cultures, both local and foreign, intermingled. The main character in this story, sea otters, are

[1] See for example Berg, 2019.

[2] The history of the fur trade as one of the engines of European expansion and the construction of great empires has been a widely studied topic. There is a vast historiography that deals with it from very different perspectives. There are, of course, the general visions of European expansion in early modernity, such as that of Eric Wolf, *Europe and the People Without History*, as well as the studies on the forging of each of the European empires. For the purposes of this chapter, it is enough here only to mention that the main historical works dedicated specifically to studying the exploitation of sea otters on the American coasts are those of Ogden, *The California Sea Otter*, 1941; Gibson, *Otter Skins,* 1992, and Ravalli, *Sea Otters,* 2018. What is important to highlight is that until now the least-worked perspectives in the exploitation of sea otters are the Spanish and the Mexican, which is why in this chapter we propose pay particular attention to it.

a species that is naturally distributed on the coasts of the North Pacific throughout a wide geographical space that goes from the Japanese islands of Hokkaido to Cedros Island in Mexican California.[3] Our purpose is to chronicle what happened to sea otters after they became prey and merchandise for European hunters and traders in the 18th century, especially on the American coasts, and particularly in California and by the Spanish. We will also be interested in studying how this vast maritime space became a scenario whose control was disputed by various foreign interests—Spanish, Russian, English, American—both to expand their territorial possession over American lands and ensure their commercial exploitation, and also to ensure its control over transpacific navigation, the last maritime stretch that was conquered to dominate global communication. Starting from global, oceanic and interconnected history, we are also interested in returning to this topic with a look that integrates different spatial scales, from the wide oceanic space of the North Pacific where several European imperial powers converged that connected with China, to the local scale on the coasts of Monterrey and San Francisco on the northwestern border of the viceroyalty of New Spain. This approach will allow us to analyse the early globalization generated by the global fur trade from Alta California, a territory on the margins between Latin and Anglo-Saxon America.[4]

The persecution and extermination of sea otters between the 18th and 19th centuries was not only a socioeconomic phenomenon that had a global impact but also, above all, a phenomenon that affected a socio-ecological system that was itself globally interconnected. Nowadays we see daily the effects of climate change due to global warming,[5] a clear sign of the interconnection that exists between the different ecological systems on the planet. Of course, not all ecological changes and transformations have been the product of human activity, but it is essential to recognize

[3] The sea otter was classified in 1758 in Linnaeus' system as *Enhydra lutris*. There are three subspecies: the *nereis*, which is the smallest of the sea otters and has reddish brown fur, which swims in the waters of California from Biscayne Bay to San Francisco. The other two subspecies, the *Enhydra lutris kenyoni* and the *Enhydra lutris lutris*, larger and with dense black fur, live northward in different regions from present-day Oregon and British Columbia to Alaska and the Aleutians.

[4] On the identification of new spaces of study as a characteristic of global and relational histories see Hausberger and Rinke, "Between Spaces," 2013, p. 1418.

[5] Stefan et al., "The Trajectory", 2015, pp. 81–98.

that many of them are, especially if we want to stop and, if possible, reverse the harm we cause to the environment through our habits and customs. With precisely this intention, on this occasion we want to make this single case historically visible, the process of devastation that the sea otters suffered to satisfy the clothing fashion of the Chinese mandarins, used in turn by Western merchants to gather fortunes that allowed them to build and consolidate their control over the world economic system.[6] Understanding this phenomenon in its historical dimension is very important, even more so because today we continue to commit the same type of eco-global crimes with many species in practically all regions of the world.[7] Excessive and vain consumption today should not only be seen as reprehensible ethical behaviour but should be recognized as an activity that generates harm and destruction to socio-ecological systems and that is contributing to producing imbalances and inequalities in our environment.

European Expansion on the Coast of the North Pacific

In less than a century, between 1740 and 1830, European hunters devastated sea otter populations in the American Northwest. Its relentless pursuit as part of a value chain that connected East Asia with Northern Europe began in the 1840s, when the Russians entered the North Pacific by advancing from the Kamchatka Peninsula towards the Commander Islands, the Aleutians and Alaska. It was not a new activity. The Russians already had a long tradition as hunters and furriers that dated back to at least the 10th century. They had extended the borders of their empire by advancing from the Urals towards Siberia in search of animals with precious fur, such as sables, stoats and foxes, which not only allowed them to appropriate the territory, but also generated considerable tax revenue for

[6] Mayorga, *Pieles*, 2020; Haller, *Balleneros*, 2023; Pérez-García, "The economic", 2021.

[7] White, Transnational Environmental, 2011; Sollund, The Crimes of Wildlife Trafficking, 2019; UNODC, 2016, 2020, 2024; van Uhm, Organized Environmental Crime, 2023; Smith and Sahramäki, 2024.

the tsar's treasury. Its importance is evident when considering that around 1644, fur came to represent 10% of the empire's total income and even, in the case of Siberia, became the very basis of its economy.[8] Originally, the Russian fur market was aimed at Europe, but in the 18th century a new trade route was opened. With the signing of the Treaty of Nerchinsk in 1689 between Russia and China, the border between both countries was drawn and there Kiakhta became the main square for the sale of Russian furs in the Chinese market.[9] As populations of sables and foxes in Siberia were depleted or dispersed into forests,[10] Russians sought new species with precious fur to sell in China, and the most sought-after became the sea otter, whose fur is extraordinarily dense and soft. This great demand for otter skins in China confronted the Russians with the Japanese in the Kuril Islands, so the Russians looked for new hunting territories and moved to the Kamchatka peninsula and from there to the islands that they systematically exploited until they reached to Alaska on the American continental coast. They first moved on to the Commander Islands where, over the course of two decades from 1741, they devastated the sea otters. In that period alone, its population decreased by 85%. Otters and seals (*Phocidae*), another of the exploited species, have a very low reproduction and recovery capacity compared to sables, and they are species that do not migrate, tending always to remain in the same place. In the open sea, the strength of the currents prevented them from moving, so each population was restricted to its own area, living in shallow waters so that they could obtain their food on the seabed (urchins, crabs, abalone, among other species). Since they had little space to retreat and flee human persecution, the devastation of marine mammals was much more brutal than that of land mammals that reproduce more quickly and can enter and disperse among forests. Furthermore, since there were few other resources in the North Pacific to support imperial expansion, exploitation was concentrated on otters and seals.[11]

[8] Wolf, *Europe and the People*, 2006, p. 197.

[9] Ravalli, *Sea Otters*, 2018, pp. 8–10.

[10] Jones, "A Havock," 2011, p. 588.

[11] Ibid., p. 589.

In California, the southernmost habitat of American otters, the Spanish were the first to spot them, when the Jesuits began the colonization of that territory. In fact, the first European description of the American otter was that of Father Sigismundo Taraval in 1733, in the Islas Dolores, the archipelago whose main island today we know as Cedros (Baja California, Mexico), as recorded by Miguel Venegas in his *Noticia de la California*:

> Lately, the strangest thing that has been found in this matter in California is a species of animals very similar to Beavers,[12] if they are not already. Found them in 1733, Father Sigismundo Taraval on his trip to the Islas *Dolores*, in a place he called *San Andrés*, four and a half days away from the Mission of *San Ignacio*. So many were discovered together that those in the procession killed more than twenty, chasing them only with sticks, and the father sent some skins to Mexico. He assumes that these animals are amphibians, like beavers; but he does not say whether they found in them the rare quality of architects, building their houses, and dwellings on the rivers, with the wonderful precautions, providence, and government, which is observed in the beavers of Canada.[13]

Although from the beginning the Spanish warned that their skins were precious, at that time they were not interested in commercially exploiting sea otters, and nor did they really promote the colonization of northern California. Things changed when they learned of the Russian advances. The first confirmed official news that Spain had of the Russian presence in America were reports from the Marquis of Almodóvar, ambassador to the tsarist court, which were sent to the Spanish court in 1761. Three years later, the new Spanish ambassador in Moscow, the count de Herrería, reported that Empress Catherine II had authorized a new expedition to American lands to verify the progress that the Russians had made. Added to this news were the arrivals from England, since, in 1768, the Marquis of Grimaldi sent from London reports that were circulating in the English

[12] The Russians also described otters as beavers: Ravalli, *Sea Otters*, 2018, p. 11.

[13] Venegas, *Noticia de la California*, 1797, t. I, pp. 45–46.

court about Russian manoeuvres in America.[14] Thus, the Spanish government became concerned, not only about the security of New Spain, but also about defending the territories that it considered its own and over which it claimed sovereignty, the coasts of the American northwest and the South Sea, as they called the Pacific Ocean, a vital ocean space for the Spanish empire because it connected New Spain with the Philippines. Various measures were taken to protect their possessions, such as the founding of the San Blas station in 1767, which would be used as a base to support the colonization of San Diego, Monterrey and San Francisco in Alta California, and defend the travel routes and the return trip between Acapulco and Manila. In 1768, King Charles III ordered the immediate and effective occupation of the territory of Alta California to contain Russian incursions.[15] Maritime expeditions were also sent north to reconnoitre the territory and establish communication with its inhabitants, such as that of Juan Pérez in 1774 aboard the *Santiago*, which reached the Queen Charlotte Archipelago and Nootka (today part of British Columbia). Indeed, the crew of the *Santiago* dealt with a group of Haida Indians, with whom they exchanged some objects. The Haida gave the Spanish otter skins, carved wooden boxes and Chilkat blankets, while the Spanish gave them abalone shells, beads, knives, hats, as well as two silver spoons that went down in history because they were the evidence material that the Spanish were the first Europeans to arrive there.[16] These were only ephemeral contacts, however. In 1775 the Spanish government sent another reconnaissance expedition composed of two ships, one under the command of Bruno Hezeta and the other captained by Juan Francisco de Bodega y Quadra, who sailed to 57 degrees latitude and claimed to have taken possession of that territory for the Spanish Crown. However, as soon as the Spanish set sail, the Indians tore down the wooden cross which the Spanish had erected. The expedition served for the Spanish to collect direct information on the state of those coasts, at

[14] The interest in knowing what that part of the world was like in the process of exploration and knowing the reach that the different empires were achieving in those territories was common to several of the European courts and lasted for many years, as is made clear in Coxe's book, *Account of the Russian*, 1780.

[15] Pinzón, *Actions and reactions*, 2012, p. 131 and Ortega, *Alta California*, 2001, p. 27.

[16] Archer, "The Transient Presence", 1973, p. 3 and Fisher, *Contact and Conflict*, 1992, p. 2.

least regarding Russian advances; even then, however, the Spaniards were not interested in the fur wealth of the region.

A few years later, in 1778, the English expedition under the command of James Cook arrived in Nootka and its crew carried out some exchanges with the Indians (or Indigenous people). The otter skins they then acquired were sold to Guangzhou and there they made an enormous profit. It was also at this time that the English discovered the famous Spanish silver spoons as ornaments hung around the neck of a Haida. News of the prices secured by Cook's crew selling furs in China spread in 1784, when James King published the official chronicle of the expeditions carried out on the ships *Resolution* and *Discovery*. The news of such a promising market had immediate consequences, because the following year the first English merchant ship arrived in Nootka in search of otter skins.[17] It was the *Harmon*, commanded by Captain James Hanna, a ship that was significantly renamed *Sea Otter*. In that first hunting season alone, Hanna collected 560 skins in Nootka which he sold in Guangzhou for 20,600 pesos.[18]

It was in 1784, the same year that King published the news of Cook's expedition, when the Spanish finally launched a project to exploit otter skins in California. It was an initiative presented to the Spanish government by Vicente Basadre y Varela in 1784. His proposal was that Californian skins be sold in China in exchange for mercury to refine silver in New Spain. His plan was for the Indians of the Franciscan and Dominican missions of California to hunt otters and sea lions (Otariidae). Then the fathers in charge of the missions would send the skins to San Blas and from there they would be sent to Acapulco, then on to Manila and from there to Guangzou. Basadre's plan was sent for approval to the king of Spain and, after several studies carried out in 1785, it was approved. The Royal Treasury would oversee financing the first operations; Basadre would organize the collection of the skins in the missions and send them to Acapulco to load them on the Philippine ship and transport them to Asia. The exploitation of Californian furs was designed

---

[17] Fisher, *Contact and Conflict*, 1992, p. 2.

[18] Gibson, *Otter Skins*, 1992, p. 21 y Fisher, *Contact and Conflict*, 1992, p. 3, the currency with which the sale price is indicated is "Spanish dollars", that is, the pesos of eight *reales*, a currency minted in Mexico with great esteem in China.

as a seat (or monopoly) of the Spanish Crown, so all those who would intervene in this operation were insisted on seeking the "absolute prohibition of their traffic because of how much it interests the good of the State and the service of the King" to protect the royal monopoly.[19] Outside of Basadre, the missionary fathers and the captains of the San Blas ships, no one else could participate in this enterprise, neither the soldiers of the presidios nor the settlers of California.

Basadre arrived in August 1786 in Monterrey Bay, where he was received by Governor Pedro Fages and the Franciscan Father Lasuén.[20] A month later, the French expedition arrived there under the command of Jean-François de Galup, count of La Pérouse, sent by his government to explore the commercial potential of the area. Basadre and Fages received him and explained in detail the Spanish plan to exploit the Californian fur industry, which La Pérouse made public in his *Voyage author du monde*. According to the French commander, Governor Fages assured him that 20,000 skins could be collected annually in Monterrey alone, in addition to the same number in two or three establishments located north of San Francisco.[21] The figure was clearly an exaggeration, since in the first three months Basadre obtained 1060 skins in the Franciscan missions of San Carlos de Monterrey, San Antonio, San Luis Obispo, San Buenaventura and San Diego, and in the Dominican missions of Rosario and San Fernando from Velicatá.[22]

The fourth group of traders involved in the exploitation of sea otters was from the United States. John Ledyard, a native of Connecticut, was part of the crew of the *Resolution*, one of the ships in James Cook's expedition. He was present when the English obtained the skins from the Indians and when the English sold them in China at a magnificent price. Since then, Ledyard tried to set up an expedition to obtain furs, but he

---

[19] Letters to Don Francisco Trillo Bermúdez, to the governor of Californias, to the president of the missions of New California and Old California. Archivo General de la Nación, Indifferent Viceregal (Consulate), box 2779, exp. 020, pages. 6–8v.

[20] Ogden, "The Californias, 1932, *published* in *Gazeta de México*, 1786–1821: II, p.162; Ogden, *The California*, 1941, p. 15.

[21] La Pérouse, *Voyages*, 1799, t. II, p. 310.

[22] Letter from the Royal Governing Court to Vicente Basadre, February 2, 1787, Archivo General de la Nación, Indifferent Vicereinal, box 3508, exp. 044, pages. 5–5v.

failed immediately. It was a little later, in 1788, that the first New England ship arrived in the northwest, the *Columbia Rediviva*, under the command of Robert Gray. He inaugurated one of the most successful commercial circuits of the time which connected Boston with the American Northwest, the Hawaiian Islands, the Marianas and southern China—*The Golden Round* as Gibson called it[23]—a circumnavigation that would cross the Atlantic, Pacific and Indian Oceans, crossing Cape Horn and Cape Good Hope. A world circuit that would generate enormous profits by adding the benefits of different businesses at each scale: English merchandise, otter skins, sandalwood, silks and Chinese teas.[24]

Methods of Obtaining and Marketing Skins: Companies and Seats

Obtaining otter skins in America and extracting them for the Chinese market was done in different ways. The Europeans oversaw the marketing the skins, but not always the hunting of the animals, a task they left to the Indians. Regarding the marketing of skins, there were three principal methods: at the private initiative of merchants in small companies that embarked on their own; through large companies established through shares and with government backing; and, finally, the royal seat or monopoly that the Spanish attempted. The Russians also followed the system they had used in Siberia and by which they first obtained the skins as an annual tribute collected as they took possession of the territories. Although this tribute, called *yasak*, was officially abolished in 1788, it continued to be collected until the end of the century.[25] In addition to this tax imposition, the Russians obtained furs through barter with the Indians until 1799, when they founded the Russian American Company, which would, from then onward, administer the monopoly of fur exploitation in the Aleutian Islands and Alaska. This company would employ

[23] Gibson, *Otter Skins*, 1992, pp. II–III and 39.
[24] Ibid., pp. 56–58.
[25] Jones, "A `Havock", 2011, p. 92.

the Aleuts, the best otter hunters in the world, under extraordinarily harsh labour regimes.[26]

For their part, the Dutch, English and French began their fur companies in North America from the Atlantic coast, chasing beavers. They were looking for the layer of soft, curly hair that beavers have over the hide and under the layer of long, stiff guard hair. With this hair processed into wool, they made felt for hats that, with their different shapes, styles and sizes, were fashionable in Europe until the beginning of the 19th century, when other materials, such as silk, replaced beaver hair.[27] The French controlled this trade from Quebec, in the territory from the St. Lawrence River to the Great Lakes, and the Dutch from the Hudson to Albany and Lake Ontario until 1644, when they were replaced by the English, who launched an intensive search for furs. This caused the Europeans to compete to extract them, which in turn impacted the relations with the original peoples, who began to compete bitterly with each other. European companies were supplied by Indigenous intermediaries who, despite everything, managed to maintain relative autonomy until the middle of the 18th century by at least preventing the companies from directly interfering in the internal social and political organization of their communities and in their alliances and conflicts with other Indigenous communities. However, in the second half of the century, as the companies moved deeper into American territory, they tried to contact hunters directly in the places where they trapped the animals without having to negotiate with the chiefs of the indigenous groups. The middlemen were thus displaced, leading to major rebellions. This led to a direct relationship between the fur traders and the trappers and hunters, perhaps organized in small bands, who acted outside the large Indigenous groups. The impact was enormous; some original populations disappeared, others were displaced and transformed according to European demands, not only for beaver pelts, but also for food and other products. After the Thirteen Colonies gained independence from England and formed the new republic of the United States, several traders moved west overland until they reached the Pacific coast. There they would join the

[26] Gibson, "Russian Expansion", 1980, p. 135.

[27] Wolf, *Europe and the people*, 2006, pp. 197–198.

otter hunt that Europeans had already been carrying out from the sea for some time. These Boston merchants came to control the business through companies such as the Northwest Company, which in 1821 merged with the Hudson Bay Company,[28] and their association with the Russian American Company in 1804.[29]

For their part, the Spanish established that the animals would only be hunted by the Indians of the missions, prohibiting the settlers and soldiers of the presidios from intervening in any way. To ensure that the skins had been obtained legitimately, the missions would issue certificates, without which the skins would be confiscated when attempting to load them onto the ships that would take them to San Blas.[30] To set their prices, the skins would be classified by size and colour, just as the Chinese and Russians did. For example, first-class ones would cost ten pesos and had to measure around a yard, be black and cured; the third-class ones would be worth only two pesos and would measure three-quarters of a *vara*, they would be brown and would not be cured.[31]

In March 1787, Basadre left for Manila with the first shipment of Californian otter skins.[32] He took up residence in Guangzhou and there he was able to observe for 13 months the progress of the California fur industry, as he said, "extracted by the English, Portuguese, and Anglo-Americans," despite the monopoly that the Spanish tried to impose. In the end things did not turn out as well as he expected. Years later, Basadre wrote bitterly that "nothing was approved, perhaps because at that time it was proposed to extract it from Germany in better proportions [he was referring to mercury]."[33] In 1789, Basadre returned to Madrid and ran into serious problems because he was accused of having abandoned his mission in China, so much so that he had to go to the king and file

[28] Ibid., pp. 199, 20, 213 and 217.

[29] Morison, *The maritime*, 1921, p. 60.

[30] Letters to Don Francisco Trillo Bermúdez, to the governor of Californias, to the president of the missions of New California and Old California, March 7, 1786: Archivo General de la Nación, Indifferent Vicereinal (Consulate), box 2779, exp. 020, pages. 6–8v.

[31] Ogden, "The Californias," 1932, p. 447.

[32] Basadre to the Royal Governing Court, March 28, 1787, Archivo General de la Nación, Indifferent Vicereinal, box 1561, exp. 002, f. 102.

[33] Vicente y Basadre y Varela to the Prince of Peace, Veracruz, May 1, 1797, Archivo General de Indias, State, 40, n. 37.

a lawsuit that reached the Council of the Indies. In the end the lawsuit was resolved in their favour, but not until June 1791. Despite everything, Basadre summed up the success of his mission like this: "the first three consignments of furs that were taken to China in the years of 87, 88 and 89, they gave three million, one hundred and twenty thousand real fleeces to the Royal Treasury." Of these, one consignment was entrusted to Basadre himself and another to the factors of the Royal Company of the Philippines in Canton (Guangzhou); the third, according to Basadre, "was carried out through a separate file by virtue of a commission that the Captain General of Manila conferred on Don Vicente Memije [...]; I do not know the result of the subsequent remittances."[34] Without displaying any modesty, Basadre wrote that "a fact was obscured by which the country should place me next to those who discover objects of use and benefit. I proposed and carried out a new branch of commerce, and with it, I have secured for the Crown large sums that have successively entered the Royal Treasury."[35] The third of the shipments referred to by Basadre was 7127 skins of otters, foxes and sea lions collected in the Californian missions and in Alaska during the expedition of Esteban José Martínez y López de Haro. He arrived in Manila in 1790 and the person in charge of his negotiation in Macau was Vicente Memije. 106,025 pesos were obtained for these skins, which were invested in quicksilver. In 1791, disagreements between China and Russia caused the Chinese government to ban the import of otter pelts. Consignments of California pelts sent had to be stored in Manila until 1793, when the emperor again authorized the introduction of otter pelts to China.[36] It has been calculated that between 1783 and 1791—a period of eight years—14,483 skins were sent to Asia on behalf of the Royal Treasury of New Spain, a figure very far from the 20,000 pieces per year that La Pérouse recorded that could be collected in the Californian missions.

---

[34] Ogden, "The Californias", 1932, pp. 457–48.

[35] Vicente y Basadre y Varela to the Prince of Peace, Veracruz, May 1, 1797, Archivo General de Indias, State, 40, n. 37.

[36] Bernabeu, *The Illustrated Pacific*, 1992, pp. 271–272.

Impact and Transformation of Original Peoples

Sea otter hunting was certainly a very traditional activity. The original peoples of the North Pacific coasts, from the Hokkaido Islands to California, had lived with them and hunted them since prehistoric times. There is evidence from 10,000 BC in the archaeological remains of the Jomon culture of the western Pacific and in those of the ancestors of the Haida Gwaii on the eastern coast from 8000 BC. Even in some places, particularly where land resources were very scarce and agriculture practically non-existent, as in the Aleutian Islands,[37] local communities had to look for their food and clothing in the sea and otters and other marine mammals were hunted intensively, to the point that localized processes of extinction occurred, but without serious imbalances in the system.[38] It was very different from what happened between 1740 and 1830, when hunting brutally intensified to take the skins to foreign markets to make luxury clothing. The Aleuts were the most skilled otter hunters in the world, so in the 18th and 19th centuries Russian and American fur companies hired them to trap and kill otters along the entire American Northwest coast, from Alaska to California. This completely disrupted Aleut culture and destroyed several of their communities. His men were first subjected to a regime of servitude by forcing them to pay the *yasak*, the annual tribute in furs that implied the recognition of Russian sovereignty,[39] and later, when the Russian American Company was established, to a kind of distribution of labour, similar to the New Spanish *tequio* or the Peruvian *mita*, by which half of the men over 18 years of age in each community had to be employed as hunters for the Company. Every December, the elders of each town chose which young men should go hunting for the Russians between March and April. The season lasted until September, so they spent a semester away from their homes.[40] Such a prolonged absence of young men on the islands certainly caused serious disruption in the communities, so humans, as well as sea otters, suffered

---

[37] Jones, *Empire of Extinction*, 2014, pp. 93–95.

[38] Ibid., pp. 95–96; Ravalli, *Sea Otters*, 2018, pp. xx–xxi.

[39] Reedy-Maschner, "Where did all", 2010, p. 591.

[40] Gibson, *Otter Skins*, 1992, pp. 13–15.

from the fur fever that broke out in the last quarter of the 18th century. The impact suffered by the Aleutian population has been described as genocide, although there is no consensus on the matter, since the number of Aleuts who lived on the islands before contact with the Russians is unknown. The figures vary enormously—from 20,000 to 8000 people—but whatever the number, there is no doubt that serious depopulation did occur, since by 1790 only 2000 individuals were recorded as living on the islands.[41]

In general, all the Indigenous communities of the North Pacific territories were affected by the European invasion and experienced the transformation of their ancient cultures, if not their disappearance. In the territories colonized by the Spanish, the system applied was very different from the Russian one. The Indians were forced to settle and live in communities ruled by Franciscan and Dominican missionaries. There they were converted and instructed in Catholicism and forced to work in the fields harvesting plants and raising domestic animals according to European techniques. The original population of Alta California was made up of twenty different ethnic groups belonging to six different linguistic families composed of various subgroups. Between 1769 and 1834, the Franciscans reorganized them into 21 missionary groups that brought together a native population of approximately 21,000 neophytes. In the territory populated by Shoshonean groups in southern California, the missions of San Juan Capistrano, San Luis Rey, San Gabriel and San Fernando Rey were established. Among the Yuma, the San Diego mission. The missions of Santa Bárbara, Santa Inés, San Buenaventura, La Purísima Concepción and San Luis Obispo were founded in the territory of the Chumash. In that of the Costanoa, the missions of Nuestra Señora de la Soledad, Santa Cruz, Santa Clara, San Juan Bautista, San Francisco de Asís, San Carlos Borromeo and San José were established. The missions of San Rafael and San Francisco Solano in the territory of the Miwok while those of San Antonio de Padua and San Miguel Arcángel were founded in that of the Salinan. The Spaniards changed the proper names of the indigenous ethnic groups and called them according to the mission they had: *luiseños*, *diegueños*, *ventureños*, *gabrielinos*, etc. In

[41] Reedy-Maschner, "Where did all", 2010, p. 589.

addition to the missions, presidios were also established, and the formation of towns was promoted with settlers brought from the south, whose transfer and settlement was financed by the royal treasury. In fact, the costs of founding the missions, presidios and towns were covered with money from the Royal Treasury, with the idea that the communities would become self-sufficient when the settlements became stabilized, and the local economy developed following the European model of agriculture and livestock. Thus, little by little, the Californian lands were occupied with missionaries, soldiers and settlers coming from outside, combined with the natives of the different cultures of the region. The first mission was established in San Diego; from there expeditions were sent by land and sea to continue with the foundations and maintenance of missions, presidios and towns until reaching what was the northernmost Spanish establishment, in San Francisco Bay.[42]

The plan that Vicente Basadre proposed to the Spanish government to exchange the skins of California sea otters for quicksilver in China was based on organizing the Indians of the missions, under the supervision of the parents in charge of each community, to hunt the otters and prepared their skins to send them first to San Blas and from there to Acapulco, from where they would be sent to the Philippines on the Manila galleon. They would then move to Guangzhou, where Basadre himself would take care of exchanging them for mercury. Everything would be managed as a royal monopoly, so free trade in Californian furs was prohibited. Outside the missions, no one could trade with them, neither the soldiers in the presidios nor the settlers in the towns. Under this scheme, this company was not very successful and lasted a short time. Intrigues, Basadre's friction with the governor-intendant of Manila and competition with the Philippine Company made the first initiative fail and although later there were more attempts to develop other similar companies, the fur business

[42] Catherine R. Ettinger, "A new domesticity. The indigenous people of Alta California and missionary housing", electronic version paragraph 8 and final note 18 New World. Mundos Nuevos, [Online], Débats, put online on January 5, 2010. URL: <http://journals.openedition.org/nuevomundo/58264>; DOI: <https://doi.org/10.4000/nuevomundo.58264> (Accessed: November 17, 2020); Bernabeu and Ortega, "Indians and Franciscans", 2011, p. 412; Ortega, *Alta California*, 2001, pp. 29–33, 42. A map of the Spanish populations of Alta California can be seen in Ortega, *Alta California*, 2001, p. 39.

between Spanish California and China did not really prosper.[43] Another factor that contributed to the lack of success that the Spanish had in the fur business was that the California Indians were not as skilled sea otter hunters as the Aleuts. Unlike them, the Californian peoples had a greater variety of resources to feed and clothe themselves inland, so they were not as specialized in hunting marine mammals. Furthermore, limiting this trade to the missions to control it as a real monopoly did not work either. Although at first the missionary fathers collaborated with the plan and considered that it could be a good source of financing for the missions, in the long run they greatly disliked it and ended up opposing it, since otter hunting distracted the Indians from agriculture and livestock farming, distancing them from the missions and their community obligations. Another factor that affected the royal seat was that the soldiers of the presidios and the settlers wanted to increase their participation in the fur trade, and they succeeded, even collaborating with Russian and American traders despite it being a prohibited trade. Indeed, when the otters of the Aleutian Islands and those of Alaska became scarce, the Russians looked for new hunting territories to the south and reached the port of Monterrey. Even a little further north of this port, the Russians established a fixed settlement, Fort Ross. For their part, Boston merchants made an entire system of their clandestine trade in Spanish California. Knowing that they were prohibited from entering Spanish ports, the masters of Boston ships used to carry a "letter of friendship" issued by the Spanish consul in Boston in which they were authorized to enter under the argument that it was necessary to save the ship due to bad weather or lack of drinking water or other inconvenience of that nature. Furthermore, in the early years of the 19th century, Russians and Americans agreed to work together to hunt in Californian waters and were quite successful, a hunt in which they employed the Aleuts, thus forced to work far from their places of origin.[44]

The heyday of the otter pelt trade in the Chinese market was from 1785 to 1817, although otter hunting continued well into the 19th

---

[43] Ogden, "The Californias", 1932, pp. 457–458, 463 and Ogden, *The California*, 1941, p. 18; Bernabeu, *The* Illustrated Pacific, 1992, pp. 271–276.

[44] Ogden, *The California*, 1941, pp. 45–57 and Morison, *The Maritime*, 1921, p. 60.

century. The main traders were the Russians and the Americans, although the English also had a prominent participation. The Spanish tried, although with less success. The truth is that together they devastated the otter populations of the American northwest. By 1830, their numbers had declined brutally, to the point of being practically extinct in some regions. Those who lived on the islands Pribilof, in the Bering Sea were completely eradicated very soon, shortly after the arrival of the Russians in 1786. In other places they became so rare that the task of hunting them was economically unaffordable. In 1827, veteran hunters from California, the so-called "Nor'westman", complained that it was a business that was worth less and less; they continued to compete among themselves, many boats continued to search for otters, but it was increasingly difficult to find them. The Indians no longer made any effort to hunt them. The goods they received in exchange were no longer as attractive to them and the hunt had become much riskier because as the use of firearms had spread, the otters had become much more fearful and moved as far away from the coasts as possible. In the hunting season of 1827, there were ships that did not get one.[45] The harm had already been done and there were fewer and fewer specimens. Added to human persecution there were other disasters destroying sea otter populations. In Canada, the last time a wild otter was recorded was in 1929, near Kyuquot, in present-day British Columbia.[46]

From Devastation to Harm Repair

From the first moment Europeans saw sea otters, they all agreed on the extraordinary beauty of their fur, and no one resisted hunting them. The Spanish were the first Europeans to see them in America; the Russians were the first to exploit their great commercial value in China; the English publicly spread the news of the commercial success of this merchandise in the Asian market and the Americans took full advantage of its commercialization as part of a global circuit. All contributed to the species

[45] Ibid., pp. 65–66.

[46] Kenyon, "The Sea Otter", 1969, pp. 183–184.

practically becoming extinct on American coasts in less than a century. By 1969, it was estimated that the critical point had been reached as there were fewer than 2000 specimens in the world in remaining and dispersed communities.[47] In the 20th century, uncontrolled hunting was accompanied by other anthropogenic pressures that contributed to the decline of sea otters. Oil spills have proven the most serious, but they have also been affected by the acidification of the seas, the displacement of toxins from human waste into the sea, the disappearance of the species that serve as food for otters, which are in turn exploited for commercial purposes, such as sea cucumbers, crabs and urchins. In short, a long chain of disturbances caused by humans are being added to the natural ones, which has prevented sea otter populations from thriving in a large part of their distribution range.[48] However, important efforts have been made to reinsert and conserve them since 1970,[49] with relative success in the effective reestablishment of some populations. It has been largely achieved in certain places thanks to the protection system called "extraction without waste", by which native populations can hunt animals for subsistence and craft production purposes, if they do so in a controlled manner and under the monitoring by authorities. Joint work between authorities and communities has apparently been the best way to restore ecosystems, build resilient communities and restore relationships between the original peoples and these animals.[50]

The story dramatically illustrates that the commercial value of sea otter pelts drove the species to near extinction during the 18th and 19th

---

[47] Duplaix and Savage, *The Global Otter*, 2018, p. 67 *apud* Kenyon, "The Sea Otter", 1969.

[48] Estes, "Growth and equilibrium", 1990, pp. 385–401; A. Doroff and A. Burdin, Enhydra lutris. The IUCN Red List of Threatened Species 2015, <https://doi.org/10.2305/IUCN.UK.2015-2.RLTS.T7750A21939518.en.>. (Accessed: November 23, 2020); Fisheries and Oceans Canada, "Management Plan", 2014; Conrad et al., "Transmission of Toxoplasma", 2005, pp. 1155–1168; Johnson et al., "Prey choice", 2009, pp. 2242–2247.

[49] Starting in the 1970s, sea otter reintroduction programs began, and the first Otter Specialist Group was founded within the framework of the International Union for Conservation of Nature (IUCN), to provide leadership for the recovery of populations not only of the sea otter but of all otter species distributed throughout the world, both marine and adapted to freshwater habitats. In addition to these efforts, in 1975 the sea otter subspecies *Enhydra lutris nereis* was listed in Appendix I of the Convention on International Trade in Endangered Species of Wild Fauna and Flora (CITES), which arose with the objective of ensuring that international trade in specimens of wild animals and plants does not constitute a threat to their survival.

[50] Salomon et al., "First Nations", 2015, pp. 301–331.

centuries. Without a doubt, this type of trade that commodifies an animal species for abusive profit can be classified as a green crime on a global scale[51] due to its *transnational nature*, involving several countries in a distribution chain, either as centres of origin, places transit or wildlife destinations; by the *interconnection of ecosystems* through otter distribution patterns; the *presence of harm* caused by violence, mistreatment and overexploitation of otters, and therefore, deterioration in ecosystems; due to *injustice* in value chains since profits are concentrated in a few economic groups and sectors; due to the very high rates of extraction and illegal trafficking, which undoubtedly qualify as *crimes*, and due to the *differences and variations at the local and regional level* in relation to the management of otter populations.

The way we have viewed sea otters has been as commodities that can be appropriated by humans and whose value is measured based on their economic utility, and not as living beings with intrinsic value. It was impossible for them to consider it any other way in the 18th century, but even today in the 21st century, there continues to be a global market for sea otters, although completely different from the which flourished during the 18th and 19th centuries, both in purpose and in magnitude and type of articles demanded. The reports and accusations of specimens coming from the wild or classified as of unknown origin to evade responsibilities continue, as well as the personal use transactions and seizures, and today it is still the northern countries that report a greater volume of imports, mainly the United States, Canada, Japan, Germany and Italy.[52]

Green criminology asserts that justice systems must extend beyond the anthropocentric notions present in criminal justice, which until now is concerned only with human interests and does not reflect the reality of harm and abuse inflicted on animals.[53] Green criminology advocates for a broader concept of what constitutes a crime in order to incorporate the impact and consequences of environmental harm as a basis for defining crimes that should be the subject of criminological investigation.[54] In

[51] White, *Crimes against nature*, 2013 and White, "Green victimology", 2018.

[52] UNEP-WCMC, 2020. CITES Trade Database available (version 2020.1). https://trade.cites.org/es/cites_trade (Accessed: October 14, 2021).

[53] Nurse, "The damage", 2017, pp. 258–275; Nurse and Wyatt, 2021, pp. 19–65.

[54] Ibid.

sum, we believe in the need to develop ecological justice that surpasses human beings;[55] a justice that includes environmental rights as an extension of social rights, with a collective and intergenerational nature and that also considers justice for other species, with the recognition that humans are only part of a complex ecosystem in which non-human animals and plants also live.[56] With these concerns in mind, in the case of the global overexploitation and extermination of sea otters, we turn to Rob White and identify these actions as socio-environmental harms, in which there were both human and non-human victims and long-term impacts.[57]

Acknowledgments Authors acknowledge Instituto de Investigaciones Dr. José María Luis Mora and Centro Regional de Investigaciones Multidisciplinarias (CRIM, UNAM) for their continuous support and, Viridiana Mercedes Reyes Hernández for her assistance in the search and transcription of historical documents.

References

Literature

Archer, C. I. (1973). The transient presence: A re-appraisal of Spanish attitudes toward the northwest coast in the eighteenth century. *BC Studies*, University of British Columbia, *18*, 3–32.

Berg, M. (2019). Sea otters and iron: A global microhistory of value and exchange at Nootka sound, 1774–1792. *Past & present, 242*(Supplement 14), 50–82. Oxford Academic.

Bernabeu, S., & Ortega, M. (2011). Indios y franciscanos en la construcción de la Alta California. En E. G. Cruzado (coords.), *Actas de las Jornadas de Historia sobre el Descubrimiento de América Jornadas IV, V y VI (2008, 2009 y 2010)* (pp. 406–434). Universidad Internacional de Andalucía.

[55] Hall and Varona, "Victimology", 2018, pp. 107–128.

[56] White, *Crimes against nature*, 2013 and White, "Green victimology", 2018, pp. 1–7.

[57] Ibid.

Bernabeu, S. (1992). *El Pacífico ilustrado: del lago español a las grandes expediciones*. Mapfre.

Conrad, P. A., Miller, M. A., Kreuder, C., James, E. R., Mazet, J., Dabritz, H., Jessup, D. A., Gulland, F., & Grigg, M. E. (2005). Transmission of toxoplasma: Clues from the study of sea otter as sentinels of *toxoplasma gondii* flow into the marine environment. *International Journal for Parasitology*, Sociedad Australiana de Parasitología, Elsevier Ltd, núm., *35*, 1155–1168.

Coxe, W. (1780). *Account of the Russian discoveries between Asia and America. To which are added the conquest of Siberia and the history of the transactions and commerce between Russia and China* (2nd ed.). Printed by J. Nichols for T. Cadell.

de Galaup La Perousse, J. F. (1799). Observations on the trade of sea otter skin, &c. In J. F. de Galaup La Perousse (Ed.), *Voyage round the world performed in the years 1785, 1786, 1787 and 1788 by the Bouselle and Astrolabe*. A. Hamilton, t. II.

Doroff, A., & Burdin, A. (2020, November 23). *Enhydra lutris. The IUCN Red List of Threatened Species 2015*. https://doi.org/10.2305/IUCN.UK.2015-2.RLTS.T7750A21939518.en

Duplaix, N., & Savage, M. (2018). *The global otter conservation strategy*. IUCN /SSC Otter Specialist Group, Four Corners Institute.

Estes, J. A. (1990). Growth and equilibrium in sea otter populations. *Journal of Animal Ecology*, British Ecological Society, *59*, 385–401.

Ettinger, C. R. (2020, November 17). Una nueva domesticidad. Los indígenas de la Alta California y la vivienda misional. *Nuevo Mundo. Mundos Nuevos*, [Online], Débats, puesto en línea el 5 de enero de 2010. http://journals.openedition.org/nuevomundo/58264; https://doi.org/10.4000/nuevomundo.58264

Fisher, R. (1992). *Contact and conflict. Indian-European relations in British Columbia, 1774–1890*. University of British Columbia.

Fisheries and Oceans Canada. (2014). Management plan for the sea otter (*Enhydra lutris*) in Canada. In *Species at risk act management plan series*.

Gibson, J. R. (1980). Russian expansion in Siberia and America. *Geographical Review*, American Geographical Society, *70*(2), 127–136.

Gibson, J. R., Skins, O., Ships, B., & Goods, C. (1992). *The maritime fur trade of the northwest coast, 1785–1841*. McGill-Queen's University Press.

Hall, M., & Varona, G. (2018). La victimología verde como espacio de encuentro para repensar la otredad más allá de la posesión. *Revista de Victimología | Journal of Victimology*, Hugeyns, núm. 7, 107–128.

Haller, S. C. (2023). *Balleneros, loberos y guaneros en Patagonia y Malvinas. Una historia socioambiental del mar, 1800–1914*, Buenos Aires, SB.

Hausberger, B., & Rinke, S. (2013). Entre espacios: México en la historia global. *Historia Mexicana*, El Colegio de México, núm. 4, *LXII*, 1415–1420.

Johnson, C. K., Tinker, M. T., Estes, J. A., Conrad, P. A., Staedler, M. S., Miller, M. A., Jessup, D. A., & Mazet, J. A. (2009). Prey choice and habitat use drive sea otter pathogen exposure in a resource-limited coastal system. *PNAS Proceedings of the National Academy of Sciences,* University of California Press, *106*(7), 2242–2247.

Jones, R. T. (2011). A 'Havock made among them': Animals, Empire, and Extinction in the Russian North Pacific, 1741–1810. *Environmental History, 16*(4)., University Press Oxford, 585–609.

Jones, R. T. (2014). *Empire of extinction. Russians and the North Pacific's strange beasts of the sea, 1741–1867*. Oxford University Press.

Kenyon, K. W. (1969). *The sea otter in the eastern pacific ocean*. Bureau of Sport Fisheries and Wildlife, Division of Wildlife Research, (North American Fauna, num. 68).

Mayorga, M. (2020). *Pieles, Tabaco y quillangos. Relaciones entre loberos angloestadounidenses y aborígenes australes en la Patagonia (1780–1850)*. Ediciones de la Subdirección de Investigación. (Colección Cultura y Naturaleza, vol. II).

Morison Samuel Eliot. (1921). *The maritime history of Massachusetts, 1783–1860*. Houghton Mifflin Company/The University Press Cambridge.

Nurse, A. (2017). El daño a los animales y la criminología verde, cuestiones de derecho y justicia. En H. Mol, D. R. Goyes, N. South, A. Brisman (Eds.), *Introducción a la Criminología Verde: Conceptos para nuevos horizontes y diálogos socioambientales* (pp. 258–275). Editorial Temis, Colombia/Universidad Antonio Nariño.

Nurse, A., & Wyatt, T. (2021). *Wildlife Criminology. New Horizons in criminology*. Bristol University Press.

Ogden, A. (1932). The Californias in Spain's Pacific otter trade, 1775-1795. *Pacific Historical Review*, Universidad de California Press, *1*(4), 444–469.

Ogden, A. (1941). *The California Sea otter trade, 1784–1848*. University of California Press.

Ortega Soto, M. (2001). *Alta California. Una frontera olvidada del noroeste de México, 1769–1846*. Universidad Autónoma Metropolitana/Plaza y Valdés.

Pérez-García, M., & Jin, L. (2021). The economic "micro-cosmos" of Canton as a global entrepôt: Overseas trade, consumption and the Canton system from

Kangxi to Qianlong eras (1683–1795). *Atlantic Studies. Global Currentes*. https://doi.org/10.1080/14788810.2021.1895037

Pinzón Ríos, G. (2012). *Acciones y reacciones en los puertos del Mar del Sur. Desarrollo portuario del Pacífico novohispano a partir de sus políticas defensivas, 1713–1789*. IIH-UNAM/Instituto Mora.

Ravalli, R. (2018). *Sea otters. A history*. University of Nebraska Press.

Reedy-Maschner, K. (2010). Where did all Aleut men go? Historical demography and social organization. *Human Biology*, Wayne State University Press, nums. 5–6, *82*, 583–611.

Salomon, A. K., Kii'iljuus, B. J. W., White, X. E., Tanape, N., Sr., & Happynook, T. M. (2015). First nations perspectives on sea otter conservation in British Columbia and Alaska: Insights into coupled Human Ocean systems. In S. E. Larson, J. L. Bodkin, & G. R. VanBlaricom (Eds.), *Sea otter conservation*, Academic Press (pp. 301–331).

Smith, L. S., & Sahramäki, L. (2024). *Research handbook on environmental crimes and criminal enforcement*. Research Handbooks in Environmental Law. Edward Elgar Publishing.

Sollund, R. A. (2019). *The crimes of wildlife trafficking: Issues of justice, legality and morality* (1st ed.). Routledge. https://doi.org/10.4324/9781315550428

Stefen, W., Broadgate, W., Deutsch, L., Gaffney, O., & Ludwig, C. (2015). The trajectory of the Anthropocene: The great acceleration. *The Anthropocene Review*, Australia National University y Stanford University, *2*(1), 81–98.

UNEP-WCMC. (2020). CITES trade database (version 2020.1). Retrieved October 14, 2021, from https://trade.cites.org/es/cites_trade

UNODC. (2016). *World wildlife crime report 2016: Trafficking in protected species*. United Nations Office on Drugs and Crime, United Nations.

UNODC. (2020). *World wildlife crime report 2020: Trafficking in protected species*. United Nations Office on Drugs and Crime, United Nations.

UNODC. (2024). *World wildlife crime report 2024: Trafficking in protected species*. United Nations Office on Drugs and Crime, United Nations.

Van Uhm, F. (2023). *Organized environmental crime. Black Markets in Gold, wildlife, and timber*. Bloomsbury Academic.

Venegas, M. (1797). *Noticia de la California, y de su conquista temporal, y espiritual hasta el tiempo presente, sacada de la historia manuscrita, formada en México en el año de 1739, por el padre Miguel Venegas, de la Compañía de Jesús; y de otras noticias, y relaciones antiguas, y modernas: Añadida de algunos mapas particulares, y uno general de la América Septentrional, Asia Oriental, y Mar del Sur intermedio, formados sobre las memorias más recientes, y exactas que se*

publican juntamente: Dedicada al Rey Nuestro Señor por la provincia de Nueva España, de la Compañía de Jesús. En la Imprenta de la Viuda de Manuel Fernández, y del Supremo Consejo de la Inquisición, III tt.
White, R. (2011). *Transnational environmental crime: Toward an eco-global criminology*. Routledge Taylor & Francis Group.
White, R. (2013). *Crimes against nature: Environmental criminology and ecological justice*. Routledge.
White, R. (2018). Green victimology and non-human victims. *International Review of Victimology*, Sage Publications, *24*(2), 239–255.
Wolf, E. R. (2006). *Europa y la gente sin historia.*

What Mercy? A Case Study of Mercy Release in Hong Kong

Rebecca W. Y. Wong

Mercy Release in Hong Kong

Mercy release, commonly known as 'prayer animal release' or live release', is the act of releasing captured animals into the wild to prevent them from being consumed by human beings. The practice is deeply embedded in the religious context of Buddhism or Taoism since it is believed to be rooted in the concept of compassion; liberating captured wildlife will not only generate positive karma for the participants but also grant a second chance of survival for the animals released (Thach, 2024). Yet the actual implications of this practice are a far cry from positive with animals subjected to immense pain and suffering.

The practice of mercy release is carried out throughout the year but most commonly on festivals such as Buddha's birthday (which normally falls in May) and Lunar New Year. In May 2021 (the month of the

R. W. Y. Wong (✉)
City University of Hong Kong, Kowloon, Hong Kong
e-mail: rebecca.wywong@cityu.edu.hk

D. van Uhm, D. Siegel (eds.), *Global Green Crime and Ecojustice*, Palgrave Studies in Green Criminology, https://doi.org/10.1007/978-3-031-91557-4_13

Buddha's birthday), dozens of red-eared slider turtles (*Trachemys scripta elegans*) were discovered in a drainage pool in Tai Po, all with 'mercy release' written on their shells in Chinese characters. Many of the turtles had already died, and those that were still alive were found to be suffering from skin infections and injured shells (Anon, 2021). Non-followers of the Buddhism and Taoism faiths also participated in this ritual in the hope of generating good karma. Birds were once the most popular animal for mercy release, but this changed after the outbreak of the avian flu in Asia. Reptiles, amphibians and fishes are now commonly used in mercy release. It is a misconception to assume that mercy release is constrained by geographical borders since its practice has been documented in other regions such as mainland China, New Zealand, France and the United States (Tan, 2019; Hall, 2017).

The situational context of this chapter is based on Hong Kong, a coastal city based in the southern region of China. At present, there are no laws or regulations in Hong Kong governing this practice and any associated harm caused to the environment. Instead of enforcement and prosecution, the Hong Kong government is working to strengthen public awareness of the dangers of mercy release through environmental organizations such as Kadoorie Farm and the Society for the Prevention of Cruelty to Animals (Hong Kong Special Administrative Region Press Release, 2021).

This study is theoretically conceptualized against the approach of environmental harm which examines harmful, yet legally permitted actions towards wildlife and the environment. Mercy release is a prime example of environmental harm. This study examines whether facilitators (including those who supply the fishes for release and participants of the ritual) understood the harm in their actions. Interviews and observations were conducted in various mercy release hot spots across Hong Kong during 2022–2023 to collect empirical data. Based on our data, interviewees were quick to dismiss any harmful implications of their actions. They felt their good intentions in releasing captured animals were misinterpreted and will continue their practice. Further, the popularity of releasing captured animals has turned into a lucrative business opportunity for fish sellers.

Mercy Release: Environmental Harm

The Hong Kong government relies heavily on public education campaigns to inform the general public about the negative implications of mercy release yet the practice remains popular. As of today, the only legal redress for mercy release is found under the Prevention of Cruelty to Animals Ordinance (Chapter 169) for causing unnecessary suffering to the released animals. Yet the government received 16 complaints under this ordinance between 2014–2021, none of which led to prosecutions (Leung, 2021).

Theoretically, the practice of mercy release is a prime example of environmental harm. The environmental harm approach in the study of green criminology captures the 'awful but lawful' practices that cause harm to wildlife and the environment. Under a strictly legal framework, misconduct towards the environment is punishable by law if written in the legal codes, this is known as the legal procedural approach and the corporate perspective in the discipline of green criminology (Brisman, 2007; Lynch & Stretsky, 2003, 2010).

Although this approach provides an established and clear definition of the law and allows for predictability as to the kind of behaviour expected by the courts, it remains an extremely narrow approach and could 'overlook the possibility of the most harmful activities, does not allow for furtherance of the discussion of what constitutes harm and does not challenge the State's role in the continuance of this injurious behaviour' (Wyatt, 2014). A strict legal construction of environmental crime will neglect harmful actions that will result in damaging consequences for the environment such as the unregulated trade in freshwater turtles and tortoises mentioned in the earlier sections. Another example of a lawful, yet awful act against the environment is the tolerated and continuing use of plastic which would lead to plastic pollution and the untimely death of marine species since they could mistake plastic debris for food. The dire situation at sea brought upon by plastic pollution is described as 'The ingestion of plastic debris by small fish and seabirds for instance, can reduce food uptake, cause internal injury and death following blockage of intestinal tract' (Derraik, 2002, p. 845).

Departing from this narrow definition of environmental crime, we turn to the socio-legal approach, often known as the environmental harm discourse which places the concept of harm as the central consideration. Harm towards the environment is defined as 'an actual danger or adverse effect, stemming from direct and indirect social processes, that negatively impinges upon the health and wellbeing and ecological integrity of humans, specific biospheres and non human animals' (White, 2013, p. 29). The concept of harm is a much broader descriptor than crime; thus, a much wider range of environmental-related misdeeds would be captured and as such the study of green criminology could then 'recognize the fact that many environmental disruptions are legal and take place with the consent of society' (South, 2014, p. 374). Further, the environmental harm approach is a welcomed angle in understanding environmental crimes since it moves away from a human-centred interpretation of misdeeds against the environment and accepts that nonhuman beings can experience pain and suffering (Bernie, 1999; Flynn & Hall, 2017).

This approach is best suited theoretically in narrating the study of mercy release since it asks not if the action is unlawful but rather if it is harmful to society. Actions that the environmental harm approach captures include excessive consumerism on a global scale, the continuing degradation and mining of natural resources, and, of course, the practice of mercy release which is at the forefront of this chapter.

Implications of Mercy Release

The implications of mercy release can be addressed in terms of: (1) animal suffering; (2) the environment; and (3) human health.

1. Animal Suffering

The act of mercy release is motivated by the belief that the released animal will be spared from their untimely death and live freely in the wild. Yet this is very often far from the truth with regard to the welfare of the released animals. Indeed, the process of mercy release could have a significant impact on the suffering of the animal involved. The pain and suffering endured by the wildlife occurs well before it is released, beginning when it is purchased and during its transportation. Before the animals are released into the wild, one must ask: where did these animals come from? Based on existing studies, there are three principal avenues.

First, the animals are purchased from pet stores, or fresh and alive from food markets. This is evident, for example, in the release of tortoises and freshwater turtles in Hong Kong where most of the animals are purchased directly from fish stalls where the stall owners are fully aware that the purpose of the purchase is for mercy release (Wong, 2024). Between 2015 and 2019, the recorded number of turtles imported into Hong Kong (as food and non-food use) amounted to 3,573,189 while the number of CITES-listed turtles and tortoises were recorded at 2,543,497 (Agriculture, Fisheries and Conservation Department, 2024; Census and Statistics Department, 2024).

These official figures do not necessarily capture the entirety of the trade in turtles since there are also unrecorded and illegal cases of turtles being smuggled into Hong Kong. For example, Burmese starred tortoises and radiated tortoises listed on CITES Appendix I and the Asian leaf turtle and black tree monitor listed on CITES Appendix II were smuggled into Hong Kong from Thailand in December 2023 (The Government of the Hong Kong Special Administrative Region, 2023). Yet there is no empirical evidence to show that the smuggled tortoises are used for mercy release. The vast number of turtles and their unsupervised entry into Hong Kong brings us to the issue of invasive species (discussed in greater depth in the next section).

The ill effects of mercy release continue. It has been documented that animals are captured in the wild to be sold to mercy release participants and that their well-being is significantly undermined during the process of capture and transportation. Their welfare is neglected since they are treated as an economic commodity rather being seen as living beings. One depressing description of this situation is described concisely by Hong Kong Kadoorie Farm and Botanic Garden, a local botanical park that aims to promote biodiversity and general environmental awareness: '[o]ne very cruel fact surrounding this activity is that many wild animals are captured specifically to be released and for the one bird which reaches the marketplace, 10 or more birds may already have died during transportation in crowded conditions. By the time they are once again released back into the wild, many are too weak to survive. In addition, some released animals are captive-bred species that lack skills to survive in nature. Effectively the action is similar to releasing a friendly pet dog in the middle of the wilderness to fend for itself and find food. Also, the traders have turned what was considered a compassionate activity into a commercial enterprise, reaping large sums of money from the suffering of animals' (Kadoorie Farm & Botanic Garden, n.d.).

Further, there is empirical evidence to show that the released animal is often recaptured immediately after they have been released so they can be exploited again by different parties. 'There are boats that take people to release freshwater carp, and we've seen [the released carp] be fished by different boats and taken back to the market to re-sell' (Hong Kong Animal Law and Protection Organisation, 2021). In a report published by the Environment and Animal Society of Taiwan, 63 out of 155 pet stores supplied birds for mercy release and the birds were recaptured and released again. The process of repeated release and capture leads to extreme stress for the animals of resulting, eventually, in their deaths (Shiu & Stokes, 2008). This tormenting process leads to unnecessary animal pain and suffering.

In this vein, it is important to remind readers that the animals, even if they are alive when released, are unlikely to survive since they are being released into a foreign environment in which they must compete with other species within established ecosystems. On the other hand, the released animal could well be an invasive species and introducing it to foreign environments could destroy existing ecosystems (Wong, 2023).

2. Environment

Unregulated mercy release involves the releasing of vast amounts of animals into nature without any proper scientific identification as to whether they are native or foreign species. This leaves open the possibility that invasive species are being released into the environment and, in turn, resulting in the disruption of established ecosystems. In theory, mercy release is a well-intended practice that aims to liberate captured wildlife from being slaughtered as food. The reality, however, is very different.

Participants in mercy release will not make such distinctions and rarely consider whether the species they are releasing into the wild belongs to that environment. Accordingly, non-indigenous and invasive species are released into established ecosystems, causing significant disruption to the border environment (Everard et al., 2019). In the long run, unwanted biological invasion could lead to population decline in indigenous species and drastic changes in ecosystems. The far-reaching ecological harm resulting from biological invasion is well documented (Simberloff & Vitule, 2014).

The Sabah grouper is a classic example of an invasive species causing mass disruption to established ecosystems. This fish is a carnivorous hybrid species of the tiger and giant grouper, bred in a laboratory in Malaysia, which has found its way to Hong Kong via the food trade. In 2014, the *Guardian* newspaper reported on the hybrid groupers in their Environment section, describing the fish, which was to be found in Southeast Asia, as a 'Frankenfish' . In this article, the author warned of the dire consequences if the hybrid grouper was left free to breed in the wild since they were fed a mix of protein-rich pellets and do not need to consume as much as their wild counterparts to grow. Thus, in the long run they would be able to outcompete the native species for resources (Langenheim, 2014). Of particular relevance with regard to the subject of this chapter is the use of Sabah grouper in mercy release in Hong Kong. It is estimated that 440,000 kg of Sabah grouper and live tortoises were released in 2019 (Hong Kong Legislative Council, 2021). This hybrid grouper is commonly found and sold in the local fish markets in Hong Kong. Their relatively low prices and large size make the Sabah grouper a popular choice for mercy release and, consequently, it is also commonly found in local mercy release ceremonies (Cheang et al., 2020). The role played by mercy release in the introduction, spread, and establishment of invasive species in Hong Kong is acknowledged in the Environment Bureau's Hong Kong Biodiversity Strategy and Action Plan 2016–2021. In this plan, the government emphasized the urgent need to control and combat invasive species in the city (Hong Kong Biodiversity System).

The red ear slider turtle is another classic example of an invasive species released into established ecosystems as a result of mercy release. The red ear slider is native to the Mississippi River waterways in the United States, yet this species can now be found all over the world, outcompeting native species after being abandoned. The red ear slider turtle is one of the most traded turtle species in the world. Between 1989 and 2009, the US exported more than 100 million red eared sliders, amounting to over 80% of their total reptile exports. They are also a popular and impulsive pet purchase because of their low prices, small size and accessibility. Yet they are frequently abandoned by their lawful owner because they are more aggressive with age and long life span, thus making them one of the most invasive species in the wild (Valdez, 2021). Within the Hong Kong context, there was a recorded number of 60,200 red eared sliders being imported to the city between 2015 and 2019. Given their vast availability and cheap prices, the red eared sliders are one of the most popular animals used in mercy release. When released into foreign ecosystems, the red eared sliders will outcompete native turtle species for resources and disrupt established food chains by consuming aquatic vegetation, invertebrates and vertebrates. This could lead to the long-term persistent decline in the population of native turtle species (Prevot-Julliard et al., 2007; Pearson et al., 2015).

Another example of non-indigenous, foreign species being released into the wild is the vast amount of Asian swamp eels introduced to the aquatic environments in North America. The Asian swamp eel is native to the Oriental region around Southeast Asia but has been introduced to North America and Europe in the food trade. Mercy release is one of the key pathways for the introduction of Asian swamp eels into North America. This situation is made worse because individuals could easily visit the food market and purchase the eels for mercy release instead of food. The situation is described as follows: 'live-food markets are a source of aquatic animals used in prayer release…such markets are common in certain metropolitan areas, especially those with large Asian populations and many offer a variety of aquatic animals, often at relatively low prices. Similar events take place in other non-Asian countries, in England prayer release practitioners purchased hundreds of live, non-native lobsters and crabs from a wholesale supplier and then illegally released them into coastal water' (Nico et al., 2019, p. 804).

The biological invasion of non-indigenous species is one of the main drivers of the loss of biodiversity and the destruction of ecosystems. Yet, this damage extends beyond that of marine life, also affecting aquatic plants. The introduction of the common carp to North America a century ago has resulted in a severe decline (of around 90%) in aquatic plant cover and richness in the Eastern Temperate Forests lakes in Minnesota. The common carp is estimated to be 70% excessively abundant in this area and its impact is likely to be felt from Canada down to the Gulf of Mexico since it has been observed across these areas (Zambrano et al., 2006; Bajer et al., 2016).

3. Human Health

In addition to animal suffering and the disruption of established ecosystems, mercy release could have a direct impact on the human population given that the released animal could potentially be carrying viruses and bacteria which could be spread to human communities, so-called zoonotic diseases. Unregulated mercy release has great potential for disease transmission given that animals are part of an unregulated trade and not quarantined.

One prime example of a zoonotic outbreak is the bird flu. Gutierrez et al.'s study (2011) shows that 50 out of 94 birds tested from Buddhist temples in Phnom Penh, Cambodia destined for mercy release tested positive for H5N1, leading them to conclude that 'merit release birds represent a risk for human contamination in countries where avian influenza virus is circulating and where this religious ritual is practiced' (Gutierrez et al., 2011). Further, 10% of the birds sold in an open market in Cambodia for mercy release were also found to be positive for influenza A virus while other viruses, such as *C. psittaci* (a bacteria that infects birds but could also infect humans, resulting in a disease called psittacosis) and *M. genavense* (the most common mycobacteriosis found in birds and considered as a potential zoonotic agent), were also found in the birds (Gilbert et al., 2012). The capture, handling, transportation and trading of the released wildlife to the mercy release site could increase the risks of zoonotic spillover, the process of transmitting pathogen from a living animal to the human population (Plowright et al., 2017).

Methods

This study aims to find out whether the facilitators of mercy release understood the harmful implications of their involvement. During 2021–2023, I conducted 35 interviews and on-site observations in mercy release hot spots across Hong Kong. I also joined a local organization in their monthly mercy release outing at sea. My interviewees included fish stall owners operating in wet markets that sell marine species for mercy release, organizers of mercy release excursions, and individuals involved in the actual mercy release process. Fortunately, the majority of the interviewees were happy to share with me their experience in mercy release while some of them even invited me to join them in their mercy release activities, which I declined.

The structure of our interviews was semi-structured and causal. I would begin our conversation by asking about the marine species they were selling/about to release to see if they had time to converse. If they agreed to talk to me, I would then ask them about their experience in releasing captured animals and I followed by asking whether they have reflected on their involvement.

Discussion

Three general conclusions could be drawn from my empirical data.

First, none of the interviewees felt that their involvement in mercy release was harmful to the released animal or environment. All the interviewees denied any wrongdoing on their part and redirected their responses with a heavy emphasis on the good they felt they were doing. The following response from a fish supplier demonstrates this: 'I am a registered business owner and I have been selling fish for decades. They [customers] come to me and I provide for them, there is nothing unlawful about this. The animals go back into their ocean, where they belong and my customers are happy.' When asked how they would feel if the fish died as a result of the release, they saw it as destiny at play, feeling it was not an outcome they could have controlled. This narrative sits comfortably with the concept of interpretive denial, where the facts of the case are not being denied but given a different interpretation from what appears clear to other people (Cohen, 2001).

Further, there is a profound sense of charity and generosity among my interviewees when they speak of their involvement in mercy release. They genuinely believed they were doing a noble act in releasing marine species into the ocean. They found it offensive that people would misinterpret their intentions. This finding is similar to the views of deer hunters, who found it offensive to be labelled poachers since the term carries a negative meaning (Eliason & Dodder, 1999). Another theme identified in my empirical data is the diffusion of responsibility, the understanding that because everybody else is doing it so it is not harmful. As exemplified by the following remark from a female participant in her mid-50s, 'every month we come as a group to release marine species and over the years, more and more people came to join us. What could be wrong if so many of us are doing it? We cannot all be wrong.' This speaks to Martin et al's analysis of the written responses to sanctions for environmental crime in which perpetrators simply pointed out that "everybody does it" before moving on to question the legitimacy of the law (Martin et al., 2008, p. 632)

Second, the released animals are subjected to immense pain and suffering. Before they are released, the marine species are transported to various locations across Hong Kong in plastic containers and overcrowding was a common phenomenon in my observations. I have seen fish crammed into small containers and rapidly grasping at the water's surface, a clear sign that they are not getting enough oxygen because of the shared compact space. Overcrowding is one of the biggest concerns in aquaculture since it can lead to poor water quality (due to excessive fish waste), higher chances of injuries, and increased stress levels among the fish. In essence, 'fishes' physical and mental well-being, and freedom to express their natural behaviour can be severely compromised by overcrowding in poor conditions. Overcrowded fishes, like any animals, suffer greater levels of stress and injury, and have a higher susceptibility to disease' (Brown & Dorey, 2019, p. 189). When I asked the fish supplier why the fish were jumping up and down in the container, he replied that the fish were excited to be released. On many occasions, I have also seen injured fish and stingrays being released into the ocean. They will probably die once released into the water not only because they are adapting to a new environment but also physically disadvantaged to vend for themselves. Here, I would like to remind readers that there are established scientific evidence that demonstrates that fishes are sentient beings, capable of experiencing pain, fear, and psychological stress (Conte, 2004; Chandroo et al., 2004; Culum, 2014) Thus, the concept of animal welfare should be applied to fish. This wider discussion of fish as sentient beings capable of experiencing pain and suffering prompts us to consider the emerging study of blue criminology, which examines maritime security, ocean governance and environmental crime/ harm against the aquatic environment (Ruiz Garcia et al., 2022)

Lastly, the influence of cultural norms in shaping human behaviour is evident in the practice of mercy release. The practice of mercy release is rooted in Buddhist teachings, yet my data suggest that many of the participants are non-Buddhists. Their participation in mercy release is shaped by cultural beliefs. Some of the younger interviewees revealed that their parents have always believed in the karmic effect of mercy release, so they continue to follow this tradition. I have seen young children, under the encouragement of their parents, take part in mercy release. The extent to which cultural beliefs can influence behaviour should not be overlooked, especially in the context of environmental harm and crime where special attention must be placed on cultural beliefs and norms (South, 1998). While social media could be a platform to promote and monitor environmental issues, it could also backfire and facilitate the promotion of harmful behaviour (Stringham et al., 2021; Shaw et al., 2022). There are numerous mercy release-related social groups across the Internet openly advertising such release excursions; the vast scale of these operations suggests that mercy release has turned into a lucrative business opportunity.

Conclusion

This chapter examines the practice of mercy release, a distinct type of harm towards nonhuman animals. This unregulated practice endangers not only the released animals but also the environment in which the animal is being introduced since invasive species are being released into established ecosystems. Participants fail to see the harmful implications of their involvement. The pain and suffering inflicted on the animals or the damage it is causing to the environment remains unacknowledged. Further, this practice has gained popularity due to promotion across social media platforms.

The level of harm brought upon by mercy release should not be overlooked and it should be outlawed. The harmful implications of this practice must be communicated to the public while a change in attitude and human behaviour is crucial to protect the environment. A key motivation for individuals to engage in mercy release is because they genuinely believe that their action is saving lives and that they are engaged in a meaningful action. This is a misconception that must be clarified to stop the public from engaging in mercy release. Unnecessary suffering endured by animals and disruption to the environment will be continued until such is achieved.

Acknowledgement This work was supported by the Environment Conservation Fund (33/2021) funded by the Government of Hong Kong Special Administrative Region.

References

Agriculture, Fisheries and Conservation Department, Hong Kong Special administrative Region. (2024). *Import of reptiles, rodents and miscellaneous pet mammals*. Retrieved from Agriculture, Fisheries and Conservation Department: https://www.afcd.gov.hk/english/quarantine/qua_ie/qua_ie_ipab/qua_ie_ipab_irrmpm/qua_ie_ipab_irrmpm.html

Anon. (2021, May 13). *Dead and dying non-native turtles set free for merit fished from drains*. Retrieved from The Standard: https://www.thestandard.com.hk/breaking-news/section/4/172101/Dead-and-dying-non-native-turtles-set-free-for-merit-fished-from-drains

Bajer, P., Beck, M., Cross, T., Koch, J., Bartodzej, W., & Sorensen, P. (2016). Biological invasion by a benthivorous fish reduced the cover and species richness of aquatic plants in most lakes of a large North American ecoregion. *Global Change Biology*, 3937–3947.

Bernie, P. (1999). For a nonspeciesist criminology: animal abuse as an object of study. *Criminology*, 29–59.

Brisman, A. (2007). Crime–Environment relationships and environmental justice crime-envrionment relationships. *Seattle Journal for Social Justice, 6.*

Brown, C., & Dorey, C. (2019). Pain and emotion in fishes—Fish Welfare Implications for Fisheries and. *Animal Studies Journal, 8*, 175–201.

Census and Statistics Department, Hong Kong Special Adminstrative Region. (2024). *Interactive Data Dissemination Service for Trade Statistics.* Retrieved from Census and Statistics Department: https://tradeidds.censtatd.gov.hk/Index/ea95100de50d4c169595e19fb078cec7

Chandroo, K., Duncan, I., & Moccia, R. (2004). Can fish suffer?: Perspectives on sentience, pain, fear and stress. *Applied Animal Behavioural Science, 86*, 225–250.

Cheang, C.-C., Lee, B.-Y., Ip, B. H.-Y., Yiu, W.-H., Tsang, L.-M., & Ang, P. (2020). Fish and crustacean biodiversity in an outer maritime estuary of the Pearl River Delta revealed by environmental DNA. *Marine Pollution Bulletin, 111707.*

Cohen, S. (2001). *States of denial: Knowing about atrocities and suffering.* Polity Press.

Conte, F. (2004). Stress and the welfare of cultured fish. *Applied Animal Behaviour Science, 86*, 205–223.

Culum, B. (2014). Fish intelligence, sentience and ethics. *Animal Cognition, 18*, 1–17.

Derraik, J. G. (2002). The pollution of the marine environment by plastic debris: A review. *Marine Pollution Bulletin, 44*, 842–852.

Eliason, S. L., & Dodder, R. A. (1999). Techniques of neutralisation used by deer poachers in the Western United States: A research note. *Deviant Behaviour, 20*(3), 233–252. https://doi.org/10.1080/016396299266489

Everard, M., Pinder, A. C., Raghavan, R., & Kataria, G. (2019). Are well-intended Buddhist practices an under-appreciated threat to global aquatic biodiversity? *Aquatic Conservation: Marine and Freshwater Ecosystems, 29*(1), 136–141. https://doi.org/10.1002/aqc.2997

Flynn, M., & Hall, M. (2017). The case for a victimology of nonhuman animal harms. *Contemporary Justice Review: Issues in Criminal, Social and Restorative Justice, 20*, 299–318.

Gilbert, M., Sokha, C., Joyner, P. H., Thomson, R. L., & Poole, C. (2012). Characterising the trade of wilde birds for merit release in Phnom Penh, Cambodia and associated risks to health and ecology. *Biological Conservation, 153*, 10–16.

Gutierrez, R. A., Sorn, S., Nicholls, J. M., & Buchy, P. (2011). Eurasian Tree Sparrows, risk for H5N1 virus spread and human contaimination through Buddhist ritual: An experimental approach. *PLOS One, e28609.*

Hall, J. (2017, January 24). *A Buddhist tradition to save animals has taken an ugly turn.* Retrieved from National Geograhpic: https://www.nationalgeographic.com/animals/article/wildlife-watch-mercy-release-buddhist-china-illegal-trade

Hong Kong Animal Law and Protetcion Organisation. (2021, May 28). Retrieved from Tell me more: why there is no mercy in release: https://www.hkalpo.com/newsandblogs/tellmemore-why-there-is-no-mercy-in-release

Hong Kong Special Administrative Region Press Release. (2021). Regulation of Improper Release of Animals in Taiwan. Available at: https://www.legco.gov.hk/research-publications/english/essentials-2021ise26-regulation-of-improper-release-of-animals-in-taiwan.htm (Last accessed July 10th 2025).

Hong Kong Legislative Council. (2021). *Regulation of improper release of animals in Taiwan.* Hong Kong Legislative Council.

Kadoorie Farm & Botanic Garden. (n.d.). Retrieved from "Mercy Release: brings Misfortune to the Animals: https://www.kfbg.org/en/press-release/article/Mercy-Release-brings-Misfortune-to-the-Animals

Langenheim, J. (2014, June 20). *Why we should be worried about 'Frakenfish' in South East Asia.* Retrieved from *The Guardian*: https://www.theguardian.com/environment/blog/2014/jun/20/why-worried-frankenfish-south-east-asia

Leung, C. (2021). *Regulation of improper release of animals in Taiwan.* Hong Kong Legislative Council.

Lynch, M., & Stretsky, P. (2003). The meaning of green: Contrasting criminological perspectives. *Theoretical Criminology, 7*, 217–238.

Lynch, M., & Stretsky, P. (2010). *Global environmental harm.* Taylor & Francis.

Martin, A. M., Salazar-Laplace, M.-E., Ruiz, C., & Kaplan, M. F. (2008). Individual breachers of environmental laws in cases from public administration files. *Deviant Behaviour, 29*, 611–639.

Nico, L. G., Ropicki, A. J., Kilian, J. V., & Harper, M. (2019). Asian swamp eels in North America linked to the live-food trade and prayer-release rituals. *Aquatic Invasions, 14*(4), 775–814. https://doi.org/10.3391/ai.2019.14.4.14

Pearson, S. H., Avery, H. W., & Spotila, J. R. (2015). Juvenile invasive red-eared slider turtles negatively impact the growth of native turtles: Implications for global freshwater turtle populations. *Biological Conservation*, 115–121.

Plowright, R. K., Parrish, C. R., McCallum, H., Hudson, P. J., Ko, A. I., Graham, A. L., & Lloyd-Smith, J. O. (2017). Pathways to zoonotic spillover. *Nature Reviews Microbiology, 15*, 502–510.

Prevot-Julliard, A., Gousset, E., Archinard, C., Cadi, A., & Girondot, M. (2007). Pets and invasion risks: Is the slider turtle strictly carnivorous? *Amphibia-reptilia, 28*, 139–143.

Ruiz Garcia, A., South, N., & Brisman, A. (2022). Eco-crimes and ecocide at sea: Toward a New Blue Criminology. *International Journal of Offender Therapy and Comparative Criminology, 66*, 407–429.

Shaw, M. N., Borrie, W. T., McLeod, E. M., & Miller, K. K. (2022). Wildlife photos on social media: A quantitative content analysis of conservation organisations' Instagram images. *Animals, 12*, 1787.

Shiu, H., & Stokes, L. (2008). Buddhist animal release practices: Historic, environmental, public health and economic concerns. *Contemporary Buddhism, 9*, 181–196.

Simberloff, D., & Vitule, J. R. S. (2014). A call for an end to calls for the end of invasion biology. *Oikos, 123*(4), 408–413. https://doi.org/10.1111/j.1600-0706.2013.01228.x

South, N. (1998). Corporate and state crimes against the environment: Foundations for a green perspective in European criminology. In V. Ruggiero & I. Taylor (Eds.), *The New European Criminology* (pp. 443–461). Routledge.

South, N. (2014). Green criminology, enviromemtal crime prevention and the gaps bewteen law, legitimacy and justice. *Revija Za Kriminalistiko in Kriminologijo, 65*, 373–381.

Stringham, O. C., Toomes, A., Kanishka, A. M., Mitchell, L., Heinrich, S., Ross, J. V., & Cassey, P. (2021). A guide to using the internet to monitor and quantify the wildlife trade. *Conservation Biology, 35*, 1130–1139.

Tan, L. (2019, January 13). *Buddhist animal release pratice: Saving lives or contributing to an ecological disaster?* Retrieved from New Zealand Herald: https://www.nzherald.co.nz/nz/buddhist-animal-release-practice-saving-lives-or-contributing-to-an-ecological-disaster/WOMMCGC3RXK6FLZGBO3M3LFSWU/

Thach, M. (2024). Animal mercy release, enironmental conservation, and the media in Vietnam. In I. U. K. Heise (Ed.), *Environment and narrative in Vietnam. Literatures, culture and the environment*. Palgrave Macmillian.

The Government of the Hong Kong Special Administrative Region. (2023, December 27). *Man jailed for smuggling live endangered turtles and lizards.*

Retrieved from GovHK Press Release: https://www.info.gov.hk/gia/general/202312/27/P2023122700556.htm

Valdez, J. W. (2021). Using Google trends to detemine current, past and future trends in the reptile pet trade. *Animals, 11*, 676.

White, R. (2013). *Environmental harm: An eco-justice perspective*. Bristol University Press.

Wong, R. W. (2023). Animal abuse: Beyond companion animals and domestic households. In O. H. Chan & R. W. Wong (Eds.), *Animal abuse and interpresonal violence: A psycho-criminological understanding* (pp. 11–18). John Wiley & Sons Limited.

Wong, R. W. (2024). The harms and crimes against terrestrial wildlife (Nonhuman animals). In *Oxford research encyclopedia of criminology*. Oxford University Press.

Wyatt, T. (2014). Non-human animal abuse and widlife trade: Harm in the fur and falcon trades. *Society and Animals, 22*, 194–210.

Zambrano, L., Martinez-Meyer, E., Menezes, N., & Peterson, A. (2006). Invasive potential of common carp (*Cyprinus carpio*) and Nile tilapia (Oreochromis *niloticus*) in American freshwater systems. *Canadian Journal of Fisheries and Aquatic Sciences, 63*, 1903–1910.

Photovoice for Ecojustice: Following the European Eel from the Sargasso Sea to the River Severn

Alison Hutchinson

Introduction

The European eel, *Anguilla anguilla*, is a fish with many contradictions. Seemingly populous—in their early life stages—whilst simultaneously also severely depleted and critically endangered (Pike et al., 2018). Their bodies, both delicacies and drivers of the globalised sushi economy, are also phantoms of the night, writhing and slimy, and a low contender for conservation interventions. Their presence in both legal and illegal international wildlife trade speaks to their high market desirability. Here, their bodies are transformed from their murky riverine origins and become 'white gold' (Sage, 2007). Yet, the rush to extract and exploit the eel is just one thread in the web of human-made threats towards them. Notably, Gutiérrez and Duffy (2023, p. 73) have demonstrated how a focus on illegal eel trade 'renders wider non-fisheries related harms invisible'. To bring light to these issues, this chapter follows the eel's migratory

A. Hutchinson (✉)
Newcastle University, Newcastle Upon Tyne, UK
e-mail: Alison.Hutchinson@newcastle.ac.uk

D. van Uhm, D. Siegel (eds.), *Global Green Crime and Ecojustice*, Palgrave Studies in Green Criminology, https://doi.org/10.1007/978-3-031-91557-4_14

journey; using photovoice narratives to illustrate the intersections of (non-)human and environmental harms and injustices to tease out the otherwise overlooked and socially accepted harms towards the eel, their ecosystem, and our shared entanglements with both.

To focus attention towards the largely invisible, legal, and state-authorised harms committed towards European eels, this study draws from a green criminological ecojustice framework (White, 2013). This holistic view of justice—encompassing environmental, ecological, and species justice—focuses attention on the harms and victimhood of all living beings as well as the natural environment; thus, recognising the violation of individual eels who are captured, killed, or otherwise victimised by human activities, as well as the harms towards the non-living community (rivers and oceanographic ecosystems). This position contrasts with established governance measures for the eel which are couched in anthropocentrism. Whilst fishing without or in breach of licences, failure to keep and submit catch records, and trading without appropriate permissions are subject to civil sanctions and criminal regulatory systems,[1] neither the eel nor their environment are considered victims of harm. Integrating an ecojustice framework expands upon the gaps in existing governance, to include harms which are not criminalised or sanctioned. For example, actions which impact population and genetic diversity, diminish welfare, impede animal cultures, and diminish the intrinsic rights of both eels and their ecosystems. These combined ecojustice perspectives provide a strong foundation to interconnect with issues of species, ecological, and social injustice and to recognise and reflect on the joint and often parallel oppression of 'nature' and people.

Methodological Approach: Visualising Harm

Visual methods are increasingly adopted within green-cultural criminological spheres and build on visual criminological attentiveness to the power and meaning of images (McClanahan, 2021). For example,

[1] See: EU Council Regulation (1224/2009) and the EU 'Wildlife Trade Regulations' (Council Regulation 338/97, 1996).

photo-elicitation methods make use of images as conversational cues (Berger, 2013). This approach is less interested in the *truth* of narratives, focusing instead on the symbolic and emotive *meanings* ascribed to images by those who are familiar with the issues presented (see works by Coleman, 2016; Natali & McClanahan, 2020; Natali, 2023). In a similar vein, photovoice engagement requests that participants present and describe their own images (Sutton-Brown, 2014). This approach advances social and epistemological justice by deconstructing researcher–participant power dynamics and providing space for those whose voices may otherwise be overlooked by traditional research methods (Breny & McMorrow, 2021). The use of photos for giving a richer meaning to environmental and wildlife conflicts is well established amongst more-than-human geographers and environmental scholars (see Bexell et al., 2023; Margulies, 2019). Whilst the use of this approach has been limited within green criminology, a recent and innovative addition from Pons-Hernandez (2024) demonstrates the use of photos to illustrate and evidence harms towards the European eel.

The photo-voice gallery presented in this chapter arose from the 'Beastly Business: Examining the Illegal Wildlife Trade in Europe' project undertaken at the University of Sheffield, UK. Data collection took place between November 2022 and April 2023 and was approved by the University of Sheffield's ethical review board (reference: 049587). The research involved collaboration with 14 individuals, including academics, conservationists, commercial and recreational fishers, fish traders, government representatives, and non-governmental organisation (NGO) authorities. Two primary sites of interest were chosen to focus the photo-voice narrative. The first took place at the start of the eels' transatlantic spawning migration in the Sargasso Sea and involved joining a fishing trip with a group of recreational and commercial fishers. Three photo-voice interviews with a fisher, a maritime surveillance officer, and a marine conservationist were held on board the fishing vessel. A further three interviews with a government official, an NGO representative, and a marine researcher took place in Bermuda—the closest land mass to the Sargasso Sea. The second site for interviews followed the eels' juvenile migration to the River Severn in the UK. Eight interviews and numerous observations with fishers, fish traders, enforcement authorities, and NGO representatives were carried out during the glass eel fishing season in 2023. All participants have given permission for their responses and photographs to be used. To ensure confidentiality and safeguarding commitments, all names given in this chapter are pseudonyms.

A thematic analysis of the interviews led to the categorisation of five core-conflicts across both study sites (a condensed thematic analysis is shown in Fig. 1). The photo-voice journey is structured by the level of prevalence given to these conflicts at both locations.

Core conflicts:

1. Threats from negligence

   Negligent acts are those that have unintended impacts on the eels' health and survival (e.g., polluting activities and habitat degradation).
2. Threats from ignorance

   Ignorant impacts arise when harms to the eel are not adequately considered (e.g., when uncertainties surrounding scientific knowledge, or a lack of collaboration, leads to ineffective policies, governance, or management measures).

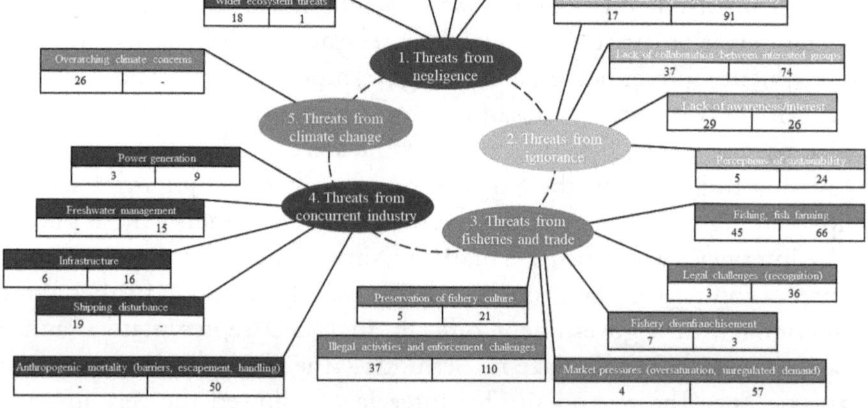

Fig. 1 Overview of thematic analysis. The prevalence of common concerns underpinning the five conflicts are shown (Sargasso Sea group on the left, and the River Severn group on the right)

3. Threats from fisheries and trade
   Threats are inclusive of direct harms to the eel (e.g., il/legal fishing, farming, and trade) as well as socio-cultural harms from shifting fishery dynamics (e.g., unregulated demand, oversaturated markets, fishery disenfranchisement, and loss of fishery culture).
4. Threats from concurrent industries
   Threats are inclusive of broader threats from wider industries (e.g., infrastructure development, waterway management, power generation, and global shipping).
5. Threats from climate change
   Inclusive of the overarching climate crisis (note, this only emerged as a core-concern amongst the Sargasso Sea group).

Photo-voice Gallery: Eels in the Sargasso Sea—Core-Conflicts in Focus

Threats from Negligence

The first image in the gallery (Image 1) was selected by Megan, a marine scientist, who has conducted long-term studies of Bermuda's coral reefs, coastline, and the open ocean. During a walking interview and community litter pick, one participant described how marine debris becomes congregated by the Sargasso Sea gyre, leading to a near-constant cycle of litter accumulation along Bermuda's coastline. Image 1 serves as a reflection on the 'insidious challenge' that marine debris presents to both marine and terrestrial ecosystems.

Pollution and poor waste management were the most frequently discussed threats toward the eel (core conflict 1), reflecting the ubiquitous nature of the problem:

> The problem is on land, away from the ocean … [it's] not messy fishermen or lazy beachgoers, it's people who are hundreds of miles away from the ocean with poor waste management … Their litter just gets in a stream, river or estuary and it ends up here…
>
> …If you think of those developing eel leptocephali … their diet is going to be impacted by plastic. Whether they're feeding of copepods that have

Image 1 Washed-up pollution. Shared by Megan, a marine scientist. Megan described how the image shows large and small plastic debris that have become tangled up within sargassum seaweed. These smaller pieces of degraded plastic are almost imperceptible, leading to false impressions of 'pristine' beaches

> been feeding on microplastics, they're going to get that plastic ... Anything that has to eat in this ocean, that is going to be there for months to years, they will see more vulnerability'—Megan

The excerpts above demonstrate the local, national, and global dimensions of ocean-born pollution; and is indicative of the negligent, commonplace, and routine activities that produce 'ordinary harms' (Agnew, 2020). The amplification of harms through the bioaccumulation in marine food webs is further evocative of the 'slow violence' and invisible ecological consequences from delayed, dispersed, and destructive practices (Nixon, 2011).

Threats from Fishery and Trade

The next major concern amongst participants centres around the challenges within the fishing sector (Image 2, core-conflict 3). Image 2 was shared by Joe, a life-long commercial fisher in Bermuda. Joe described how limited capacity for fishery enforcement and ineffective fishery management practices means there is very little resource to prevent illegal fishing, impacting both the legitimacy of fishing efforts and the abundance of marine species in the region. A lack of monitoring also means fishers can sell their catch directly to the market without having accountability through market traders. This opens opportunities for illegal or undocumented fish to enter the legal market (Joe, pers. communication).

A combination of low enforcement capacity and challenges in monitoring trade means that legitimate fishers are often outcompeted by illegal operators, with direct implications for their personal and financial security.

Image 2 Passive security, a sign. Chosen by Joe, a commercial fisher in Bermuda. This image is intentionally anonymous, indicative of a sense of unease within the fishing community. It serves as a visual metaphor to illustrate the passive and anonymous nature of fishery enforcement

> You have to find your own markets to sell it [fish]. And it just leaves a blank area because people lie on their statistics all the time on what they are catching and not telling the honest truth. And even a licensed fishermen is not going to tell you exactly what they caught.—Joe

This fine line between honest reporting and competitive or fraudulent activities demonstrates how legal and illegal fishing have become closely intertwined. This relationship builds on the phenomenon of 'green-collar crime' (see Wolf, 2011; Iordăchescu et al. 2022) to demonstrate how fishers can opportunistically shift towards non-compliance and illegal activity under the cover of legitimate practices. The wider context of fishery disenfranchisement perhaps also creates a platform for neutralisation—a sense that everyone is operating illegally without getting caught—to rationalise fraudulent or criminal activities and absolve personal responsibility (Thiel and South, 2022).

Threats from Ignorance

The third image digs deeper into the problem of poor management to recognise the broader and often-unseen threats from inadequate fishery governance (Image 3, core conflict 2). The image was taken with Anupa (a representative for the Sargasso Sea Commission Secretariat). It shows the Sargasso Sea's golden sargassum seaweed and serves to prompt a reflection on the need for new perspectives in high-seas management.

There are two strands to this theme of ignorance. The first relates to an ignorance towards natural systems within management at the international scale. For example, Anupa described how the transitory nature of the Sargasso Sea challenges conventional international policies and high-seas management approaches:

> Right now, protection of the high seas is achieved through a patchwork of regional and sectoral organisations, which leaves a lot of gaps. There is no holistic way to approach protecting high seas ecosystems ... We want to have our little square box and that's what we're managing. But that's not how high seas ecosystems operate. The features are dynamic. And that goes against a lot of governance measures.—Anupa

Image 3 Managing the unseen. Chosen by Anupa, a representative for the Sargasso Sea Commission. Anupa described how the biodiverse landscape within the Sargassum ecosystem is frequently considered a lesser priority within management; a consequence of being hidden from view, unfamiliar, and unknown

This narrow and rigid focus within governance could leave critical developmental habitat and migratory corridors for many marine species, including the eel, unprotected (Anupa, pers. communication). This collective ignorance within international governance illustrates how ignorance, whether rational or inadvertent, becomes a mechanism for harm. For example, the rigid governance of marine protected areas (opposed to adaptable and holistic governance) reflect active and strategic management choices—e.g., the costs of alternatives are too high, there are too many unknowns (rational ignorance), or the ecological significance of dynamic high-seas environments are simply not appreciated or fully understood (inadvertent ignorance) (see Somin, 2022). Whatever the cause, this state of ignorance produces harms towards unprotected species, environments, and vulnerable marine systems that fall outside of those 'little square' management boxes.

The second issue of ignorance relates to challenges at the local level. Connecting to the issue of fishery disenfranchisement (Section "Threats from Fishery and Trade"), participants described having lack of trust in the ability of local governance to protect fishery interests. Enforcement and juridical procedures were often perceived as giving lesser importance to fishery crimes:

> The court system is not set up for environmental prosecutions … in Bermuda you just get a slap on the wrist and [offenders] can just go on with business.—Joe (a commercial fisher)

And:

> There's a whole lot of wardens who have just had the wind taken out of their sails. Guys who have an airtight case, [thinking] this guy is going away, and then nothing happens. Or somebody waves a gaffe at them or an axe at them or stuff like that … It doesn't hit them … But death threat kind of things. And then they don't go to court, and nothing happens. Or they go to court, and everything gets thrown out. And that guy kinda said like "if you step on my boat, I'll kill you". And it doesn't go anywhere.—Adam (a recreational fisher)

These excerpts highlight an atmosphere of intimidation and distrust between the fishing community and management bodies, exasperated by an ignorance of the social, political, and environmental nuances surrounding fishery crime. The root of this ignorance may lie in the hierarchy of who and what are considered to be victims (see Wyatt, 2021). For example, considering the grey areas between legal and illegal fishing, we have a 'cat-in-the-box' problem where fishers may either be considered legitimate and therefore ideal victims, or illegitimate and less-worthy victims. Less-worthy victims receive less prioritisation the further down this hierarchy of recognition they appear (notably, non-humans fall even further down this hierarchy). This hierarchical view of victimhood can be understood as both anthropocentric and strategic ignorance; further marginalising both fishers and enforcement officers who are threatened or ostracised, as well as marine wildlife and ecosystems that receive little to no recognition as victims.

Threats from Climate Change

The impacts from the climate crisis were a salient issue amongst all interviewees; they touched on issues impacting Bermuda's mangrove forests, coral reef habitats, and the wider marine community. Image 4 was shared by Megan, a marine scientist, to illustrate how changing climates and oceanographic dynamics are already impacting the Sargasso community by slowing the growth of calcifying animals like corals, bryozoans, and barnacles (Image 4, core conflict 5).

Despite a variety of local impacts, harm was often discussed at the global and ecosystem level, for example the impact on whole ecosystem dynamics:

> We have striking evidence that the Sargasso Sea is becoming warmer, saltier, and more acidic. These changing ocean variables can affect the lifestyles of many high-seas organisms who have evolved to live under different conditions. Long-term persistent warming, as well as discrete extreme periods of warming like marine heatwaves can alter the structure and functioning of entire ecosystems.—Anupa

Recognising the impacts on marine wildlife and environments connects to concerns for species and ecological justice and highlights the 'universal victimisation' of climate change (White, 2018).

Threats from Concurrent Industries

The final image for the Sargasso Sea gallery was shared by Anupa and shows a container ship to illustrate concerns over concurrent ocean industries (Image 5, core conflict 4). Increases in ship traffic and their associated noise and pollution impacts were raised in numerous interviews; and large container or cruise ships often dominated the horizon.

Image 4 Consequences for ecosystems. Shared by Megan, a marine scientist. This image shows a brain coral, *Pseudo-diploria*, with black band disease—an indicator and consequence of a warming climate and increasing oceanic acidity

This focus on concurrent industries reveals a complex relationship between shipping and environmental governance. On one hand, Anupa described how shipping and noise pollution produces harms towards vulnerable marine species:

> Ship traffic has increased on the Sargasso Sea in recent years, paths become entrenched, they [ships] break up the sargassum mats and who knows what that's doing to the ecosystem … it's probably better for a lot of the things developing to have a big mat rather than a tiny clump.—Anupa

Image 5 Increasing disturbance. Shared by Anupa, a representative for the Sargasso Sea Commission. Anupa described how increased shipping traffic is physically disruptive to the Sargassum ecosystem, as well as contributing to greater levels of noise pollution, bio-fouling, and polluting discharges

In contrast, Elliot (a waste-reduction and community engagement representative) described how rising ship traffic is closely tied to Bermuda's economy and island logistics. Most of the islands' waste is shipped overseas for processing, a system that makes use of empty container ships from imported goods (Elliott, pers. communication). This waste management process is viewed as an appropriate solution to keep the local environment clean and healthy, whilst providing Bermudian residents with goods and services. However, we can also understand this relationship through the treadmill of production framework (see Long et al., 2016); wherein the globalisation of production and consumption, along with waste management, recycling, and consumer industries, has become fundamentally intertwined with environmental harms and 'ecological disorganisation'.

Photovoice Gallery: Eels in the River Severn—Core Conflicts in Focus

Threats from Fishery and Trade

Following the juvenile eels' migration, the photovoice gallery now moves to the River Severn in the UK. Discussions here overwhelmingly focussed on the enforcement challenges surrounding fishing and trade. These conflicts are aptly symbolised in Image 6, shared by Sheena, a fishery researcher, which depicts the divisions of land and access that underpin elver fishing (Image 6, core conflict 3).

John, an elver fisher, described how the allocation of fishing rights is interweaved with land ownership and controlled by trade interests. Unlike the direct-to-market process in Bermuda, trade operators dictate to fishers the size of the market order, closing once the order has been fulfilled. Following Brexit, this system has become increasingly challenging as the number of fishers has begun to outweigh the UK market capacity. Increasing rates for tumps (spots on the river to fish) in prime locations additionally means that elver fishers need to catch increasing quantities of glass eels before they get a return on their financial outlay.

The competitive nature of the glass eel trade has led to an increasing atmosphere of distrust and disconnect between fishers, market dealers (who control the glass eel trade), and land owners (who control the tump trade). Numerous fishers recounted instances of aggression and sabotage within the fishing community (anonymous correspondence) and ample speculation surrounded the size and destination of the 2023 UK order, particularly whether the glass eels were destined for Kaliningrad, Russia (this has since been confirmed, see APHA, 2023). Elver fishers who had already purchased their tumps and licences described their dilemma over supplying the market to a country actively engaged in war (anonymous correspondence). To avoid selling to the Russian market, some fishers

Image 6 Controlling fishers. Chosen by Sheena, a fishery researcher. During a walking interview, Sheena pointed out this sign indicating restricted access for elver fishers. Signs like these are commonplace in the area

instead planned to donate their catch to restocking efforts. However, permissions for restocking were not granted during the 2023 fishing season and it is not clear what happened to those eels which had been caught for restocking.

As with the Sargasso Sea case, networks and practices within the eel fishery provide an overlap between legal and illegal fishing activities and the potential for green-collar crime. Alex, a fishery representative who works closely with fishers, fish traders, enforcement officers, and researchers, cautioned that restocking glass eels effectively perpetuates the fishery without providing long-term solutions. Restocking efforts essentially seek to translocate and repopulate eels in areas where their numbers have declined. This practice could potentially be exploited to navigate illegal trade through legal channels; for example, by declaring the movement and trade of glass eels for restocking, when they are actually intended for the market. Such overlaps between the restocking and trafficking of eels have been evidenced by Guiterrez and Duffy (2023) and discrepancies in numbers and misdeclared shipments have also been noted by Ibáñez Alonso and van Uhm (2023).

Whilst this wider context of distrust and competition in the fishing community meant that harm was primarily framed towards the fishers, e.g., via a loss of livelihood, social pastimes, and interpersonal connections within the fishing community; many fishers and fishery representatives were also keen to note how maintaining an active fishery could be central to eel protection:

> If you stop eating eel, you then break the final thread of a once close relationship between humans and eels ... We have to connect humans to nature, and fishing for them in a responsible way is part of keeping the human relationship with nature alive.—Alex

During observations, fishers were frequently sensitive to issues of sustainability and welfare. Whilst this attentiveness may be framed as the protection of a highly valuable product, fishers showed a strong sense of care and a duty of protection towards the glass eels. For example, they demonstrated how the timing of the dip net needs to be precise to prevent crushing the glass eels under the weight of river debris. Handling was also kept to a minimum to prevent thermal shock to the eel's delicate body, and bycatch was removed and returned to the river using a system of sieves. Sam, an elver fisher, additionally showed how elvers in storage tanks need specific water and aeration conditions and fare better with careful feeding regimes. Sam described these elvers as if they were his pets and was strongly supportive of efforts to protect wild populations, including both restocking (the movement off eels to increase population abundance in depleted areas) and translocation (assisted migration beyond barriers) efforts.

Threats from Ignorance

The next image, shared by Matt an eel business owner and trader, builds further on this connection between the fishery and the eel. Alongside their eel trading business, Matt also conducts population monitoring research and spoke of serious uncertainties in the accuracy and legitimacy of population research which has been used to determine fishery and trade restrictions (Image 7, core conflict 2).

Many interviewees questioned the scientific rigour of population assessments and discussed the need for better, research-informed policy. For instance, Matt cautioned that the measures used to estimate

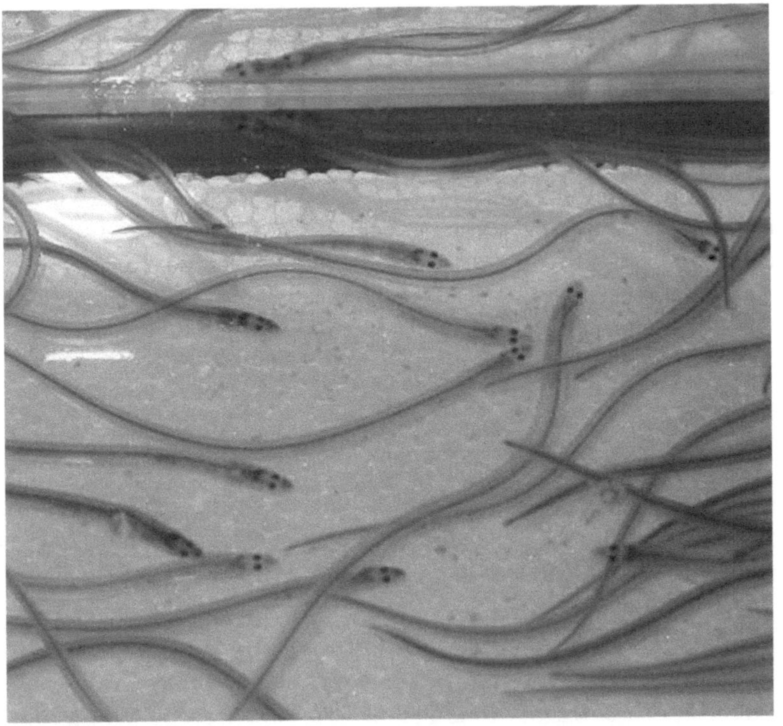

Image 7 How to protect the invisible? Shared by Matt, an eel business owner and trader. This image shows glass eels that have been dyed pink using Rhodamine B as part of a study into their upstream survival

population abundance (e.g., using glass eels as a proxy for the whole population) leaves considerable scientific uncertainty around the species' global population dynamics. Alex additionally described how population data for adult European eels is severely limited:

> Very little measurement is done and there's very little confidence in what those measurements mean.—Alex

The oversight of mature eel populations and uncertainties around the eels' population status connect with Birkenholtz and Simon's (2022) description of ignorance arising from the interaction between scientific practice, political rationality, and agenda setting. These uncertainties feed into policy responses that place an emphasis on glass eel recruitment (believed to be quite strong in the UK) rather than the escapement of adult eels (which is significantly reduced). Such responses are highly contested, with lasting impacts on how the fishery, monitoring, and restocking efforts function—as Matt noted:

> You've got these bureaucrats in the Environment Agency who are just pen pushers, that think they're fishery scientists … I think, well, what sort of qualifications have they got to make these sorts of scientific judgements? … Some have absolutely no science qualifications, others have a very limited knowledge with reference to eels' science—and the overall knowledge is spread very thinly…

And:

> …We don't do any translocation in the UK … We've tried to, we've done a little bit. But it's really difficult to get permission … We're not allowed to do any stocking this year.—Matt

Matt claimed that population-level uncertainties are tolerated by policy makers and used to defend strict fishery regulations that prevent restocking (notably, these regulations are in place to prevent laundering and illegal activities). Although restocking in the UK may have some conservation value (see Fleming et al., 2023), its overall value remains highly contested (ICES, 2022). These disagreements between fishery research and policy have produced considerable conflict; particularly, as noted above, Matt understands the problem to be with mature eel escapement rather than the glass eel population on which the fishery depends.

Threats from Concurrent Industries

Another prominent concern centred on threats from concurrent industries, particularly infrastructure development and freshwater management (Image 8, core conflict 4).

Image 8 Insurmountable barriers. Chosen by Alex, a fishery representative. Alex emphasised the threat posed by the industrialisation of freshwater river systems, where barriers such as this three-gate flood defence profoundly alter the natural functioning of riverways and prevent the migration of eels

Building on previous discussions around the parallels of increasing production and increasing 'ecological disorganisation' (Long et al. 2016), fishery representatives also emphasised wider environmental harms and anthropogenic mortality, listing the development of hydropower stations, water engineering, and floodgates as key concerns:

> As society has become an industrial one, our relationship with eel has changed. And now from eating and respecting it, it's often just collateral damage to water engineering ... Technology moved from wood [gates] and banks of grass that eels could crawl through, to concrete walls, steel doors, and barriers that are just too efficient ... That's suffocated the whole life cycle. Gradually breaking a life pattern of 100 million years, so slowly that we went to sleep watching it...—Alex

Fishers further described how barriers from weirs and floodgates force glass eels to congregate in large numbers. This makes them highly vulnerable to predation from other fish, birds, and opportunistic fishers. These harsh barriers impede the eels' migration, contribute to a loss of critical eel habitat, and significantly reduce the juvenile eels' habitable range as they transition from marine to freshwater environments.

Threats from Negligence

The final image in the gallery demonstrates concerns over the loss and degradation of the eels' habitable range (Image 9, core conflict 1).

When sharing the image above, Lelie described how the historic draining of marshland and the development of flood defences to improve agricultural land (particularly in the Somerset levels) has drastically reduced the habitable range of the eel. Similarly, Matt also emphasised how habitat loss is a potentially insurmountable issue:

> It's obviously an impossible task to get back to the original situation because just look at the amount of wetlands that have been lost. Hundreds and hundreds of acres, thousands, millions of acres that have been lost.—Matt

This recognition introduces a complex problem for how we envisage nature conservation and the aesthetics of wilderness in an environment

Image 9 Habitat deterioration. Shared by Lelie, a conservation scientist. Lelie described wanting to draw attention the carelessness surrounding the preservation and protection of freshwater habitats

that has been historically and continually reshaped and manipulated by human activities.

Discussion

The breadth of issues raised throughout the photovoice gallery demonstrates the complexity and interwoven nature of threats facing the eel. The question for ecojustice now rests on how differing viewpoints produce or obscure the visibility of harm and what this means for the eel and our entangled (or distant) lives with them. The heavy weighting towards fishery and trade management issues is perhaps not surprising considering a heightened awareness of criminality following numerous high-visibility police investigations (European Anti-Fraud Office, 2023; EUROPOL, 2022), and the recognised potential for criminal actors to embed within, or interact with, legal operators (Ibáñez Alonso & Van Uhm, 2023). However, the photovoice gallery has also highlighted how this threat is only part of a very precarious picture, with significant concurrent anthropogenic pressures under-represented within social discourse.

Factors Influencing Perceptions of Harm

Despite the Sargasso Sea's critical importance to the European eel as a breeding ground and spawning site, there was very little connection with the species in Bermuda. Two external elements may have influenced this. First, a major concern amongst interviewees related to pollution. During the interview period, a single-use plastics ban was being introduced into legislation and beginning to take effect (see Government of Bermuda, 2021). This may explain why plastic pollution was central to many discussions at the time. The second major focus on fishery disenfranchisement and enforcement capacity was also timely, as the 'Bermuda Ocean Prosperity Programme', an ocean management scheme, was also undergoing a public consultation (Government of Bermuda, 2022). This consultation was particularly contentious as proposals centred around the expansion of 'no-take' marine protected areas.

A different set of social and political issues may have informed the direction of discussion in the UK. The international European eel trade rests within an increasingly fractious geopolitical landscape. Following Brexit, glass eel trade within the UK has fallen into a state of flux, with—as yet—no clear resolution for establishing international markets in compliance with CITES (Stein & Bunt, 2021). Speculation also surrounds the UK–Russia eel trade relationship (Whipple, 2023), which has reignited political and moral conflicts. These pressures on the continuation of the UK fishery, combined with an increased awareness of wildlife crime, underscored much of the discussion with fishery and conservation groups alike.

Invisibility of Harm

Whilst the prioritisation given to conflicts can reflect how narratives around harm are shaped, absences in discussion are also relevant. Pollution and climate change were leading concerns amongst the Sargasso Sea interviewees, yet neither issue was meaningfully described by the UK group; perhaps demonstrating how these issues are disproportionately realised across the globe (see White, 2018). Further absences were also seen between the two galleries; oceanographic threats were not raised within the UK group, inland threats were not discussed in the Sargasso Sea group, and nether group raised harms from aquaculture. There appears to be a disconnect between distinguishing known and understood *threats* (e.g., fisheries, pollution, habitat loss, climate change) from more commonplace and accepted *harms* (e.g., the normalcy of capturing and farming animals for food). As such, the broader social, environmental, and geopolitical context of globalised fishery markets, supply chains, and industries become obscured within localised views.

Threats towards the eel were primarily framed at a population level (concerning the species), with additional threats recognised towards financially and socially invested human groups (fisheries and industry). This leaves minimal recognition towards broader ecosystem harms, and a complete lack of visibility or discussion of harms towards individual eels, or indeed humans, whose health and well-being are impacted both by the direct exploitation of the eel and by the numerous forms of negligence and damage towards their natural environment. With the above in mind, the next section expands on potential avenues for species and eco-justice.

Harm Recognition for Ecojustice

Throughout the photovoice gallery the eel was firmly understood as a fishery species with harms primarily understood at the economic and personal level. To expand the parameters of harm towards the eel, I am mindful of Margulies' (2022, p. 242) reflection that: '...matters of care are always inherently also political matters intersecting the governing of nonhuman life and transformations wrought by global capitalism'. In this sense, the production and recognition of harms towards the eel can be viewed as a matter of how (and where) a balance is struck between commodification, conservation, and care for the species. Following this guide, each of these elements are discussed in turn.

Harms from Commodification: Exploitation

Fishery-related harms were often understood in terms of economic losses. In the UK, the focus turned to social issues around access and market control. Here, the purchasing of licences and tumps is common practice; however, when combined with limited market demand, this process creates a system of financial control that incentivises fishery activities. Fishers become increasingly motivated to make back their multiple downpayments, and this creates a highly competitive environment that can potentially encourage non-compliance with regulations. Similar conflicts emerged in the Sargasso Sea case, where legal and illegal fishers operate within the same competitive environment.

The photovoice gallery also showcased how the eels' habitat is increasingly reconstructed, fragmented, and polluted. These environmental harms are intimately tied to the commodification of nature, exploiting concepts of natural capital and ecosystem services. Evidence of the anthropogenic impacts on river systems are ubiquitous. Up to 99% of UK rivers contain barriers (Amber International, 2020), including around 1600 hydro-power plants (Spry, 2022) which severely impact both the up- and down-stream migration of the eel. The scale of this problem is amplified throughout Europe, with an estimated 1.2 million barriers fragmenting river systems (Amber International, 2020). Beyond this, the Atlantic Ocean and Sargasso Sea are experiencing unprecedented and unabated increases in ocean pollution, warming, acidification, overfishing, and habitat destruction (IOC-UNESCO, 2022). Despite the significance of these environmental harms, pollution and habitat loss were often painted passively and outwardly by interviewees. Pollution (which was only actively problematised within the Sargasso Sea gallery) was viewed mostly as an external threat inundating shores and coastlines.

And although habitat loss and degradation were recognised as significant issues, they too were often described as insurmountable, revealing a fundamental conflict between land designated for human-use and land left for the eel.

Harms from Conservation: Uncertainties

A lack of scientific engagement with the eel across their lifespan makes understanding the harms they experience impossible to gauge. The strength of scientific monitoring and assessments were repeatedly challenged by fishery interest groups, who contend that a focus on glass eels in population assessments does not adequately capture regional differences and could potentially obscure harms to mature eels (where monitoring is limited). It is simply not known how populations, particularly of mature eels, are impacted by the threats highlighted in this chapter.

Although research involving glass eels can close gaps in knowledge and make for stronger management interventions, there is also a need to be sensitive to the impact of such investigations. For instance, although Rhodamine B (the dye used for population analysis shown in Image 7) is often used as a wildlife biomarker (Fry & Dunbar, 2007), it is recognised as a carcinogenic and genotoxin which can cause heritable damage to DNA (Du et al., 2022). Impacts around water quality and ecosystem health are unclear, however, concerns have been raised over potential toxicity for freshwater and marine organisms (Skjolding et al., 2021; Sudarshan et al., 2022). Whilst the concentrations used in the monitoring study pictured did not result in any immediate mortalities (Aprahamian & Wood, 2021), it does not appear that threshold toxicity limits have been carried out for glass eels. The impact on the wider environment and ecosystem is also not known.

Harms from a Lack of Care: The Invisible Eel

The photovoice gallery has shone a light on issues towards the wider environment (pollution and habitat loss), population-level impacts (fishery industry), and towards people (cultural and trade connections). However, there has been little reflection on how the eel personally experiences harm, or how harms towards the species may produce cascading harms to the ecosystem or wider environment. This oversight speaks to Beirne's (2018) warning that much of the killing and abuse of wildlife is legal and invisible. For instance, whilst the traditional method for fishing glass eels (described in Section "Threats from Fishery and Trade") was connected to perceptions of higher welfare, this framing obscures from view the reality of harm throughout the supply chain. Specifically, these 'high welfare' glass eels will inevitably be transferred to farming facilities, where their welfare and lifespan will be reduced. Ultimately, the killing of eels for food production is largely normalised and mortalities from industry and habitat loss are also largely regarded as secondary losses to developmental needs.

To focus in on the individual eel and emphasise with the harms they face, we may additionally draw from Pons-Hernandez's (2024, p. 92) study on eels; noting how visual exposure to eel suffering can a begin a process of addressing the violence towards the species. This photovoice study has shown how fishers can demonstrate a deep affection and intimate knowledge towards eels. For example, by understanding patterns in migration and behavioural responses. These close interspecies connections potentially motivate an attitude of care, albeit one that is tied to exploitation. Perhaps, just as exposure to images of eel suffering can motivate empathy and compassion amongst wider groups (Pons-Hernandez, 2024), the fishers' personal exposure to the eel as both a living being and a commodity can instigate both an ethic of compassion as well as desensitisation to harm. These contradictions between instrumentally and intrinsically motivated care reflect a conceptual impasse in the recognition of harm and victimhood, and make conceptualising harms to individuals, populations, and wider ecosystems considerably challenging.

Final Remarks

By following the eel across their migratory journey and punctuating this with visual examples of the threats they face in their environment, it is possible to recognise how narratives surrounding threats are used to legitimise action and shape management responses; with often-harmful repercussions towards the species, wider ecosystems, and humans. The predominant focus levied towards the fishing industry demonstrates the need to expand definitions of harm within and beyond these groups, where the needs of (and care towards) the eel frequently remain secondary considerations balanced against human demands—for power generation, waterway management, agriculture, industry development, and food provisioning. Species and ecological justice can only be seriously championed when harms to the individual, and their entangled relationships with others, are meaningfully engaged with and prioritised. Only then, with a holistic and ecocentric awareness of harm and victimhood, can policy and conservation decision-making be challenged to respond more dynamically to institutionalised, socially normalised, and legally permitted harms.

References

Agnew, R. (2020). The ordinary acts that contribute to ecocide: A criminological analysis. In *Routledge International Handbook of Green Criminology* (pp. 52–67). Routledge.

Amber International. (2020). D4.2 report of case studies demonstrating the effects of barrier removal. *Mitigation and Installation*. p. 330.

APHA. (2023). Request for information: Glass eels exports to Russian Federation. Retrieved July 19, 2024, from https://assets.publishing.service.gov.uk/media/652d3f73697260000dccf88f/FOI2023_06947.pdf

Aprahamian, M., & Wood, P. (2021). Estimation of glass eel (*Anguilla Anguilla*) exploitation in the Severn Estuary, England. *Fisheries Management and Ecology, 28*(1), 65–75.

Beirne, P. (2018). *Murdering animals*. Palgrave Macmillan UK.

Berger, J. (2013). *Understanding a photograph*. Penguin UK.

Bexell, S. M., Nikitina, K. V., & Manna, L. E. (2023). 'This image haunts me to this day': Exploring the psychological impacts of wildlife loss on conservation professionals through Photovoice. *Ecopsychology, 15*(3), 294–304.

Birkenholtz, T., & Simon, G. (2022). Introduction to themed issue: Ignorance and uncertainty in environmental decision-making. *Geoforum, 132*, 154–161.

Breny, J. M., & McMorrow, S. L. (2021). *Photovoice for social justice: Visual representation in action*. SAGE Publishing.

Coleman, T. (2016). Photo elicitation as method: A participatory approach. In *Practicing qualitative methods in health geographies* (pp. 34–53). Routledge.
Du, H., Zhang, Y., Jiang, H., & Wang, H. (2022). Adsorption of rhodamine B on polyvinyl chloride, polystyrene, and polyethylene terephthalate microplastics in aqueous environments. *Environmental Technology & Innovation*, 27.
European Anti-Fraud Office. (2023). *OLAF in operation against illegal trafficking of eels*. OLAF European Anti-Fraud Office.
EUROPOL. (2022). *49 individuals across Europe arrested in major blow to eels trafficking*. Europol Press and Media.
Fleming, L. V., Walker, A., Evans, D., Aprahamian, M., James, M., Bašić, T., Watts, S., Horsburgh, G., Allin, R., Blake, K., Connor, S., McAlpine, J., & Littlewood, A. H. L. (2023). *Non-detriment finding assessment for the export from the United Kingdom of CITES-listed European eel Anguilla anguilla* (2023–26). JNCC Report 745, JNCC, Peterborough, ISSN 0963-8091.
Fry, T. L., & Dunbar, M. R. (2007). *A review of biomarkers used for wildlife damage and disease management*. USDA Wildlife Services Staff Publications.
Government of Bermuda. (2021). *Regulating single use plastics in Bermuda*. p. 3.
Government of Bermuda. (2022). *Draft marine spatial plan*. p. 126.
Gutiérrez, L., & Duffy, R. (2023). Harms and the illegal wildlife trade: Political ecology, green criminology and the European eel. *Critical Criminology*, 1–16.
Ibáñez Alonso, A. I., & van Uhm, D. P. (2023). The illegal trade in European eels: outsourcing, funding, and complex symbiotic-antithetical relationships. In *Trends in organized crime, 26* (pp. 293–307). https://doi.org/10.1007/s12117-023-09490-5
ICES. (2022). ICES Advice on fishing opportunities and conservation Ecoregions in the Northeast Atlantic. Retrieved July 19, 2024, from https://www.fishsec.org/app/uploads/2023/07/221103-ICES-Advice-on-fishing-opportunities-for-eel-2023.pdf
IOC-UNESCO. (2022). *State of the Ocean Report 2022 Pilot Edition*. IOC. Technical series, 173. https://unesdoc.unesco.org/ark:/48223/pf0000381921
Iordăchescu, G., Lappe-Osthege, T., Dickinson, H., Duffy, R., & Burns, C. (2022). Political ecologies of green-collar crime: Understanding illegal trades in European wildlife. *Environmental Politics, 32*(5), 923–930. https://doi.org/10.1080/09644016.2022.2156173
Pike, C., Crook, V., & Gollock, M. (2018). *Anguilla anguilla, The IUCN Red List of Threatened Species 2020:* e.T60344A152845178. https://doi.org/10.2305/IUCN.UK.2020-2.RLTS.T60344A152845178.en

Long, M., Stretesky, P., & Lynch, M. (2016). The treadmill of production, planetary boundaries and green criminology. In *Environmental crime and its victims* (pp. 263–275). Routledge.

Margulies, J. D. (2019). On coming into animal presence with photovoice. *Environment and Planning E: Nature and Space, 2*(4), 850–873.

Margulies, J. D. (2022). A political ecology of desire. *Environmental Humanities, 14*(2), 241–264.

McClanahan, B. (2021). *Visual criminology*. Policy Press.

Natali, L. (2023). The social perception of environmental victimization. A visual and sensory methodological proposal. *Crime, law and social change, 81* (pp. 561–580). https://doi.org/10.1007/s10611-023-10099-6

Natali, L., & McClanahan, B. (2020). The visual dimensions of green criminology. In A. Brisman & N. South (Eds.), *Routledge international handbook of green criminology* (Routledge international handbooks) (2nd ed., pp. 95–109). Routledge.

Nixon, R. (2011). *Slow Violence and the Environmentalism of the Poor*. Harvard University Press.

Pons-Hernandez, M. (2024). 'Missing the trees for the Forest?' An analysis of the harms to European eels caused by their trafficking and trade. *Critical Criminology, 32*(1), 77–95. https://doi.org/10.1007/s10612-024-09765-5

Sage, A. (2007). Eel poachers put 'white gold' on critical list. *The Times*. Available at: https://www.thetimes.com/uk/crime/article/eel-poachers-put-white-gold-on-critical-list-763f603rhz6#:~:text=The%20dark%20world%20of%20eel,delicacy%20known%20as%20white%20gold.

Skjolding, L. M., Jørgensen, L. G., Dyhr, K. S., Köppl, C. J., McKnight, U. S., Bauer-Gottwein, P., Mayer, P., Bjerg, P. L., & Baun, A. (2021). Assessing the aquatic toxicity and environmental safety of tracer compounds rhodamine B and rhodamine WT. *Water Research*, 197. https://doi.org/10.1016/j.watres.2021.117109

Somin, I. (2022). Rational Ignorance. In M. Gross, and L. McGoey, (Eds.), *Routledge international handbook of ignorance studies* (Vol. 2, pp. 313–322). London: Routledge.

Spry, W. (2022). *Regional renewable electricity in 2021*. Department for Energy Security & Net Zero, p. 13.

Stein, F. M., & Bunt, D. (2021). *Brexit and the European Eel*. Institute of Fisheries Management.

Sudarshan, S., Bharti, V. S., Harikrishnan, S., Shukla, S. P., & Bhuvaneswari Govindarajan, R. (2022). Eco-toxicological effect of a commercial dye rhodamine B on freshwater microalgae chlorella vulgaris. *Archives of Microbiology, 204*(10), 658.

Sutton-Brown, C. A. (2014). Photovoice: A methodological guide. *Photography and Culture, 7*(2), 169–185.

Thiel, D., & South, N. (2022). Criminal ignorance, environmental harms and processes of denial. In *Routledge international handbook of ignorance studies* (pp. 334–344). Routledge.

Whipple, T. (2023). Fears British eel conservation in Russia could help fund Ukraine war. *The Times.*

White, R. (2013). Green criminology and the pursuit of social and ecological justice. In *Issues in green criminology* (pp. 32–54). Willan.

White, R. (2018). *Climate change criminology* (1st ed.). Bristol University Press.

Wolf, B. (2011). 'Green-collar crime': Environmental crime and justice in the sociological perspective. *Sociology Compass, 5*(7), 499–511. https://doi.org/10.1111/j.1751-9020.2011.00381.x

Wyatt, T. (2021). *Wildlife trafficking: A deconstruction of the crime, victims and offenders*. Springer Nature.

A Source of Disease, Health, or Both? Wildlife Consumption in China after the Emergence of COVID-19

Jessica Bell Rizzolo, Annah Zhu, and Ruishan Chen

Introduction

Globally, the illegal wildlife trade, which includes the poaching, trafficking, and consumption of live animals and animal parts for food, medicine, decoration and entertainment, is one of the most prevalent global criminal enterprises (Rosen & Smith, 2010). While scholars and policymakers distinguish between the illegal and legal wildlife trades both in terms of international treaties such as CITES (The Convention on International Trade in Endangered Species) and in the research literature (Reeve, 2006; Warchol, 2004; Biggs et al., 2017), the line between these

J. B. Rizzolo (✉)
Independent Researcher, Chicagoo, IL, USA

A. Zhu
Wageningen University, Wageningen, Netherlands
e-mail: annah.zhu@wur.nl

R. Chen
Shanghai Jiao Tong University, Shanghai, China

D. van Uhm, D. Siegel (eds.), *Global Green Crime and Ecojustice*, Palgrave Studies in Green Criminology, https://doi.org/10.1007/978-3-031-91557-4_15

two forms of trade is often ambiguous, complicated by laundering, corruption, and inadequate enforcement (Bennett, 2015; Lyons & Natusch, 2011). In addition, the legal trade and consumption of wildlife is not necessarily sustainable; it can threaten species with extinction both within and outside the geographic area(s) in which it is legalized (Zhang et al., 2008; Livingstone & Shepherd, 2016). In an attempt to stem poaching from the wild, some countries (China included) have supported the legal trade of products from farmed wildlife. However, this practice can increase the demand for wildlife products through increasing the acceptability of wildlife consumption (Rizzolo, 2021).

Since the emergence of COVID-19, which is widely viewed to have zoonotic origins linked to wildlife consumption (Shivaprakash et al., 2021; Kumar et al., 2020), there has been an increased interest in wildlife consumption bans as a potential tool not only to reduce illegal trade, but also to mitigate the potential public health detriments of wildlife consumption. The illegal trade of wildlife within China occurs both within and across its borders (Liang et al., 2023; Chow et al., 2014). China plays a pivotal role in the illegal wildlife trade due to a plethora of factors, including technological development, demographic and economic changes, and demand trends for wildlife products (Esmail et al., 2020). In an effort to stem poaching (and to generate income), China has promoted the intensive farming of numerous wildlife taxa, including badgers, bears, bamboo rats, civets, crocodiles, deer, ferrets, foxes, frogs, macaques, mink, muskrats, pheasants, pangolins, raccoon dogs, snakes, tigers, and turtles (Whitfort, 2021). China also imports a large number of farmed wildlife. Over half of wildlife imported into China from 1997 to 2016 was captive-bred, and this percentage was even higher for birds and fish (Jiao & Lee, 2022). However, farmed wildlife has not saturated the demand of Chinese customers for (often illegally caught) wild-sourced products. Consumers are often willing to pay more for wild-sourced products (Dutton et al., 2011; Coals et al., 2020) and may not be willing to switch from wild to farmed products (Hinsley et al., 2022); this can create dual markets: one for legal farmed products and one for illegal wild-caught products.

While some scholars contend that broad wildlife consumption bans are not feasible given cultural and livelihood attachments to wildlife products (Zhu & Zhu, 2020), others argue that wildlife consumption bans are an integral first step towards meaningful, effective protection of both wildlife and human health (Xiao et al., 2021a). In China, the government responded to the spread of COVID-19 with a wildlife consumption ban issued in February 2020 (Koh et al., 2021; Sardana & Fischer, 2021). This ban prohibited the farming, hunting, selling, purchasing, and trafficking of terrestrial wild animals consumed for food (Huang et al., 2021). However, this ban did not cover aquatic species and also made exceptions for the use of wildlife for medicinal, ornamental, or scientific purposes (White, 2021; Rizzolo, 2021). Furthermore, for species utilized for various purposes such as civets and racoon dogs, consumed both as food and fur, only production and consumption for food was included in the ban (Whitfort, 2021).

In this chapter, we discuss our research in China on attitudes towards wildlife consumption, wildlife consumption bans, and zoonotic disease in the wake of COVID-19 (see Zhu et al., 2023a, 2023b; Rizzolo et al., 2023a, 2023b) We discuss how support for wildlife consumption bans intersects with perceptions of zoonotic risk, the source of wildlife products (farmed or wild-caught), and the use of wildlife supplements for Traditional Chinese Medicine (TCM). With the use of China as a case study, we outline how (in the wake of COVID-19), researchers and practitioners can and should analyze attitudes towards zoonotic risk and health when instituting bans or other regulatory mechanisms on wildlife consumption.

Sampling and Respondents

The research discussed in this chapter is primarily based on an online survey of adult Chinese respondents we conducted in June 2021. The sample (N = 974) of adult respondents (aged 20 to 79 years old) from Mainland China was recruited through the research firm Qualtrics, which has expertise in China-based sampling (e.g., Rizzolo, 2021; Ammann et al., 2020) and offers a more representative sampling process than other online survey tools (Boas et al., 2020). The survey was originally written in English and then translated into Chinese by a bilingual member of the research team. Before commencement of the full survey, two pretests in Chinese were conducted with a convenience sample in order to assess translation clarity, data quality, and survey validity. A soft launch procedure was also used; this involves testing the survey with a small number of respondents ($n = 50$, recruited from the Qualtrics panel) and then pausing the data collection to analyze response quality before proceeding.

The final survey consisted of several sections: (a) demographic information; (b) beliefs and behaviours related to the origins of COVID-19 and wildlife consumption; (c) beliefs and behaviours that pertain to health and TCM; and (d) attitudes about the regulation of wildlife consumption and the prevention of potential zoonotic transfer. We included a wide range of taxa and focused on species that, in China, are commonly consumed (Zhang & Yin, 2014), farmed (Rizzolo, 2021; Wang et al., 2019), and/or associated with past zoonotic outbreaks (Wang & Eaton, 2007). Qualtrics sampling for this study drew from a China-based panel of individuals recruited from diverse sources, which include website intercept recruitment, member referrals, targeted email lists, gaming sites, customer loyalty web portals, permission-based networks, and social media. Panel members' names, addresses, and dates of birth are validated via third-party verification measures prior to their inclusion in the panel. The final survey used multiple quality control measures. This involved tracking the respondent's time to completion and excluding respondents who took less than one-third the median time to completion (as calculated during the soft launch). This allows variability in time to survey completion but excludes respondents who speed through the survey too quickly to thoughtfully answer the questions. In addition to this standard Qualtrics protocol, additional quality control measures included a total of ten screens, excluding respondents who answered in an internally contradictory manner. Our analysis involved descriptive statistics and binary logistic models. Prior to analysis, we checked for collinearity between the independent variables, and correlations indicated that collinearity was not present. We controlled for gender, geographic location (urban/rural), income, education, and age by including those variables in our regression models.

The sample included gender and age quotas representative of China's adult population and was diverse in terms of gender, age, income, education, and occupation (Fig. 1). The sample was predominantly urban (90.7% of respondents lived in an urban area). In comparison, in 2019, 60.6% of Chinese people were urban residents whereas 39.4% were rural residents. In 2019, the average wage in China of employed persons in urban private units was RMB ¥ 53604 annually (or RMB ¥ 4467 per month); for employed persons in urban non-private units, the average wage was RMB ¥ 90501 annually (or RMB ¥ 7542 per month). Although our survey measured income in a categorical manner, the most frequent category was RMB ¥ 5000–10000 per month. Our sample had more college-educated people than China's national average of 14.6% of adults but included respondents from all regions of China, although the number of respondents per region varied (Fig. 2). We had more respondents

Gender (in %)	
Male	52.3% (n=509)
Female	47.7% (n=465)

Monthly Person Income (in %)	
Under RMB ¥ 1,000	02.0% (n=19)
RMB ¥ 1,000-5,000	23.6% (n=230)
RMB ¥ 5,000-10,000	37.3% (n=363)
RMB ¥ 10,000-20,000	21.6% (n=210)
Over RMB ¥ 20,000	15.6% (n=152)

Age (in %)	
20-29	17.6% (n=171)
30-39	16.5% (n=161)
40-49	23.4% (n=228)
50-59	20.9% (n=204)
60-69	14.9% (n=145)
70-79	06.7% (n=65)

Highest Level of Education (in %)	
Middle school or below	01.8% (n=18)
High school	15.5% (n=151)
College	74.2% (n=723)
Graduate degree	08.4% (n=82)

*Occupation Sector (in %) **	
Agriculture	10.2% (n=99)
Construction	13.3% (n=130)
Education	12.6% (n=123)
Food service	09.2% (n=90)
Government	05.7% (n=56)
Healthcare	04.1% (n=40)
Technology	19.5% (n=190)
Other	25.3% (n-246)

Fig. 1 Characteristics of respondents (n = 974) as self-reported answers to single-choice demographic questions (from Rizzolo et al., 2023b). *The listed sectors are the primary eight occupational categories in China (https://www.diyifanwen.com/zhichangzhinan/renshizhengce/599954.html)

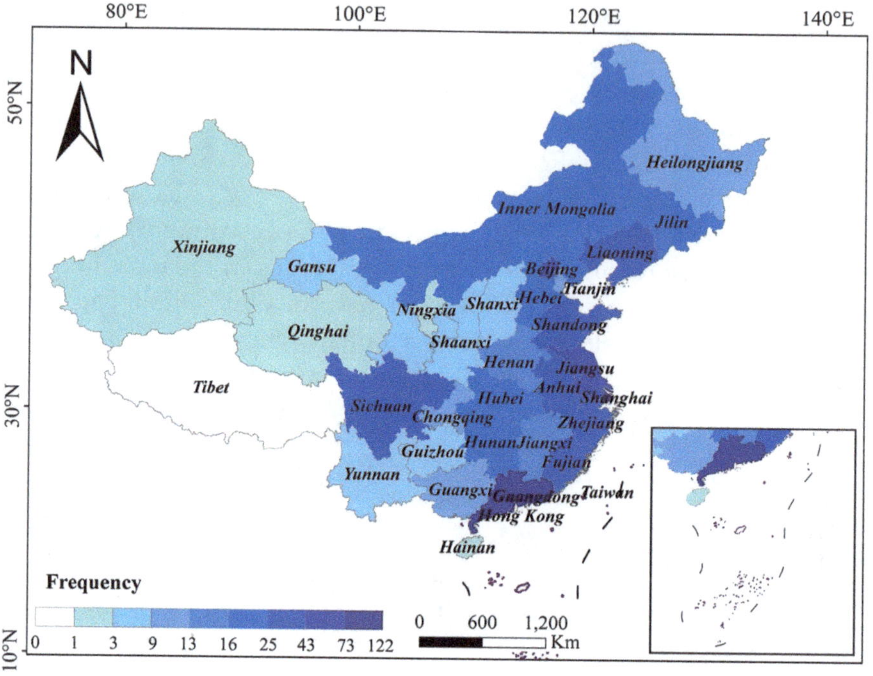

Fig. 2 Geographic distribution of respondents ($n = 974$) (from Rizzolo et al., 2023b)

in the Eastern part of the country, particularly in the southeast coastal areas; this is similar to the distribution of China's population as a whole.

Wildlife Consumption in China: A Source of Disease or Health?

Zoonotic diseases refer to infectious diseases that are transmitted from nonhuman animals to humans. In addition to COVID-19, examples include Ebola Virus Disease (Buceta & Johnson, 2017), Severe Acute Respiratory Syndrome or SARS (Shi & Hu, 2008), Human Immunodeficiency Virus or HIV (Hemelaar, 2012), and Influenza A (Bengis et al., 2004). It is estimated that 60% of human pathogen species are zoonotic and that, of these, 75% originated in wild rather than domesticated species (Ferreira et al., 2021). The increase in zoonotic transfer from animals to humans has been linked to wildlife consumption (Murray et al., 2016; Shi & Hu, 2008) and the wildlife trade (Bell et al., 2004) in addition to larger trends of defaunation, deforestation, and environmental exploitation (van Uhm & Zaitch, 2021; Gatti, 2020). Potential sites of zoonotic transfer include wildlife markets (Mizumoto et al., 2020; Naguib et al., 2021) as well as wildlife farms (Liu, 2003).

In addition, the source of zoonotic disease is often contested and influenced by sociopolitical factors (Zhu et al., 2023a, 2023b; Gatti, 2020). Within China, the notion that COVID-19 emerged from a Chinese wildlife market is not universally accepted. Our work (Zhu et al., 2023a, 2023b) found that more than half of Chinese respondents (53%) believed that COVID-19 originated in the United States, while only 17% believed it originated in China. Further, the perceived *country* of zoonotic origin influenced the perceived *source* of origin as well. For respondents who indicated China as the most likely origin of COVID-19, the majority (75%) selected wet markets as an origin source. In contrast, of those respondents who noted that the United States was the most likely origin of COVID-19, the majority (51%) selected laboratories as an origin source (Zhu et al., 2023b).

Although wildlife consumption is a potential source of zoonotic infection, China also has a complex history of wildlife consumption for perceived health benefits through Traditional Chinese Medicine, or TCM (Xu & Yang, 2009; Chee, 2021). TCM is an integral part of China's health care system as well as everyday cultural practices (Xu & Yang, 2009; Zhu & Zhu, 2020; Cheung et al., 2020). There are TCM departments in approximately 90% of the hospitals in China (Xu & Yang, 2009). Over 10,000 plant and animal species are utilized in TCM for a plethora of disorders, with gastrointestinal and neurological syndromes among the most commonly addressed (Williamson et al., 2013). In TCM, wildlife products are often combined with other ingredients to address particular bodily systems and imbalances. For example, although there is an absence of scientific evidence for these effectiveness of these treatments (Fears et al., 2020), rhino horn is used for heat imbalances in kidneys and the liver, bear bile to target the gall bladder, liver and heart, and tiger bone for bone health and rheumatic pain (Cheung et al., 2020). While wildlife consumption is one component of TCM, other practices that do not use wildlife, such as plant-based supplements, acupuncture, massage, and cupping, are also widespread in China (Chen, 2019). In our work (Zhu et al., 2023a, 2023b; Rizzolo et al., 2023a, 2023b), we found that 11% of our respondents had consumed wildlife as part of TCM, although other forms of TCM use were more common (Fig. 3).

Further, 42% of our respondents believed that it was beneficial to their health to consume meat from animals that have been slaughtered recently. A smaller, but still significant percentage of respondents (16%) thought it was beneficial to their health to consume wild animals. These beliefs are connected; respondents who endorsed the benefits of fresh slaughter were more likely to find the consumption of wild animals beneficial. In addition, people in their forties, fifties and seventies were significantly more likely than respondents in their twenties to find the consumption of wild animals beneficial, as were rural respondents (compared to urban

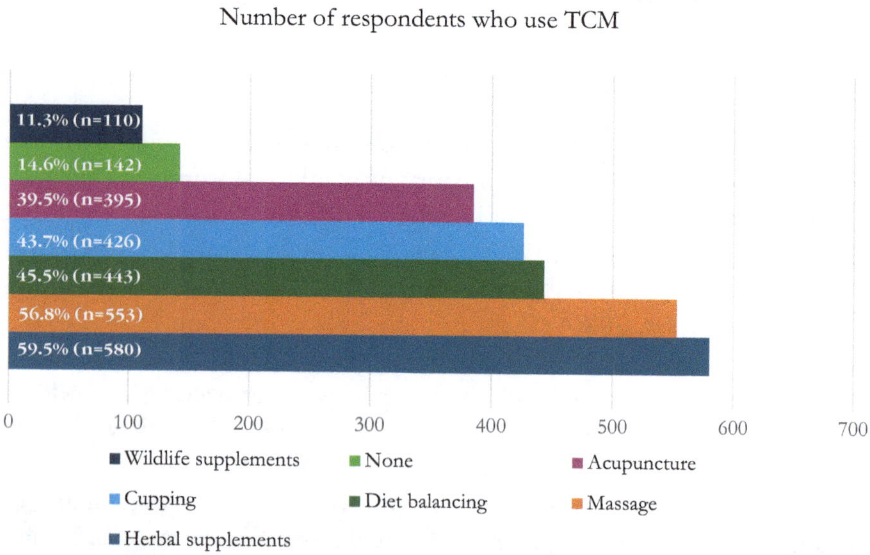

Fig. 3 Responses in the sample to a multiple choice question about types of TCM use (e.g., "Which of the following Traditional Chinese Medicine products and techniques do you use? Select all that apply") (from Rizzolo et al., 2023b)

respondents). TCM consumer preference for wild animals (over farmed or domesticated animals) is linked to beliefs about the increased effectiveness of wild animal products (Liu et al., 2016), as well as the traditional use of wildlife products to achieve bodily balance (Cheung et al., 2020).

It is important to note that, even if people believe that wildlife can be a source of zoonotic disease, they may still consume wildlife (Li et al., 2021). This may be because of a dearth of knowledge about how diseases are transmitted (e.g., through which species and/or transmission methods) and/or due to varied risk perceptions based on where the wildlife is sourced from or used for (Li et al., 2021). For example, one survey (Li et al., 2021) found that approximately 80% of their Chinese respondents believed that the wildlife trade is likely to transmit disease from wildlife to humans. However, in this same survey, almost half of respondents viewed the consumption of "only farmed animals" as a protective measure against zoonotic disease (Li et al., 2021).

Ban Support, TCM Use, and Zoonotic Risk Perceptions

Wild-caught and Farmed Wildlife Consumption

Overall, there was a high level of support for wildlife consumption bans. The majority of respondents (70.7%) supported a ban on the trade of all wild-caught wildlife species. A lesser majority (57.4%) endorsed a ban on the trade of all farmed wildlife species. In almost all species (Table 1), support for bans of wild-caught wildlife was higher than support for bans of farmed wildlife.

In addition, we found that the demographic correlates of ban support varied for wild-caught versus farmed wildlife (Tables 2 and 3). For example, we found that respondents who were included in a higher-income bracket were more likely to support a ban on farmed wildlife (for various species). Although there are variations by species and products, overall, farmed wildlife products are often perceived as less valuable or effective, whereas wild-caught wildlife products are often considered potent and socially desirable (Coals et al., 2020; Zhang et al., 2008). If higher

Table 1 Support for wildlife bans (from Rizzolo et al., 2023a)

Wildlife product	Percentage of sample who support wild-caught ban (%)	Percentage of sample who support farmed ban (%)
All wild animals	70.7	57.4
Bamboo rat	77.5	72.6
Bat	85.7	87.8
Bear bile	83.4	82.1
Civet	84.1	83.8
Frog	75.2	64.8
Mink	80.5	68.4
Pangolin	85.1	73.8
Snake	77.0	64.9
Tiger bone	83.7	80.0

earners can afford wild-caught wildlife products, which are often pricier than their farmed equivalents, bans on farmed wildlife may not threaten their consumption patterns. However, it is also possible that respondents who are higher earners have objections to wildlife farms that are not shared by lower-income groups. In addition, respondent age had opposite effects for these two types of bans. Our results suggested that, while younger respondents may be more supportive of wild-caught bans, older respondents may be more supportive of wildlife farming bans. This may be because older respondents view wild-caught species as more effective in the treatment of health ailments and thus not object to wildlife farming bans. Overall, the indication of these results is that efforts to amplify support for wildlife consumption bans should be tailored to *different portions of the Chinese population depending upon whether the wildlife is farmed or wild-caught.* Failing to differentiate between wild-caught and farmed risks misidentification both of the support for wildlife consumption bans and the segments of the populace most receptive to or antagonistic to a ban.

In addition, we found that differential support for wild-caught and farmed wildlife bans are linked to the perceived zoonotic risk of wildlife consumption. For farmed wildlife, the perceived zoonotic risk of the species (e.g., the perception that consuming that species could lead to disease) only increased ban support for two species: civets and frogs. For civets, this may be because of their role in the zoonotic outbreak of SARS as well as the detection of COVID-19 in civets in wet markets in Wuhan

Table 2 Support for wild-caught wildlife bans of various species (from Rizzolo et al., 2023a)

	Civet	Frog	Tiger	Bear	Bamboo Rat	Mink	Snake	Bat	Pangolin
Zoonotic Risk	Exp(B)	Exp(B)	Exp(B)	Exp(B)	Exp(B)	Exp(B)	Exp(B)	Exp(B)	Exp(B)
Believe species likely to transmit COVID-19	**7.098***	**2.718***	**12.657***	**8.366***	**6.391***	**14.284***	**2.662***	1.595	1.073
Consumption Patterns									
Consume reptiles	0.781	**0.470***	0.506	0.574	0.728	0.721	0.570	0.720	0.705
Consume amphibians	0.950	0.706	0.908	0.848	0.623	0.526	0.508	0.880	0.663
Consume mammals	1.346	0.773	**4.321***	**3.894***	0.462	0.909	0.684	3.389	2.703
Ban attitude									
Current ban adequate	**2.189***	1.612	**2.419***	**2.176***	1.586	**2.086***	**1.932***	**2.622***	**2.405***
Gender									
Male	0.810	1.016	0.654	0.637	0.870	0.943	1.034	0.743	0.814
Education									
High school	0.587	4.802	1.555	1.733	1.733	0.401	3.286	0.829	0.935
College	0.507	2.888	0.881	1.002	1.312	0.536	1.677	0.511	0.591
Graduate degree	0.304	2.535	0.440	0.457	0.843	0.359	1.168	0.268	0.320
Age									
20 to 29	0.258	**0.136***	0.162	0.154	0.162	1.039	0.145	0.222	0.207
30 to 39	0.603	**0.154***	0.209	0.197	0.220	1.437	0.166	0.432	0.319
40 to 49	0.659	0.274	0.513	0.441	0.200	1.116	0.222	1.368	0.739
50 to 59	0.552	0.331	0.299	0.257	0.271	1.199	0.233	0.245	0.35
60 to 69	**0.057***	**0.034***	**0.035***	**0.026***	**0.023***	0.139	**0.029***	**0.041***	**0.037***
Monthly income									
RMB ¥ 1,000–5,000	2.2	2.183	1.286	1.288	4.491	2.289	2.299	1.744	2.055
RMB ¥ 5,000–10,000	2.227	1.866	1.535	1.515	3.709	1.621	2.325	1.658	1.798
RMB ¥ 10,000–20,000	6.331	2.976	4.066	4.082	7.030	3.300	4.567	6.165	4.938
Over RMB ¥ 20,000	3.220	0.955	1.975	2.316	1.844	0.913	1.183	2.270	2.679
Constant	5.420	3.501	14.527	19.103	3.673	3.521	5.197	11.759	15.823

***p* < 0.001**

For categorical variables, the reference categories are: female (for gender), middle school or below (for education), 70–79 (for age), and < CNY 1,000 (for income)

Table 3 Support for farmed wildlife bans of various species (from Rizzolo et al., 2023a)

	Civet	Frog	Tiger	Bear	Bamboo Rat	Mink	Snake	Bat	Pangolin
Zoonotic risk	Exp(B)	Exp(B)	Exp(B)	Exp(B)	Exp(B)	Exp(B)	Exp(B)	Exp(B)	Exp(B)
Believe species likely to transmit COVID-19	**1.928***	**2.477***	2.966	1.926	1.329	0.864	1.165	1.370	0.918
Consumption patterns									
Consume reptiles	0.848	**0.430***	**0.436***	**0.444***	0.54	**0.34***	**0.453***	0.636	**0.471***
Consume amphibians	0.583	**0.481***	0.833	0.818	0.664	**0.469***	**0.432***	0.825	0.62
Consume mammals	1.165	**0.377***	0.701	1.056	0.706	0.576	0.521	1.268	0.641
Ban Attitude									
Current ban adequate	0.958	0.897	0.792	0.988	1.229	0.975	1.131	0.96	0.822
Gender									
Male	0.729	0.655	0.713	0.708	0.856	0.702	0.633	0.917	0.662
Education									
High school	0.693	0.596	1.010	2.276	1.835	0.396	0.622	0	0.558
College	0.625	0.309	0.691	1.824	1.109	0.257	0.304	0	0.32
Graduate degree	0.840	0.369	0.768	1.797	1.425	0.422	0.394	0	0.221
Age									
20 to 29	**0.162***	1.947	0.308	**0.044***	0.503	1.668	1.902	**0.074***	1.888
30 to 39	**0.159***	1.137	0.270	**0.040***	0.609	1.418	1.458	0.103	1.083
40 to 49	0.272	1.488	0.485	**0.083***	0.548	1.607	1.793	0.313	1.370
50 to 59	1.093	2.816	0.593	0.124	1.096	2.373	3.083	0.467	2.570
60 to 69	1.306	**4.622***	1.286	0.603	2.202	4.073	**4.907***	3.622	**4.371***
Monthly income									
RMB ¥ 1,000–5,000	3.193	3.085	2.332	2.350	2.823	2.330	1.975	2.551	1.524
RMB ¥ 5,000–10,000	3.883	**5.555***	4.434	3.288	2.732	4.736	3.398	2.171	3.586
RMB ¥ 10,000–20,000	5.961	**8.512***	**7.165***	4.983	3.467	**7.765***	5.455	4.088	**6.663***
Over RMB ¥ 20,000	5.589	**6.692***	**12.58***	6.796	2.826	5.361	3.39	3.169	**13.263***
Constant	5.823	1.141	4.715	12.18	1.247	2.131	1.496	1.419E+9	2.848

***$p < 0.001$**

For categorical variables, the reference categories are: female (for gender), middle school or below (for education), 70–79 (for age), and < CNY 1,000 (for income) at the beginning of COVID-19 (Wang & Eaton, 2007; Hassanin et al., 2021). While the finding on frogs requires more research, this may be due to debate over frog farming in the wake of COVID-19 (Lee, 2020), which has brought attention to the zoonotic risks and livelihood issues related to this practice.

In contrast, for wild-caught wildlife, zoonotic risk was significant (and by larger factors) for almost all the species examined. When the broad term "wildlife" is used in discussions of COVID-19-related risks and regulations, people may conceptualize this term as referring only to wild-caught wildlife, even though most wildlife sources (such as wet markets) include both wild-caught and farmed species (Xiao et al., 2021b). In turn, this generalized frame might yield differential support for farmed wildlife versus wild-caught wildlife bans.

Finally, there are important species differences in zoonotic risk perceptions (Fig. 4) that intersect with attitudes towards wildlife consumption bans.

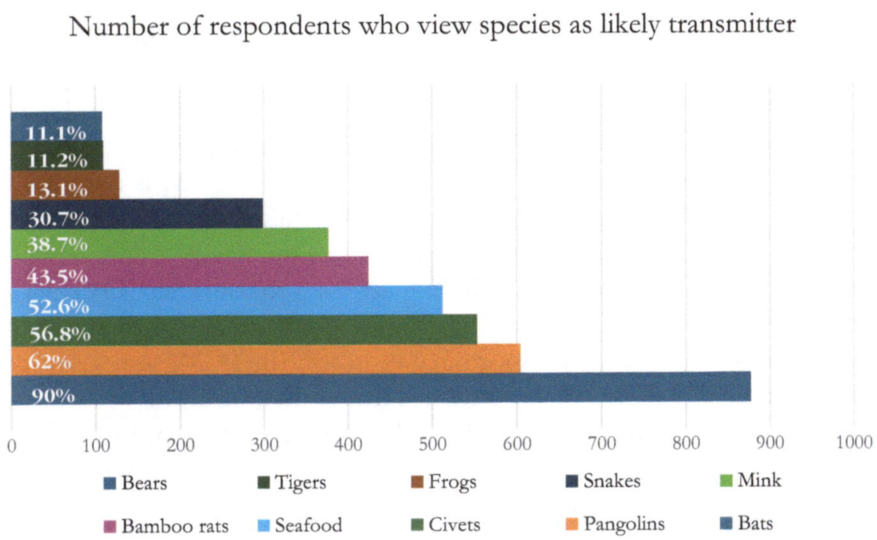

Fig. 4 Responses in the sample (n = 974) to a multiple choice question about which species are viewed as likely transmitters of COVID-19 (e.g., "Which of the following species do you think are likely transmitters of COVID-19? Select all that apply") (from Rizzolo et al., 2023b)

Both in our work (Rizzolo et al., 2023a, 2023b) and in the larger literature (Giri et al., 2021; Gatti, 2020), pangolins and bats are likely to be perceived as potential sources of COVID-19. Even though bats and pangolins were considered likely zoonotic transmitters overall, in our research their status as a zoonotic source was not a factor in predicting ban support. However, what did impact support for bans of wild-caught pangolins and bats was one's attitude towards China's *current ban.* Namely, if respondents thought that China's current wildlife consumption ban was adequate, they were more likely to support bans on wild-caught pangolins and bats. Since pangolins and bats are species that have received a substantial attention as zoonotic vectors in both the policy and scientific realms, it follows that support for bans of these species would be connected with attitudes towards the effectiveness of the government's regulatory response.

Attitudes Towards Health and Disease

Attitudes towards health (and TCM in particular) and zoonotic risk perceptions are pivotal factors in support for bans on wildlife consumption in China. Most of our respondents (66.9%) considered TCM to be an effective form of medicine, and only 14.6% of the respondents did not practice any aspect of TCM. This strong support for TCM is consistent with the prevalence and integration of TCM into the health care system within China (Xu & Yang, 2009). Further, in our sample, people who used wildlife in TCM were four times *more likely* to consider COVID-19 very likely zoonotic. This indicates that a person may simultaneously view wildlife consumption as a source of health (in TCM) and a source of disease (of COVID-19). In fact, one form of use (for health) may actually *increase* the perceived zoonotic risks of that wildlife product. This may be because use of wildlife for TCM increases knowledge of the wildlife product and/or yield more interactions with potential zoonotic reservoirs.

In fact, 80% of our sampled Chinese respondents said that protection against future pandemics and protection of the environment was a motivator for ban support (Rizzolo et al., 2023a, 2023b). When asked to select effective measures for the prevention of infectious disease (overall, not just in terms of COVID-19), most respondents endorsed regulations on wild animal farms (76%), wet markets (75.4%), and the wildlife trade (69.8%) as effective prevention tools. The most frequently selected preventative measure was the monitoring and restricting of imported products (78.3%). This may be because of doubts that COVID-19 originated in China and/or because species that are viewed as higher risk for zoonotic disease are imported from countries outside China (Zhu et al., 2023a, 2023b). However, most respondents (82%) viewed COVID-19 as very or somewhat likely to be zoonotic, even if they didn't see it as originating in China.

Wildlife Consumption Bans post-COVID-19

Overall, we suggest that there are several important variables to consider when analyzing how attitudes towards health and disease impact support for wildlife consumption bans.

Perceived Origins

First, it is important to understand where people perceive the zoonotic disease as originating from. This includes both the country and the source (e.g., wildlife farm, laboratory, etc.). These two aspects of origin intersect (Fig. 5). For example, we found that considering the United States as the origin region of COVID-19 increased the probability of selecting imported frozen foods or laboratories/research as the origin source (Zhu et al., 2023a, 2023b). In contrast, viewing the United States as the origin region decreased the probability of thinking that COVID-19 emerged from wild animals in a wet market. Importantly, our results also demonstrated that respondents who selected wild animals in a wet market as a likely source of the pandemic had higher odds of supporting a trade ban on all wild-caught animals and all farmed wildlife. Therefore, attitudes towards origins can influence perceived source, which in turn impacts ban support.

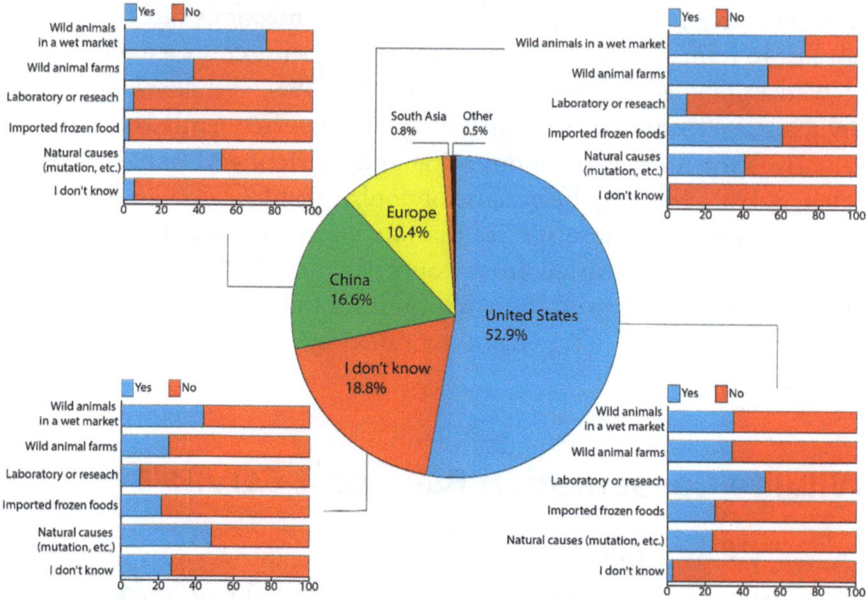

Fig. 5 Perceived source of COVID-19 based on perceived country of origin (from Zhu et al., 2023a)

Attitudes towards Traditional Chinese Medicine (TCM)

Attitudes towards Traditional Chinese Medicine exerts a substantial impact on wildlife consumption. On both a legal and cultural level, TCM-based wildlife consumption in China is often granted particular exemptions not given to other types of wildlife use, such as the consumption of wildlife for food (Rizzolo, 2021). Prior work has found that agreement with the use of wildlife products for TCM (a pro-TCM orientation) can spillover to and increase non-TCM uses of wildlife (Rizzolo, 2021). However, TCM use might also increase knowledge of zoonotic risk, which in turn can increase support for wildlife bans. We found that respondents who used wildlife for TCM were four times more likely than those who didn't to believe that COVID-19 was very likely zoonotic.

Differences Between Farmed and Wild-caught Wildlife

Finally, researchers should distinguish between farmed and wild-caught wildlife in terms of consumption patterns, perceived zoonotic risk, and support for wildlife bans. Our research suggests that this is a primary distinction in the minds of Chinese respondents. A sizable portion of our sample (Table 4) believed that, while the hunting of wild animals should be prohibited, the farming of wildlife is allowable.

Further, we found that a respondent's attitude towards China's current ban (Fig. 6) was only a significant factor for wild-caught wildlife consumption. For numerous species, a person's belief that the current ban was adequate increased the likelihood that a person would support a wild-caught ban. This may be because respondents consider the workability of bans based on their beliefs about the effects of China's current ban. In other words, if they think that China's ban has been effective at safeguarding wildlife and human health, it follows that they would develop an overall pro-ban attitude. In contrast, a person's attitude towards China's current wildlife consumption ban had no effect on support for farmed wildlife bans. It may be that respondents view China's current ban as only applicable to wild-caught animals and/or do not consider China's current ban when appraising the utility of farmed wildlife bans. This has important implications for the wildlife trade because China's ban as originally written applied to *both* wild-caught and farmed wildlife (You, 2020). However, our respondents do not seem to consider this current ban when evaluating hypothetical farmed wildlife bans. Even

Table 4 Attitudes towards regulations of wild versus domesticated animals

How should wild animals be regulated differently than domesticated? (select as many as apply)	Percentage of sample who endorse this regulatory framework (%)
No hunting or farming of wild animals should be allowed	48.8
More regulations should be put in place for wild animal farming than for domesticated animal farming	45.2
No hunting of wild animals should be allowed, but farming wild animals is allowable	32.1

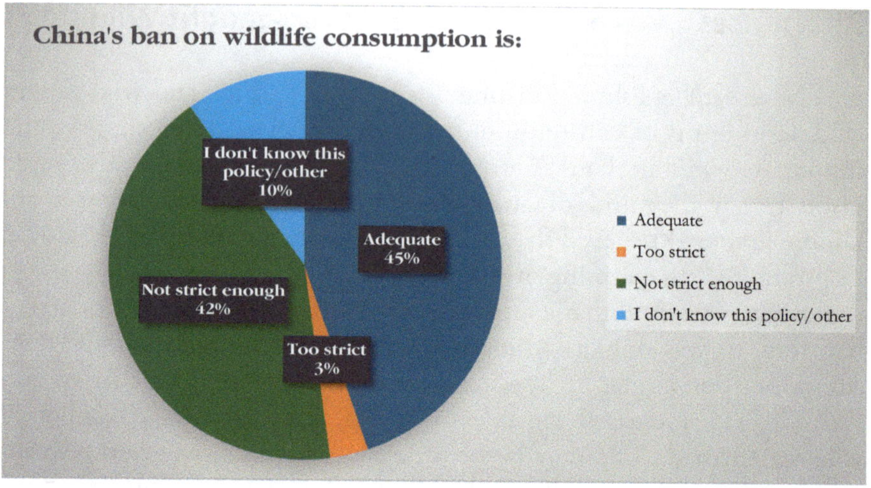

Fig. 6 Attitudes in sample (N = 974) towards China's current ban on wildlife consumption

though China's ban has altered the legal circumstances of wildlife farms (You, 2020), it is necessary to consider that *consumers* are not necessarily aware that farmed wildlife are included in China's ban.

Conclusion

Since the emergence of COVID-19, which is widely considered to have zoonotic origins in wildlife consumption (Shivaprakash et al., 2021; Kumar et al., 2020), there has been an amplification of interest in wildlife consumption bans as a potential method to both address illegal trade and to alleviate the potential public health repercussions of this consumption. Focusing on China in particular, in this chapter we have described how attitudes towards health and disease impact support for wildlife consumption bans. Namely, we have outlined how the perceived origins of zoonotic disease, attitudes towards Traditional Chinese Medicine (TCM), and differences between farmed and wild-caught wildlife impact support for China's current wildlife consumption ban as well as hypothetical additional bans. The inclusion of these factors can help conservationists and policymakers develop a nuanced understanding of, and tailored interventions to amplify, support for wildlife consumption bans, with potential benefits for both wildlife conservation and public health.

References

Ammann, J., Egolf, A., Hartmann, C., & Siegrist, M. (2020). Cross-national comparison of the Food Disgust Picture Scale between Switzerland and China using confirmatory factor analysis. *Food Quality and Preference, 79*, 103756.

Bell, D., Roberton, S., & Hunter, P. R. (2004). Animal origins of SARS coronavirus: Possible links with the international trade in small carnivores. *Philosophical Transactions of the Royal Society of London. Series B: Biological Sciences, 359*(1447), 1107–1114.

Bengis, R. G., Leighton, F. A., Fischer, J. R., Artois, M., Morner, T., & Tate, C. M. (2004). The role of wildlife in emerging and re-emerging zoonoses. *Revue scientifique et technique-office international des epizooties, 23*(2), 497–512.

Bennett, E. L. (2015). Legal ivory trade in a corrupt world and its impact on African elephant populations. *Conservation Biology, 29*(1), 54–60.

Biggs, D., Cooney, R., Roe, D., Dublin, H. T., Allan, J. R., Challender, D. W., & Skinner, D. (2017). Developing a theory of change for a community-based response to illegal wildlife trade. *Conservation Biology, 31*(1), 5–12.

Boas, T. C., Christenson, D. P. & Glick, D. M. (2020). Recruiting large online samples in the United States and India: Facebook, Mechanical Turk, and Qualtrics. *Political Science Research and Methods, 8*(2), 232–250.

Buceta, J., & Johnson, K. (2017). Modeling the Ebola zoonotic dynamics: Interplay between enviroclimatic factors and bat ecology. *PLoS One, 12*(6), e0179559.

Chee, L. P. (2021). *Mao's bestiary: Medicinal animals and modern China*. Duke University Press.

Chen, W. (2019). Indigenous knowledge systems: Practices in modern day China. In D. R. Katerere, W. Applequist, O. M. Aboyade, & C. Togo (Eds.), *Traditional and indigenous knowledge for the modern era: A natural and applied science perspective* (pp. 117–127). CRC Press.

Cheung, H., Doughty, H., Hinsley, A., Hsu, E., Lee, T. M., Milner-Gulland, E. J., Possingham, H., & Biggs, D. (2020). Understanding traditional

Chinese medicine to strengthen conservation outcomes. *People and Nature, 3*(1), 115–128.
Chow, A. T., Cheung, S., & Yip, P. K. (2014). Wildlife markets in south China. *Human–Wildlife Interactions, 8*(1), 108–112.
Coals, P., Moorhouse, T. P., D'Cruze, N. C., Macdonald, D. W., & Loveridge, A. J. (2020). Preferences for lion and tiger bone wines amongst the urban public in China and Vietnam. *Journal for Nature Conservation, 57*, 125874.
Dutton, A. J., Hepburn, C., & Macdonald, D. W. (2011). A stated preference investigation into the Chinese demand for farmed vs. wild bear bile. *PloS One, 6*(7), e21243.
Esmail, N., Wintle, B. C., t Sas-Rolfes, M., Athanas, A., Beale, C. M., Bending, Z., et al. (2020). Emerging illegal wildlife trade issues: A global horizon scan. *Conservation Letters, 13*(4), e12715.
Fears, R., Griffin, G. E., Larhammar, D., ter Meulen, V., van der Meer, J. W. M. (2020). Globalization of Traditional Chinese Medicine: What are the issues for ensuring evidence-based diagnosis and therapy? *Journal of Internal Medicine, 287*(2), 210–213.
Ferreira, M. N., Elliott, W., Kroner, R. G., Kinnaird, M. F., Prist, P. R., Valdujo, P., & Vale, M. M. (2021). Drivers and causes of zoonotic diseases: An overview. *Parks, 27*, 15–24.
Gatti, R. C. (2020). The pangolin's revenge: SARS-CoV-2 did not emerge from a lab but from wildlife exploitation. *GAIA-Ecological Perspectives for Science and Society, 29*(2), 79–82.
Giri, R., Bhardwaj, T., Shegane, M., Gehi, B. R., Kumar, P., Gadhave, K., et al. (2021). Understanding COVID-19 via comparative analysis of dark proteomes of SARS-CoV-2, human SARS and bat SARS-like coronaviruses. *Cellular and Molecular Life Sciences, 78*(4), 1655–1688.
Hassanin, A., Grandcolas, P., & Veron, G. (2021). Covid-19: Natural or anthropic origin? *Mammalia, 85*(1), 1–7.
Hemelaar, J. (2012). The origin and diversity of the HIV-1 pandemic. *Trends in Molecular Medicine, 18*(3), 182–192.
Hinsley, A., Wan, A. K. Y., Garshelis, D., Hoffmann, M., Hu, S., Lee, T. M., Meginnis, K., Moyle, B., Qiu, Y., Ruan, X., & Milner-Gulland, E. J. (2022). Understanding why consumers in China switch between wild, farmed, and synthetic bear bile products. *Conservation Biology, 36*(3), e13895.
Huang, Q., Wang, F., Yang, H., Valitutto, M., & Songer, M. (2021). Will the COVID-19 outbreak be a turning point for China's wildlife protection: New developments and challenges of wildlife conservation in China. *Biological Conservation, 254*, 108937.

Jiao, Y., & Lee, T. M. (2022). The global magnitude and implications of legal and illegal wildlife trade in China. *Oryx, 56*(3), 404–411.

Koh, L. P., Li, Y., & Lee, J. S. H. (2021). The value of China's ban on wildlife trade and consumption. *Nature Sustainability, 4*(1), 2–4.

Kumar, V., Pruthvishree, B., Pande, T., Sinha, D., Singh, B., Dhama, K., & Malik, Y. S. (2020). SARS-CoV-2 (COVID-19): Zoonotic origin and susceptibility of domestic and wild animals. *Journal of Pure and Applied Microbiology, 4*(suppl 1), 741–747.

Lee, A. (2020). Chinese frog breeders call for help as wildlife trade ban shuts down business. *South China Morning Post.*

Li, H., Daszak, F., Chmura, A., Zhang, Y., Terry, P., & Fielder, M. (2021). Knowledge, attitude, and practice regarding zoonotic risk in wildlife trade, Southern China. *EcoHealth, 8*, 95–106.

Liang, D., Giam, X., Hu, S., Ma, L., & Wilcove, D. S. (2023). Assessing the illegal hunting of native wildlife in China. *Nature, 623*(7985), 100–105.

Liu, J. (2003). SARS, wildlife, and human health. *Science, 302*(5642), 53–54.

Liu, Z., Jiang, Z., Fang, H., Li, C., Mi, A., Chen, J., Zhang, X., Cui, S., Chen, D., Ping, X., Li, F., Li, C., Tang, S., Luo, Z., Zeng, Y., & Meng, Z. (2016). Perception, price and preference: Consumption and protection of wild animals used in traditional medicine. *PloS one, 11*(3), e0145901.

Livingstone, E., & Shepherd, C. R. (2016). Bear farms in Lao PDR expand illegally and fail to conserve wild bears. *Oryx, 50*(1), 176–184.

Lyons, J. A., & Natusch, D. J. (2011). Wildlife laundering through breeding farms: Illegal harvest, population declines and a means of regulating the trade of green pythons (*Morelia viridis*) from Indonesia. *Biological Conservation, 144*(12), 3073–3081.

Mizumoto, K., Kagaya, K., & Chowell, G. (2020). Effect of a wet market on coronavirus disease (COVID-19) transmission dynamics in China, 2019–2020. *International Journal of Infectious Diseases, 97*, 96–101.

Murray, K. A., Allen, T., Loh, E., Machalaba, C., & Daszak, P. (2016). Emerging viral zoonoses from wildlife associated with animal-based food systems: risks and opportunities. In *Food Safety Risks from Wildlife* (pp. 31–57). Springer.

Naguib, M. M., Li, R., Ling, J., Grace, D., Nguyen-Viet, H., & Lindahl, J. F. (2021). Live and wet markets: Food access versus the risk of disease emergence. *Trends in Microbiology, 29*(7), 573–581.

Reeve, R. (2006). Wildlife trade, sanctions and compliance: Lessons from the CITES regime. *International Affairs, 82*(5), 881–897.

Rizzolo, J. B. (2021). Effects of legalization and wildlife farming on conservation. *Global Ecology and Conservation, 25*, e01390.

Rizzolo, J. B., Zhu, A., & Chen, R. (2023b). Wildlife consumption, health, and zoonotic disease in China after the emergence of COVID-19. *EcoHealth, 20*(3), 323–342.

Rizzolo, J. B., Zhu, A. L., & Chen, R. (2023a). Support for wildlife consumption bans and policies in China post-Covid-19. *Oryx, 57*(6), 747–756.

Rosen, G. E., & Smith, K. F. (2010). Summarizing the evidence on the international trade in illegal wildlife. *EcoHealth, 7*, 24–32.

Sardana, S., & Fischer, C. (2021). *Implications of a wildlife trade ban*. Asian Development Bank.

Shi, Z., & Hu, Z. (2008). A review of studies on animal reservoirs of the SARS coronavirus. *Virus Research, 133*(1), 74–87.

Shivaprakash, K. N., Sen, S., Paul, S., Kiesecker, J. M., & Bawa, K. S. (2021). Mammals, wildlife trade, and the next global pandemic. *Current Biology, 31*(16), 3671–3677.

van Uhm, D., & Zaitch, D. (2021). Defaunation, wildlife exploitation and zoonotic diseases: A green criminological perspective. In D. Siegel (Ed.), *Notes from isolation: Global Criminological perspectives on coronavirus pandemic* (pp. 11–30). Eleven International Publishing.

Wang, L. F., & Eaton, B. T. (2007). Bats, civets and the emergence of SARS. In J. E. Childs, J. S. Mackenzie, & J. A. Richt (Eds.), *Wildlife and emerging zoonotic diseases: The Biology, circumstances and consequences of cross-species transmission* (pp. 325–344). Springer.

Wang, W., Yang, L., Wronski, T., Chen, S., Hu, Y., & Huang, S. (2019). Captive breeding of wildlife resources—China's revised supply-side approach to conservation. *Wildlife Society Bulletin, 43*(3), 425–435.

Warchol, G. L. (2004). The transnational illegal wildlife trade. *Criminal Justice Studies, 17*(1), 57–73.

White, A. (2021). How China's wildlife trade legislation permits commercial trade in protected wild animal species. *The China Quarterly, 248*, 1–14.

Whitfort, A. (2021). COVID-19 and wildlife farming in China: Legislating to protect wild animal health and welfare in the wake of a global pandemic. *Journal of Environmental Law, 33*(1), 57–84.

Williamson, E. M., Lorenc, A., Booker, A., & Robinson, N. (2013). The rise of traditional Chinese medicine and its materia medica: A comparison of the frequency and safety of materials and species used in Europe and China. *Journal of Ethnopharmacology, 149*(2), 453–462.

Xiao, L., Lu, Z., Li, X., Zhao, X., & Li, B. V. (2021a). Why do we need a wildlife consumption ban in China? *Current Biology, 31*(4), R168–R172.

Xiao, X., Newman, C., Buesching, C. D., Macdonald, D. W., & Zhou, Z. M. (2021b). Animal sales from Wuhan wet markets immediately prior to the COVID-19 pandemic. *Scientific Reports, 11*(1), 1–7.

Xu, J., & Yang, Y. (2009). Traditional Chinese medicine in the Chinese health care system. *Health Policy, 90*(2–3), 133–139.

You, M. (2020). Changes of China's regulatory regime on commercial artificial breeding of terrestrial wildlife in time of COVID-19 outbreak and impacts on the future. *Biological Conservation, 250*, 108756.

Zhang, L., Hua, N., & Sun, S. (2008). Wildlife trade, consumption and conservation awareness in southwest China. *Biodiversity and Conservation, 17*, 1493–1516.

Zhang, L., & Yin, F. (2014). Wildlife consumption and conservation awareness in China: A long way to go. *Biodiversity and Conservation, 23*, 2371–2381.

Zhu, A., & Zhu, G. (2020). Understanding China's wildlife markets: Trade and tradition in an age of pandemic. *World Development, 136*, 105108.

Zhu, A. L., Chen, R., Rizzolo, J., & Li, X. (2023b). The politicization of COVID-19 origin stories: Insights from a cross-sectional survey in China. *Societies, 13*(2), 37.

Zhu, A. L., Chen, R., Rizzolo, J., & Qian, J. (2023a). Perceptions of COVID-19 origins and China's wildlife policy reforms. *Global Ecology and Conservation, 43*, e02463.

Conflict, Organized Crime and Crime Convergence

Environmental Crimes, Illicit Economies and the Emergence of a Mekong Crime Complex

Mark Brown

Introduction

Since Myanmar's fragile experiment in democracy was derailed by a military coup in February 2021 the country has transformed into a global centre of criminality. A global organised crime index based on data collected even before the coup gave Myanmar an organised criminality ranking of 3rd in the world (out of 193 countries), as well as placing it 1st in both Asia and the Southeast Asian region (Global Initiative, 2021). In 2018 it had been ranked 82nd out of 84 economies in the world, labelled as an enabler of crime and noted for its manifest lack of commitment to countering illicit trade (behind only Libya and Iraq; Transnational Alliance to Combat Illicit Trade, 2018). Since then, the ensuing civil war,

M. Brown (✉)
School of Law, The University of Sheffield, Sheffield, UK

School ofSocial and Political Sciences, University of Melbourne, Melbourne, Australia
e-mail: mark.brown@sheffield.ac.uk

D. van Uhm, D. Siegel (eds.), *Global Green Crime and Ecojustice*, Palgrave Studies in Green Criminology, https://doi.org/10.1007/978-3-031-91557-4_16

reductions in state law enforcement capacity and indeed increased state participation as a criminal actor itself, have all pushed Myanmar into a position as the current global epicentre of transnational organised crime. Green or environmental crimes have long been an important feature of this criminal landscape. The fracturing of state authority and policing and regulatory capacity, as well as the increasing permissiveness that has accompanied the conflict, has done little to reign in the size and destructive impacts of the many criminal enterprises that continue to loot Myanmar's environmental resource inheritance.

Addressing the question of how to make sense of the rising tide of environmental crimes in Myanmar and how we should best combat them, this chapter will proceed in four stages, each developing part of the chapter's central argument that new, radical and innovative thinking is required in this area. The first section provides a brief overview of environmental crime in Myanmar both pre- and post-coup, illustrating the wide array of destructive activities and pointing towards the intersecting nature of environmental and other organised crime in Myanmar and the wider region. Following that, the chapter very briefly develops the question of crime dynamics further, drawing on the literature on complex and intersecting criminalities exhibited in most illicit economies, before highlighting the wider linkages that mark and shape organised crime in the Mekong region, a geographical area generally taken to extend from the Chinese province of Yunnan in the north, southwards through Myanmar, Laos, Vietnam, Thailand and Cambodia. So, while Myanmar might be the epicentre of this expanding ecosystem of crime, its effects spill out across the region, resulting in frequent calls for regional-level responses. Such efforts have shown little success and the third section of the chapter will propose a rather radical conceptual alternative to the standard call for greater regional integration in law enforcement and related activities. Instead, and drawing on the environmental science of wildland fire behaviour and its notion of the 'fire complex'—a multi-front, super-size, super-heated fire that exhibits unique thermal and behavioural dynamics—the idea of a Mekong Crime Complex is proposed. This crime complex operates on a regional scalar that is more than the additive sum of national parts and exhibits crime dynamics unique to its scale and challenging to conventional approaches towards crime control. Akin to wildfire complexes, the Mekong Crime Complex thus also demands integration of disciplines that today mainly reside in silos and in which disciplinary assumptions produce a kind of bounded rationality that limits capacities to understand and predict crime dynamics or model intervention impacts. Finally, then, the chapter will conclude with some thoughts of the benefits such reframings as proposed here offer.

Environmental Crimes in Myanmar

This first section of the chapter provides a highly abbreviated summary of some key forms of environmental crime in Myanmar (see also MacBeath, 2025).

Teak, Burmese rosewood and Padauk: Timber harvesting has been a major source of income and livelihood for Burmese since at least the nineteenth century. Yet between 2001 and 2020 an estimated 4 million hectares of forest was lost: an area roughly the size of Switzerland (Doherty, 2021). Timber may only be exported from Myanmar via Yangon. Yet, research by the Global Initiative Against Transnational Organised Crime has established that most, in fact, flows illegally across the porous north-eastern border with China, the value of which is estimated at US$600 million a year, with around US$200 million finding its way to the US and Europe (Mahadevan, 2021). China, India, and Thailand are the largest customers for Burmese teak. The Environmental Investigation Agency (EIA) reports that between them the three countries imported 4 million cubic metres of teak logs in the ten years to 2017, worth an estimated US$2.8 billion. Complex regulations around timber harvesting, milling, shipping, and the like can blur clear distinctions between licit and illicit and allow the infiltration of illicit timber products into the licit supply chain. Further, the mis-grading of teak logs (grading premium logs as poor-quality saw logs), in collusion with corrupt officials and other key actors (see below), is one of the key means by which harvesting embargos and other attempts to regulate the industry are continually evaded, providing the opportunity to extract vast illicit profits in subsequent downstream sales. EIA investigations have revealed long-standing and complex criminal, corporate, and government relationships spanning Myanmar,

Thailand, Singapore, China, Hong Kong, Malaysia, and Europe that sustain this trade.

Jade and precious metals: In 2014 an estimated US$31 billion of jade was mined in Myanmar, a figure representing almost half the country's official GDP and 46 times the level of government expenditure on health (RUSI, 2017). Another estimate puts the figure in 2016 as high as US$44 billion. While such high estimates remain disputed, and the trade—almost all of which is conducted illicitly—remains obscured, something like US$15billion a year would seem a reasonable estimate of the value of a trade where profits flow almost entirely into the hands of criminal networks and corrupt state and military actors. Part of the difficulty in estimating the size of this illicit market is that prices are subject to significant fluctuation, potentially doubling or trebling value from one year to the next based on fluctuations in demand or supply. As a result, attention has turned to production volumes, which remain high. Yet this should not be so. For in an effort to improve regulation to squeeze criminal and conflict actors out of the industry, in 2016 the National League for Democracy (NLD)—the governing party that ruled in an uneasy relationship with the Myanmar military between 2012 and 2021—suspended issuance of new mining licences. The last licences were due to expire in 2021, meaning that today all jade is effectively illicitly mined and traded. Due to the limited reach of the state, however, official collusion and corruption, and now the civil war, this has wholly failed to stem the industry. Its beating heart is the town of Hpakant, in Kachin State, which is, in fact, the world's largest jade extraction site. The financing of this notionally closed-down industry is Chinese, and key actors in Kachin include four ethnic armed organisations (EAOs) (the Arakan Army, the Kachin Independence Army, the United Wa State Army, and the Myanmar National Democratic Alliance Army), the Myanmar military, and thousands of artisanal miners who toil in the mine tailings and whose profits are heavily taxed by both conflict actors. Global Witness found illegal taxation of jade in 2021 to be 10–25% to the armed groups and 25% to the Myanmar military. Despite fighting 'in the field', there is evidence of cooperation and collusion among all the Kachin conflict actors in jade extraction around Hpakant, including in the recent post-coup period (Global Witness, 2021). Senior criminal figures also appear to have interests in other areas, such as drug trafficking, reflecting the breadth of activities of criminal networks (Global Witness, 2015). Overall, the jade industry stands as one of the key conflict-supporting resources sectors, with deep involvement of both junta and EAOs creating few incentives to return to the reform agenda pursued from 2016 by the NLD.

Extensive mining of precious metals, including tin by the United Wa State Army, and gold by various actors also forms part of the extractive resources picture in Myanmar. For example, a 2023 report released by the Kachin State Accountability Resource Governance group estimated a tenfold increase in unregulated gold mining in the state since the coup of 2021. Nevertheless, conflict actors on both sides tax such operations. The report estimated that a gold miner operating a 100-square-foot plot in Hpakant would owe US$71,400–95,200 to the Kachin Independence Army and US$95,200–142,800 to the junta (Radio Free Asia, 2023).

Environment: Large-scale environmental destruction and illicit profits result from criminal and EAO activities, as well as from Chinese government and Myanmar junta facilitation of so-called 'economic development' along their shared border. Crimes include not only deforestation due to illegal logging, but also destruction from more notionally 'legitimate' developments such as China's proposed Mong Tong dam on Shan State's Salween River. China is reported to be supplying EAOs in the region (including UWSA, SSPP, and the TNLA) with arms to secure the levels of security deemed necessary for developments such as Mong Tan to proceed and for commercial trade routes to operate effectively (Fawthrop, 2022). At the same time, highly destructive environmental practices support the extraction of rare earth minerals, where sales (mainly) to China have increased in value from US$1.5 million in 2014 to US$780 million in 2021. Many of the profits of this industry flow directly to EAOs who control the mountainous regions where mines are located. As such, miners fall under the permitting and excise demands of EAOs, who then provide both notional licences (to operate illicitly) and security. Oversight and visibility in such areas, such as from civil society groups, presents risks of violence which, in turn, means that not much is known of these extraction operations or their expansion. However, satellite over-flights commissioned by Global Witness in 2022 revealed more than 2,700 settlement ponds at almost 300 locations across north-eastern

Myanmar, covering an area the size of Singapore. The costs of environmental remediation are unknown, but in now-closed mining areas in China they are estimated in the many billions of US$ and may take up to 100 years to complete (Global Witness, 2022).

Wildlife trade: The trade in wildlife and wildlife products, frequently centred on Myanmar's special economic zones (see below), is recognised to be extensive, but data here too are thin. A 2008 study for the US Congress found the most frequently trafficked species were Asiatic black bears, clouded leopards, Asian elephants, reptiles, and turtles (Wyler, 2008), while more recent reportage has found a wide variety of animal parts, including pangolin (implicated in the 2020 Wuhan Covid-19 outbreak), on sale (BBC, 2014). More recently, van Uhm and Wong (2021) described the extensive involvement of Chinese organised crime groups in this trade and their role frequently as intermediaries and facilitators working between local poachers, for example, and drawing upon resources of the wider criminal economy. Frequently identified hot spots include Mong La, close to the Chinese border, and Tachilek, on the border with Thailand, more of which will be mentioned in the next section which considers the integration of criminal economies and the regional character of transnational organised crimes centred on Myanmar.

Intersecting Criminality and Regional Flows

While academic and law enforcement practice tends towards expertise-based specialisation in certain types of organised crime, criminal organisations themselves increasingly operate on a "mixed-business" model. Systematic study of this phenomenon has thus far not matched its pace of change, as Guerette and Aziani (2022, p. 9) recently observed, noting that 'macro-level (regional/national/transnational) analyses have thus far developed in a largely fragmented manner via various studies of specific transnational crime types'. Evidence of convergence nevertheless exists, and environmental crimes have been a notable aspect of this. Anagnostou and Doberstein (2022, p. 1615), for example, concluded after identifying 150 studies published between 2000 and 2020 that 'the illegal trade in many of the most frequently trafficked species have…converged with numerous other serious and organised crimes, most commonly drug trafficking'. Meanwhile van Uhm and Nijman (2022, p. 543) summarise the broader environmental crime nexus by observing of 'organized crime syndicates' that they often 'diversify into the lucrative business of tropical timber, endangered species, and natural minerals alongside their traditional activities'.

This conclusion is reinforced by studies of environmental crimes and the illicit economy in Myanmar described above. Indeed, the very locations of the myriad forms of organised crimes map across each other not only at the national level (e.g., many centring upon Kachin and Shan states in northern and north eastern Myanmar) but indeed right down to the street-address level, where casinos and criminal compounds in the armed ethnic enclaves of Myanmar's border regions host, for example, not only a burgeoning trade in wildlife species, but also illegal gambling, cyber-scam centres holding human slaves, trading centres for female brides, markets for women for forced childbearing, and other types of illicit activity (Robinson & Branchini, 2018; US Department of State, 2022). Moreover, the impact of the 2021 military coup d'état and the escalating conflict since then has been to create zones of increasing permissiveness and lawlessness (see below). The effect has been a rapid scaling-up of criminality in an area that, as evidenced above, had already been ranked 3rd globally for organised crime prior to the coup. Environmental crimes in this context become but one element in a spiralling vortex of criminality, the scale of which is unprecedented. For example, the UN Office on Drugs and Crime estimates that in 2022 alone methamphetamine production concentrated in Myanmar's Shan state generated criminal proceeds of more than US$60 billion (UNODC, 2022). At the same time, in 2023 the country leapfrogged Afghanistan to become the world's largest opium producer, after seeing a sharp rise in land returning to poppy production following the coup, and increasing sophistication in the industry (UNODC, 2023). Consequently, when the Global Initiative Against Transnational Organised Crime revised its global crime index in 2023, Myanmar had leapt to the top, now ranking number 1 globally on the basis of its intersecting criminal economies and manifest state weakness (lack of resiliency) and active participation in organised crime (Global Initiative, 2023; see also International Crisis

Group, 2023, 2024; Jesperson et al., 2023; United States Institute for Peace, 2024).

Special Economic Zones and Intersecting Criminality

The 'special economic zones' that sprang up around Asia in recent years have also, in many cases, become zones of intensive and intersecting criminality (Gore, et al., 2022). In an oft-repeated pattern, this begins with environmental crimes as a hub is established in remote forest tracts far from regulation and surveillance, while the hub itself then develops as a kind of criminal entrepôt. Crimes found in these zones, particularly those established in Myanmar close to the China border, include illegal gambling (the Chinese market for which is estimated at US$145 billion, annually); the sale of rare or endangered wildlife both from the region but also trafficked-in from as far away as Africa; the sale of wildlife parts, such as tiger skins, ivory, or pangolin scales; the laundering of proceeds of crime through the casinos themselves; online scamming and fraud operations based in casinos or other facilities, often utilising the labour of trafficked Burmese nationals as well as nationals from the wider region; human slavery; kidnapping and extortion; prostitution; drug trafficking; unregistered or illegal mining, and so on.

Casinos and the criminal enterprises they support frequently combine the operations of Chinese organised crime groups and Burmese EAOs (Tower & Clapp, 2020). Furthermore, the trafficking networks based out of these zones are closely tied with regional neighbours, such as Kings Romans, and Golden Botan City, both in Laos, and integrated into regional and global flows of illicit goods and services. For example, Mong La in Special Region 4 in far eastern Myanmar emerged as a model and has since been replicated in similar developments in other places in the region. EAOs (including the National Democratic Alliance Army (NDAA) and the United Wa State Army (UWSA)) are reported to have taken important roles in these criminal enterprises, providing security, protection, and facilitating trafficking, while linked Burmese banks have been designated by the US Department of Treasury for terror financing (US Treasury, 2021) and regional partners sanctioned, such as the so-called Zhao Wei Transnational Criminal Organization (US Treasury, 2018). Similar relationships and dynamics can be found further north in Special Region 1, the Kokang Self-Administered Zone, where illicit funds are channelled to support Border Guard Forces (BGFs) and junta officers, as well as being laundered into the formal Myanmar economy via companies such as the Fully Light Company that invests in projects in Mandalay and Yangon (Clapp & Tower, 2021). The military junta, themselves, are shareholders in companies like Myanmar Economic Holdings Limited, Myanmar Economic Corporation (jade), Myanmar Timber Enterprise (teak), and the like. Various investigations name military, government, and business figures with stakes in companies identified as being associated with jade or timber, drugs, arms and a variety of other clandestine activities (Justice for Myanmar, 2024).

A Regional Response: Same Wine, Same Old Bottles

Given the regional character of these illicit flows, alongside the collapse of any functional law enforcement or border and customs capacity in Myanmar since the coup d'état of February 2021, prescriptions for its solution fall back continually on the idea of improved regional responses. It is reflected, for example, in proposals received at the eighth ASEAN Ministerial Meeting on Drug Matters (AMMD-8) in Vientiane, Laos, in August 2023 to maintain the efficiency of existing drug control cooperation mechanisms, conduct more joint patrols, and improve cross-border information sharing (People's Army Newspaper, 2023).

The point to be made here does not so much concern the regional approach per se. Rather, it concerns the underlying analysis and its implicit assumption that, as the scale of the Myanmar-centred regional crime problem grows, it can be effectively met by law enforcers more effectively doing the same thing, grounded in their respective national systems. The idea of the region, and the regional response, in this sense is invoked for two reasons. First, it reflects the national lens through which almost all law enforcement operates. Second, and more importantly, regional or cross-border approaches are invoked reflexively because it is presumed that *crime as a phenomenon obeys certain laws*: that it grows in a linear, additive fashion; that it grows in environments that vary in nature but conform in some essential way to the elementary contours of Weberian national statehood; and that additive growth can be countered with additive law enforcement capacity, such as is achieved by recruiting and coordinating neighbouring national systems. It is a central argument of this chapter that such a view is flawed. Yet, how else might crime as a phenomenon be understood? What other conceptual model might be applied to recognise that point when the old laws break down, when they no longer fit the data?

The Problem of Complexity and Insights about the Wildland Fire Science

If these typical assumptions we use about how crime phenomena operate—such as about linearity—are wrong, what might be the alternatives? Van Calster (2010) observes that such traditional approaches are common across the social sciences. They originated in the eighteenth century and were designed to deal with *complicated* systems and processes. Their solutions to complication are characterised by various forms of reduction, ranging in scale from reductive experimental methods to isolate and concentrate on single variables, to the reduction achieved by disciplines specialising in the study of single types of individuals, social processes or phenomena (e.g., crime and criminals in criminology, learning and children in education, etc.). What these methods are particularly poor at, Van Calster argues, is dealing with *complex* phenomena. These are things that work in non-linear ways, often via feedback loops that change the dynamics of the phenomenon itself in ways that are rapid and challenge effective prediction. One of the reasons criminology has achieved such poor results in most areas where it holds putative expertise, Van Calster proposes, is that it attempts to apply traditional complication-reducing methods to phenomena, including organised crime, that are inherently complex in nature. The widespread failure of criminological models to explain more than small amounts of variance in crime or criminal justice actions (Walker, 2011), to be replicable even when it does (Reuter, 2017), or to satisfy scientific error testing (e.g., via Bayesian modelling;

Niermeyer, et al. 2022) reinforces the empirical element of this argument. So, what is the alternative?

Wildfire Science and the Wildfire Complex

Another discipline increasingly recognising the need to apply complexity theory and methods to phenomena is the environmental science of wildland fire behaviour. Recent work on fire complexes—multi-front, super-size, super-heated fires that exhibit unique thermal and behavioural dynamics—seems to have particular applicability, via analogous reasoning, to the problem of organised crime in Myanmar and the wider Mekong region. Wildfire complexes fundamentally alter the ecological structure and balance in landscapes. At the same time, they're characterised by sometimes significant spatial variability in intensity across the complex, reflecting different vulnerability/susceptibility levels of trees and other vegetation species, as well as the terrain through which the fire moves (Trouvé et al., 2021). As an example of the non-linear character of events in wildfire complexes, Trouvé et al. (2021, p. 2) observe that 'Disentangling the roles of fire intensity and tree and stand susceptibility presents a challenge because the observed pattern of fire severity is influenced by both.' Thus, fire scientists some time ago developed methods of non-linear simultaneous estimation to model the way one phenomenon—such as fire intensity—changes dynamically under recursive feedback-looping effects of other variables within the fire complex. Among these other variables are wind, temperature, humidity, and soil structure, which simultaneously influence one other as well as fire intensity in the complex system.

During a wildfire event, it is the scale of the contiguous fire complex that drives many of the forces undermining ordinary fire assumptions and models. For example, when a fire complex reaches scale, it begins to change wide-scale atmospheric conditions and, literally, to make its own weather. And the weather it produces, incidentally, predisposes higher burn rates and other factors that hasten fire spread. For example, Dowdy et al. (2017, p. 7342) studied the huge Black Saturday fires in southeast Australia in 2009, illustrating how the 'pyroconvection and fire-atmosphere feedback processes' produce pyrocumulonimbus storms and dry lightning strikes that further expand a conflagration. McRae et al. (2014, p. 417) had earlier studied fire complexes in eastern Australia's Blue Mountains, observing of these phenomena that the emergence of 'pyrocumulus or pyrocumulonimbus (pyroCb) signals a transition in fire development from what is primarily a surface phenomenon to a coupled fire–atmosphere phenomenon in which the fire is able to interact with upper levels of the atmosphere', including the upper troposphere and lower stratosphere. Studying why extant models of fire behaviour fail under such fire complex conditions, McCormick (2002, p. 1) concluded that explanatory models of ordinary fire essentially 'scale locally' the drivers of fire. While larger-scale, probabilistic, models are also available, their generality—we might call this a kind of abstract scale—provides insufficiently useful prediction within a terrain/landscape of a middle scale (e.g., a region or sub-region) such as is found in a fire complex.

Reflecting on the sometimes-tragic misjudgements made during fights against wildfires, Cunha et al. (2022, p. 2020) examined firefighter decision-making in Portugal's 2017 Pedrógão Grande fire complex, a fire they describe as 'of Dantesque proportions' in which 66 lives were lost. Drawing on real-time incident data, they conclude that lack of fire-complex-scale data (due to a data systems failure) led firefighters to default to fighting various parts of the fire complex using local data—and thus local-scale methods. The resulting catastrophe, they argue, can be seen as a case of improvisational myopia, a topic on which there is a body of literature in organisational studies. Improvisational myopia 'refers to paying primary attention to the local context both physically and temporally during improvisational action instead of to complete system level information' (Cunha et al., 2022, p. 2022). In a wildfire complex where, as McCormick (2002) observe, locally scaled data fail to capture the most meaningful drivers of the blaze, this produces in firefighters forms of what Cunha (2022, p. 2022) term 'corrosive myopia', meaning that local-level actions either have little effect, poor payoff for the effort, or, in fact, harm the wider operation. Examples might include actions across different fire teams that run counter to each other or lack the appropriate order or sequencing necessary for effectiveness, or that employ methods inappropriate to the scale of the problem.

The Mekong Crime Complex: A Proposal

Similarities between the phenomenon of wildfire complexes and the Mekong regional organised crime phenomenon centred on Myanmar should be clear. While Myanmar sits at its epicentre, the whole phenomenon of regional organised crime is better understood as a single crime complex. But why is that so? What is wrong with speaking simply of organised crime in the Mekong region? The answer, as the wildfire complex example illustrates, is that fire or crime complexes exhibit different characteristics and behave in ways that do not respond readily to traditional, locally-scaled efforts at fire or crime fighting. A fact well evidenced by the marked ineffectiveness and often counter productiveness of such efforts. This chapter will finish by addressing two aspects of the Mekong Crime Complex, which should offer a novel and fruitful new approach to addressing environmental and other crimes in this part of the world. These are: very briefly, the concept's utility for criminologists; and an outline of some key features of the Mekong Crime Complex that distinguish it from typical or 'ordinary' criminality.

Crime Complexes in Criminology

Despite its face validity the idea of a crime complex phenomenon has received remarkably little attention in criminology. Garland (2001) delivered the most influential account of a crime complex in his book *The Culture of Control*. But what he termed 'the "crime complex" of late modernity' was very different to what we're discussing here. For Garland, the complex was a 'cultural formation' 'characterized by a distinctive cluster of attitudes, beliefs, and assumptions' (Garland, 2001, p. 163) about the ubiquity, normality and intractability of crime and the new dynamics of its control in the mainly urban centres of contemporary western societies (but in reality, mainly the UK and USA) (see also Campbell, 2008; Pakes, 2005).

Thereafter, the idea of a crime complex largely disappeared from criminological writing for almost two decades. Recently it has re-emerged in three apparently unrelated studies, each of which offers convergent support for the spatial sense in which the term is being used here. First, Goodall (2023) coined the term ‘rural enterprise crime complex’ to describe a certain arrangement of cultural and spatial factors that supported crimes against non-human species in the countryside. Two further studies engage the maritime domain and transnational organised crimes. Satrial et al. (2020, p. 356) propose ‘the international fisheries crime complex’ as a useful rubric not just for describing a certain constellation of related and supporting crimes, but also for ‘reveal[ing] the predicaments which cause fisheries crime to be so underrated despite the concrete damages it causes’. Utilising a case study of Somalia-based piracy, Jacobsen and Larsen (2023, p. 83) highlight that not only do crime convergences exist – ‘hence the notion of a *maritime crime complex*’ – but that this has manifold implications. In particular, they argue that ‘that tackling a seemingly “segregated” crime—like piracy—in isolation from related types of crime in a broader crime complex entails a risk of overlooking crucial interlinkages and dynamics’ (p. 84).

This point is well made and an important assumption within the vision of crime complexes being developed here. But while each of these studies suggests new and productive ways for ‘thinking outside the box’ of ordinary crime analysis, they are marked by significant limitations that the approach to a Mekong Crime Complex aims to surmount. The idea of a crime complex in each of the ways it is used above is ultimately an intervention upon established thinking about crime convergence. The Mekong Crime Complex notion is much more than this, as reflected by its origins in the environmental science of wildfire behaviour and its approach to complex rather than merely complicated, or convergent, phenomena and to the changes that scale and intensity themselves generate.

Anatomy of the Mekong Crime Complex

What marks out transnational organised crime in the Mekong Crime Complex is thus its scale, intensity and capacity to affect virtually all other processes and systems across the affected terrain of its reach. At its centre is Myanmar. In late 2023 the country was ranked 1st globally—as the most criminal country in the world—in the authoritative global organised crime index compiled by the Global Initiative Against Transnational Organised Crime, Global Initiative, 2023). Yet the contribution of wildfire science has been to draw attention beyond this 'single fire' to view—or, nation by nation approach—to consider the conflagration as a whole and its associated multiplier effects. Thus, how does the size and scale reflected in the Mekong Crime Complex produce impacts that are more than the simple additive sum of national parts? How, across the Mekong, are social and political structures and equilibriums being changed, perhaps irreparably, by the fire of this illicit economy in its manifold dynamic parts? In what way is this fire leaping beyond its normal terrestrial parameters to 'change the weather' or fundamentally alter—through volatile interaction—the conditions across the complex of social and political order itself? And how, across the Mekong Crime Complex, do we find spatial variability in susceptibility and resilience that reflect the recursive effects of crime altering local social ecologies over time? These are just a few of the questions that this notion of a crime complex poses. Answering even these might be a monograph-length effort. Nevertheless, for illustrative purposes in this chapter it is possible to highlight one facet as a kind of exemplar of what might be possible: the impacts where organised crime reaches such scale that, like wildfire-atmosphere volatilities, it creates new atmospheric conditions that further drive the crime complex and exacerbate its devastating effects.

The Emergence of Regimes of Extreme Permission

A distinctive feature of the Mekong Crime Complex is the role played by special economic zones (SEZs), key features of which were described in the second section of the chapter. Criminology has paid little attention to their emergence and a recent review of SEZs' global rise observed that 'such economic enclaves have only made a guest appearance in criminology' (Hall et al., 2023, p. 279). To remedy that, Hall et al. (2023, p. 279) proposed what they term 'a typology of the domains in which crime and harm are evident: illicit trade, elite wealth holdings and environmental degradation'. Environmental degradations are indeed a feature of Mekong SEZs, as are a range of other environmental crimes, including illicit natural resource extraction and a wide array of wildlife crimes. Yet there is a paradox in the account these scholars give of SEZs. It is that, like McCormick's (2002) observations about probabilistic wildfire models that provide average-fire parameters and thus struggle to fit the super-scaled heat and destructive properties of fire complexes, so too this does this global typology. Accommodating together, as it does, 'celebrated…modern-day success stories (such as the Shenzhen SEZ)' and those with promises of the same like 'the new UK Teesside Freeport' alongside Laos's notorious Golden Triangle Special Economic Zone, the typology operates on the register of general theories of complicated systems. This is not simply an argument for more granularity than a global typology can reasonably achieve. Rather, it is to make two critical observations about the anatomy of the Mekong Crime Complex.

First, while Hall and colleagues present SEZs as a kind of 'outside'—spaces beyond law and sovereign control, or in some liminal state between them—in the Mekong Crime Complex SEZs are epiphenomenal, being markers or range riders alerting us to something much more significant occurring in rearrangements of power, law, authority, and the like. For across the Mekong Crime Complex, from Laos's Golden Triangle SEZ centred upon the Kings Roman casino and Boten SEZ with its Golden City Casino (Laungaramsri, 2025, van Uhm & Wong, 2021), to Myanmar's SEZs, special regions, and self-administered regions, SEZs emerge in borderlands themselves characterised by weak or negligible rule of law, contested sovereign authority, and deeply embedded criminality that bear very little resemblance to Hall et al.'s (2023) putative 'inside' with which SEZs are supposed to contrast. Instead, what we find developing, albeit in uneven patterns, across the Mekong Crime Complex are large borderland regions of embedded criminality, marginal state control and capacity for law enforcement, and indeed widespread state or state-actor participation in the illicit economy. Laungaramsri (2025, p. 195) documents the 'retreat of the Lao state at the border [which] has not only put an end to the old frontier but [also] turned various borderlands into…estranged space[s]' or 'enclave spaces'. Cioccini and Greener (2023, p. 1309) use the term 'regimes of extreme permission' to describe SEZs and related arrangements in Southeast Asia, but the term perhaps better describes the entire borderland zones of northern and eastern

Myanmar that now lie almost wholly outside any Westphalian/internationally recognised form of nation-state sovereignty or lawfulness, as witnessed for example in the case of Wa state (Ong, 2021).

Second, across the complex, these spaces of extreme permission interact, producing rapid changes in the terrestrial location of crime 'fire fronts' and producing dynamic changes in local susceptibility or resilience. Criminal actors, by no means passive in this process, engage in dynamic and reflexive strategies of risk/regulatory arbitrage, shifting criminal operations between different thermal centres—hot spots—across the complex to maximise efficiencies and profits while reducing risks, threats, or compliance obligations of various sorts (ranging from law enforcement to conflict risks or political exposure). As the environmental destruction wrought by rare earth mining in China became apparent and the government there cracked down on use of illegal and destructive methods, that industry moved into the permissive spaces of Myanmar's borderlands. In the same way, Chinese crackdowns on methamphetamine production increased risks and costs for criminal enterprises who saw arbitrage opportunities in exploitation of the borderlands' permissive environment to address the markets. The environmental and social harms that each crime type produced were not affected at all—simply shifted geographically—as traditional single-fire-front-style law enforcement approaches were applied to a problem that had by then created extensive scale factors and dependencies that are seldom recognised and that emerge as a scale effect. What might these look like?

Meehan and Goodhand (2023, p. 11) have traced the emergence of 'coercive brokerage' in these borderlands, focusing on paramilitary groups and their role in governing frontier spaces 'where public authority is contested, the means of violence fragmented and there exists a plurality of would-be statebuilders'. Two prefacing points are worth making. First, these zones of extreme permissiveness are not aberrations or reflective of state failure, but in fact a relatively stable non-Westphalian state form in which extensive criminality is an enduring feature. Rather than being a pathology, therefore, they are a core feature of this type of political formation. Second, in the absence of a state sovereign, the powers of coercion that produce order are rendered by other actors, often in competition—or at least tension—with each other across the wider crime complex and its regional political field. The Mekong Crime Complex is thus far more than a complicated mélange of crime types, criminal actors, and illicit beneficiaries. It is a phenomenon that now generates powerful effects that radiate across the region. Furthermore, while many of these effects are pernicious, crime is equally and at the same time increasingly indispensable to the wider region and its people and societies. As Meehan and Goodhand (2023, p. 16) summarise it:

> the conversion of upland borderland spaces of Myanmar, Laos and Indonesia into resource frontiers has created new flows of timber, biofuels, rubber, foodstuffs and hydropower that are central to supporting Asia's rapid industrial development, urbanisation, rising consumption, and food and energy security. Revenues from illegal drugs, border town casinos, and the licit and illicit extraction of jade, gold and rare earths in Myanmar's borderlands have provided vast pools of capital invested in urban development and infrastructure projects.
>
> Far from being marginal, borderlands are often key laboratories of change that shape the trajectory of wider development processes.

In such spaces, old divisions between inside and outside, between states, or state actors, and criminals, or between the licit and illicit, begin to lose traction. The violence entrepreneurs, ethnic armed organisations and their ethnonationalist polities, paramilitaries, regional criminal syndicates and the host of other actors who ply the permissive spaces of the Mekong Crime Complex, are actors in a reorganisation of the terms of politics, capitalism, and globalisation. Echoing Clausewitz's observations on war, Meehan and Goodhand (2023, p. 22) contend that the coercive controls and threats of 'big violence' (state-challenging armed capability) leveraged in these zones that are at once beyond, yet integrated with, the state 'is not automatically a disabler of politics; in many contexts it represents the continuation of (a particular kind of) politics by other means' (for a wider analysis, see also Goodhand et al., 2024).

Fighting a fire complex that has created effects and dependencies that run so deep into regional politics and regional economies is thus beyond the scope of any law enforcement effort. While to criminologists and criminal justice actors its features might invoke ideas for technical solutions—improved anti-money laundering capacity, regional intelligence sharing, or the like—the solutions to the problems posed by the Mekong Crime Complex are political in nature and of a scale that defies law enforcement's 'single fire' capacity, both conceptually and practically. Just how much this is so was evidenced in late 2023 when China, frustrated by the refusal of the Myanmar military junta to shut the cyberscam centres that were not only defrauding thousands of Chinese citizens but also drawing in trafficked Chinese youth for work as slaves, opted not to oppose an ethnic armed uprising. Dubbed Operation 1027, a Three Brotherhood Alliance of EAOs launched coordinated attacks on lightly held government positions in the contested borderlands of Shan State. In its manifesto of objectives, alongside taking back ethnic lands, the alliance listed also closing down scam centres, detaining key operatives, and returning victims of trafficking to their national authorities. Within a few weeks, a huge industry that had been 'beyond reach' of regional states' law enforcement and international actors' influence, while they observed with alarm its pernicious impacts (e.g., estimates of trafficked slaves in Myanmar give a figure of 120,000 at the lower bound; Stimson Centre, 2023), was rolled up. A crucial financial flow the Myanmar junta and its paramilitary Kokang Border Guard Force and ancillaries was turned off like a tap when China reshuffled its deck of political playing cards. Yet, as Clapp and Tower (2023) observe, China's strategy carries risks as violence threatens economic corridors vital to its fragile eastern Yunnan province.

Conclusion

Green crimes in Myanmar have developed into 'anchor industries' not only supporting large sectors of population but also integrating into a wider crime complex. This chapter has traced the dynamics of these crimes, but also the apparent insolubility of organised crime and conflict problems in the wider region. It has diagnosed the ills of prevention as lying in a dated view of crime dynamics, one in which crime is presumed to obey laws of linearity. This holds that increasing crime can be effectively countered by more—or more coordinated—law enforcement.

Instead, the chapter proposed a dynamic model drawn from the environmental science of wildfire complexes: super-heated, super-scale wildfires. These fires have been modelled as *complex* systems—as distinct from the traditional social science view of problems as merely *complicated* systems amenable to reductive analysis—allowing for recursive effects, multiplier effects, and the like. Applying such thinking to what has been termed here the Mekong Crime Complex, the chapter has attempted both to demonstrate the limits to current approaches to regional organised crime, of which green crimes form such an important part, and to map out a novel alternative view and strategic approach for the region.

References

Anagnostou, M., & Doberstein, B. (2022). Illegal wildlife trade and other organised crime: A scoping review. *Ambio, 51*, 1615–1631.

BBC. (2014). *Undercover in Myanmar's Sin City where anything goes*. Retrieved September 22, 2023, from https://youtu.be/xqbw_7Vo_eA.

Campbell, L. (2008). The Culture of Control in Ireland: Theorising recent developments in criminal justice. *Web Journal of Current Legal Issues, 1*, 1–19. http://webjcli.ncl.ac.uk/2008/issue1/campbell1.html

Cioccini, P., & Greener, J. (2023). Regimes of extreme permission in Southeast Asia: Theorizing state-corporate crime in the global south. *The British Journal of Criminology, 63*, 1309–1326.

Clapp, P., & Tower, J. (2021). *Myanmar regional crime webs enjoy post-coup resurgence: The Kokang Story*. USIP. Retrieved September 22, 2023, from https://www.usip.org/publications/2021/08/myanmar-regional-crime-webs-enjoy-post-coup-resurgence-kokang-story

Clapp, P., & Tower, J. (2023). *China's influence increases amid Myanmar's instability: Beijing's longer-term strategy aims to fragment the anti-junta resistance in Myanmar*. USIP. Retrieved January 24, 2024, from https://www.usip.org/publications/2023/12/chinas-influence-increases-amid-myanmars-instability.

Cunha, M. P., Clegg, S., Rego, A., Giustiniano, L., Abrantes, A., Miner, A., & Simpson, A. (2022). Myopia during emergency improvisation: Lessons from a catastrophic wildfire. *Management Decision, 60*, 2019–2041.

Department of State. (2022). 2022 Country Reports on Human Rights Practices: Burma. US Dept of State. https://www.state.gov/reports/2022-country-reports-on-human-rights-practices/burma/

Doherty, F. (2021). *Opinion: Perilous future for Myanmar's pristine forests after coup*. Retrieved September 22, 2023, from https://www.thethirdpole.net/en/nature/perilous-future-for-myanmars-pristine-forests-after-coup/.

Dowdy, A., Fromm, M., & McCarthy, N. (2017). Pyrocumulonimbus lightning and fire ignition on Black Saturday in southeast Australia. *Journal of Geophysical Research: Atmospheres, 122*, 7342–7354.

Fawthrop, T. (2022). Shan state exemplifies China's tangled Myanmar ties. *The Diplomat*. Retrieved September 22, 2023, from https://thediplomat.com/2022/05/shan-state-exemplifies-chinas-tangled-myanmar-ties.

Garland, D. (2001). *The Culture of Control*. Oxford University Press.

Global Initiative. (2021). *Global Organised Crime Index 2021*. https://globalinitiative.net/initiatives/ocindex/

Global Initiative. (2023). *Global Organised Crime Index 2023*. https://globalinitiative.net/initiatives/ocindex/

Global Witness. (2015). *Jade: Myanmar's "Big State Secret"*. Global Witness.

Global Witness. (2021). *Jade and conflict: Myanmar's vicious circle*. Global Witness.

Global Witness. (2022). *Myanmar's poisoned mountains*. Retrieved September 22, 2023, from https://www.globalwitness.org/en/campaigns/natural-resource-governance/myanmars-poisoned-mountains

Goodall, O. (2023). The rural enterprise crime complex: 'undefendable rural space' and the threat from the fortress farm. *Crime, Law and Social Change, 80*, 215–235.

Goodhand, J., Meehan, P., Acero, C., & Koehler, J. (2024). *Coercive brokerage: Paramilitaries, illicit economies and organised crime in the frontiers of Afghanistan, Colombia and Myanmar*. Working Paper II. SOC ACE Research Paper No 28. University of Birmingham.

Gore, A., Kennedy, L., Southern, N., & van Uhm, D. (2022). *Asian Roulette: Criminogenic Casinos and illicit trade in environmental commodities in South East Asia*. GITOC.

Guerette, R., & Aziani, A. (2022). The displacement and convergence of transnational crime flows. In E. Savona, R. Guerette, & A. Aziani (Eds.), *The Evolution of Illicit Flows*. Springer Nature.

Hall, A., Antonopoulos, G., Atkinson, R., & Wyatt, T. (2023). Duty free: Turning the criminological spotlight on special economic zones. *The British Journal of Criminology, 63*, 265–282.

International Crisis Group. (2023). *Transnational crime and geopolitical contestation along the Mekong*. International Crisis Group.

International Crisis Group. (2024). *Scam Centres and Ceasefires: China-Myanmar Ties Since the Coup*. Crisis Group Asia Briefing N°179. Bangkok/Brussels, 27 March 2024.

Jacobsen, K., & Larsen, L. (2023). Maritime crime in the Western Indian Ocean: Interlinkages and dynamics. In Y. Eski & M. Wright (Eds.), *Maritime crime and policing*. Routledge.

Jesperson, S., Alffram, H., Denney, L., & Domingo, P. (2023). *Trafficking for forced criminality: The rise of exploitation in scam centres in Southeast Asia*. ODI Thematic brief. ODI.

Justice for Myanmar. (2024). Cartel finance map. Retrieved July 19, 2024, from https://data.justiceformyanmar.org/.

Laungaramsri, P. (2025). Dark zomia: Myanmar frontier and the Chinese enclosure. *Journal of Borderlands Studies*. https://doi.org/10.1080/08865655.2025.2469854

MacBeath, A. (2025). *Cashing in on Conflict: Illicit Economies and the Myanmar Civil War*. Global Initiative Against Transnational Organised Crime.

Mahadevan, P. (2021). *Myanmar's illicit timber: Flows, drivers and actors*. GITOC.

McCormick, R. (2002). On developing a meso-theoretical viewpoint of complex systems by exploring the use of artificial neural networks in modeling wildfires. In *Paper Presented at ForestSAT Symposium, Heriot Watt University*, Edinburgh, August 5–9. https://citeseerx.ist.psu.edu/document?repid=rep1&type=pdf&doi=b3a3ec20b3c2a8066ab2611473b917d6b4024607.

McRae, R., Sharples, J., & Fromm, M. (2014). Linking local wildfire dynamics to pyroCb development. *Natural Hazards and Earth Systems Sciences, 15*, 417–428.

Meehan, P., & Goodhand, J. (2023). *Coercive brokerage: Examining the paramilitary–organized crime nexus in borderlands and frontiers*. Working Paper I. SOC ACE Research Paper No. 26. University of Birmingham.

Niermeyer, R., Proctor, K., Schwartz, J. & Niermeyer, R. (2022) 'Are most published criminological research findings wrong? Taking stock of criminological research using a Bayesian simulation approach', *International Journal of Offender Therapy and Comparative Criminology*, online first, *69*(5), 475–494.

Ong, A. (2021). Tactical dissonance: Insurgent autonomy on the Myanmar-China border. *American Ethnologist, 47*, 369–386.

Pakes, F. (2005). Penalization and retreat: The changing face of Dutch criminal justice. *Criminal Justice, 5*(2), 145–161.

People's Army Newspaper. (2023). *Vietnam proposes solutions to strengthen ASEAN anti-drug cooperation*. Retrieved September 19, 2023, from https://en.qdnd.vn/asean-community/vietnam-and-asean/vietnam-proposes-solutions-to-strengthen-asean-anti-drug-cooperation-555295.

Platt and Tower, J. (2023). *Myanmar's Junta Is Losing Control of Its Border with China*. United States Institute for Peace. https://www.usip.org/publications/2023/11/myanmars-junta-losing-control-its-border-china

Radio Free Asia. (2023). *Illegal gold mining up ten-fold in Kachin state since Myanmar coup*. Retrieved September 22, 2023, from https://www.rfa.org/english/news/myanmar/mining-04262023173338.html.

Reuter, P. (2017). Learning from replication failure. *Criminology and Public Policy, 16*, 781–784.

Robinson, B., & Branchini, C. (2018). *Estimating Trafficking of Myanmar Women for Forced Marriage and Child Bearing in China*. Johns Hopkins Bloomberg School of Public Health.

RUSI. (2017). *Corruption and illicit flows in Asia*. RUSI.

Satrial, R., Sinaga, O., Bainus, A., & Darmawan, W. (2020). The need of international regime support to combat fisheries crime. In *Proceedings of ICPSH 2020—Navigating Global Society in the Disruptive Era*. UNSOED Press.

Stimson Centre. (2023). *Mekong–U.S. partnership track 1.5 policy dialogue on trafficking and cyber-enabled crime*. Stimson.

Tower, J., & Clapp, P. (2020). *Myanmar's Casino cities: The role of China and transnational criminal networks*, USIP Special Report No. 471. USIP.

Transnational Alliance to Combat Illicit Trade (TRACIT). (2018). *The global illicit trade environment index*. Economist Intelligence Unit.

Trouvé, R., Osbourne, L., & Baker, P. (2021). The effect of species, size, and fire intensity on tree mortality within a catastrophic bushfire complex. *Ecological Applications, 31*(6), e02383.

United States Institute for Peace. (2024). *Transnational crime in Southeast Asia: A growing threat to global peace and security (USIP Senior Study Group Final Report)*. Washington: USIP.

UNODC. (2022). *Synthetic drugs in East and Southeast Asia: Latest developments and challenges*. UNODC.

UNODC. (2023). *Southeast Asia Opium Survey 2023: Cultivation, production, and implications*. UNODC.

US Department of State. (2022). *Trafficking in persons report 2022: Country Profile: Burma*. Retrieved February 24, 2024, from https://www.state.gov/reports/2022-trafficking-in-persons-report/burma/.

US Department of Treasury. (2018). *Treasury sanctions the Zhao Wei transnational criminal organization*. Retrieved September 22, 2023, from https://home.treasury.gov/news/press-releases/sm0272.

US Department of Treasury. (2021). *Testimony of Samuel W. Bodman, Deputy Secretary U.S. Department of the treasury before the Senate Committee on banking, housing and urban affairs*. Retrieved September 22, 2023, from https://home.treasury.gov/news/press-releases/js1501.

Van Calster, P. (2010). Studying society, safety and security: Notes on observer involvement. In R. Lippens & P. Van Calster (Eds.), *New directions for criminology: Notes from outside the field*. Antwerp.

van Uhm, D., & Wong, Y. (2021). Chinese organized crime and the illegal wildlife trade: Diversification and outsourcing in the Golden Triangle. *Trends in Organised Crime, 24*, 486–505.

van Uhm, D., & Nijman, R. (2022). The convergence of environmental crime with other serious crimes: Subtypes within the environmental crime continuum. *European Journal of Criminology, 19*(4), 542–561.

Walker, J. (2011). Advancing science and research in criminal justice/criminology: Complex systems theory and non-linear analyses. In J. Walker (Ed.), *Social, ecological and environmental theories of crime*. Routledge.

Wyler, L. S. (2008). *Burma and transnational crime*. US Congressional Research Service.

Environmental Crimes in East Siberia and Activities of Chinese Mafia

Andrey G. Anisimov

Introduction

The Irkutsk Oblast is the central region of Eastern Siberia. Currently, East Siberia it is a territory of Russia, covering over 40% of all Russian area. It refers to a part of Northeast Asia, stretching from the Yenisei River in the west to the mountain ranges along the Pacific Ocean in the east. The region shares borders with China to the southeast and Mongolia to the south (Khamina, 2017).

Initially, this area was not a part of modern Russia. Starting from the third century BCE, various tribes inhabited its territories. Among them were the Huns, Xianbei, Jurchens, and ancient Turkic peoples. Simultaneously, Mongol tribes led by Genghis Khan settled in this region, eventually establishing dominance (Weatherford, 2004). The region was populated by Mongolian ethnic groups, and trade routes

A. G. Anisimov (✉)
Irkutsk Law Institute, The All-Russian State University of Justice,
Irkutsk, Russia

D. van Uhm, D. Siegel (eds.), *Global Green Crime and Ecojustice*, Palgrave Studies in Green Criminology, https://doi.org/10.1007/978-3-031-91557-4_17

were established, notably a branch of the great Silk Road. The Russian exploration of the region occurred approximately in the seventeenth century and, despite its outwardly conquering nature, it largely took place due to the decline of Mongolian statehood at that time. With the arrival of the Russians in the seventeenth century, some local residents, in exchange for privileges and liberation from the Buryat oppressors, assisted the Russian administration in the development of Buryat lands (Okladnikov, 1937).

As a result, the settlement of the region took place under the protectorate of the Russian government, largely with the assistance of the local population. This process spanned several centuries (from the seventeenth to the twentieth century) and occurred in waves of immigration from various populations. The outcome of this process is that, at present, the region is home to more than 130 nationalities. This includes European (Russians, Belarusians, Ukrainians, Poles, Mordvins, Romanians, Bulgarians, etc.), Turkic (Tatars, Tofalars), Mongoloid (Buryats, Mongols), and indigenous (Evenki, Tungus, etc.) ethnic groups.[1]

Chinese influence had a significant impact on the development of the region, including the evolution of culture and trade, although this pertains only to the period from the ninth to the fourteenth centuries. Starting from the seventeenth century, there was a decline in Chinese influence, followed by the assimilation and integration of Mongoloid peoples into the population of Russia. Today, the Buryat population (those who are direct descendants of Chinese origin) does not exceed 4% of all nationalities in the region. However, among the population of China, there exists a myth that the Irkutsk region is inherently Chinese territory, which, of course, is far from the actual state of affairs. Nevertheless, this belief influences the convictions and behaviour of Chinese criminals, forming part of their mythology. Many such criminal groups consider this region to be unlawfully occupied Chinese territory. They firmly believe that they have full rights to unlawfully exploit

[1] The ethnic composition of the population in the Irkutsk region, according to official statistics data as of the year 2023. *Federal State Statistics Service*: https://view.officeapps.live.com/op/view.aspx?src=https%3A%2F%2F38.rosstat.gov.ru

local natural resources, viewing their actions as an act of justice restoration.

The Irkutsk region has historically been an area with a high level of crime, conditioned by historical processes (Danchevskaya, 2019). For example, in 1913, in Russia, there were, on average, 105 convicts per 100,000 population for various crimes. In the Yenisei province, this number stood at 200, and in the Irkutsk province, it was even higher, exceeding 260 individuals. At this time, Siberia was perceived in public consciousness as a land of exiles. Before the start of World War I, the region ranked among the highest in terms of crime rates among the regions of the Russian Empire (Institute of History, 2009).

In terms of economic activity, the extraction of minerals and natural resources has been taking place here since the very beginnings of Russian exploration in the seventeenth century. This was primarily facilitated by ongoing trade with China. In the seventeenth century, industrial methods began to be used in the region for the extraction of rock salt. Simultaneously, deposits of rock crystal, carnelian, emeralds, and other semi-precious and precious stones were explored, along with deposits of copper, silver, and iron ore. However, during the seventeenth and eighteenth centuries, iron ore extraction saw the greatest development (Nikitin, 1990).

Simultaneously, the local population here extracted the skins and fur of wild animals, primarily sables and ermines. Up until the mid-nineteenth century, a significant proportion of all furs in Russia was obtained from Eastern Siberia, with over half of this volume being sold to China (Greenwald, 1872). Trade between the region and China occurred similarly to trade with neighboring regions of Russia through the staging of seasonal fairs (Shakherov, 2016).

In the twentieth century, under the management of the USSR, the region entered an era of industrialization. This saw the beginning of the industrial extraction of natural resources such as gold, coal, timber, and also, to a lesser extent, jade, gas, and oil. Concern for the environment and ecology was neglected, with a consumerist attitude prevailing until recent times.

During the period of industrial growth in the USSR, the influence of China and Chinese crime was almost completely absent from the Irkutsk region. In many ways, this situation resulted from the Soviet Union's policy of forcibly deporting ethnic Chinese to China and Central Asia during the period from 1926 to 1938, as well as the strengthening of controls over migration. As a consequence of these actions, ethnic Chinese practically disappeared from the region's territory until 1991 (Bugai, 2016). During this time the attitude towards China was condescending and arrogant (Malenkova, 2014). Trade with China, including natural resources, was centralized and managed by Moscow.

Everything changed dramatically after the collapse of the USSR. Because of the dissolution of the Soviet Union in the early 1990s, there was a collapse of the existing management and control systems and a subsequent overexploitation of natural resources and the involvement of criminal networks. There was a surge in crime, primarily organized crime (Gilinsky, 2014).

Russian organized crime quickly began to exploit the illegal natural resources market, recognizing it as one of the most profitable and rapidly growing sectors in the new Russia. Initially, this involved the illicit trade of products from extractive industries (such as aluminium, coal, petroleum products, etc.). However, over time, independent illegal enterprises began to emerge, including logging, gold mining, and activities connected to the trade of flora and fauna. Today, Russian organized crime is actively involved in environmental crimes, which the nation's authors now classify as transnational in nature (Rednikova, 2018, p. 107).

The rather painful transition of the new Russia to a market economy led to the shutdown of industrial enterprises in the Irkutsk region and also the theft of their property. The formation and flourishing of local organized crime occurred, with the Bratsk mafia emerging as a particularly noteworthy participant.[2]

The opening of borders, the reduction of border control, the decline of law enforcement agencies in the 1990s, and their merging with crime all combined, leading to a gradual, but steadily increasing arrival of illegal Chinese citizens and their gradual involvement in criminal activity.

[2] I. Nadezhdin (2022, August 25) *The bodies won't be found—the Taiga will hide everything. A Russian deputy created the most brutal organized crime group in Siberia. How did it take over cities.* Lenta.ru. https://lenta.ru/articles/2022/08/25/opg/

Methodology

In this chapter I am using the following sources of information: Official statistics from the Main Directorate of the Ministry of Internal Affairs of Russia for the Irkutsk region regarding the state of crime; and official reports from the Irkutsk regional police on the results of operational activities in combating ethnic, economic, and environmental crime, including organized crime. I also refer to the results of an official sociological survey conducted by the administration of the Governor of the Irkutsk region and the Government of the Irkutsk region on law enforcement work related to the topic 'The state of the migration situation in the Irkutsk region in 2023' and also to materials from official media sources.

My research methods include interviews with local Russian entrepreneurs who have been conducting business with China for more than a decade, including those who have lived in China and engaged in illegal business (5 respondents). I also conducted interviews with Chinese entrepreneurs proficient in the Russian language (2 respondents) and analyzed surveys of Chinese students studying at universities in Irkutsk (10 respondents); and surveys of operational personnel directly involved in combating organized economic and environmental crime, as well as operational personnel from the local Interpol office (5 respondents). Additionally, I studied a selective survey of local residents from the Alarsky, Bodaybinsky, Bokhansky, Zalarinsky, Kachugsky, Olkhonsky, and Slyudyansky districts of the Irkutsk region and a survey of local residents from the Okinsky district of the Republic of Buryatia (in total 47 respondents).

General Description

In the context of Chinese organized crime involved in environmental crimes in East Siberia, we will focus primarily on the activities of the Chinese Triads.[3] The so-called Triads, or representatives of the Chinese

[3] However, the term 'Triad' is used by us with some simplification. The problem lies in the insufficient study of the phenomenon and the abundance of legends, myths, and misconceptions regarding the Chinese Triads. Nevertheless, understanding even the general logic of criminal activity by

mafia, are the descendants of the long history of secret Chinese societies that emerged around the seventeenth century (Purbrick, 2019). The exact origin of the Triads is largely unclear and shrouded in legend. In Chinese, they are often referred to as "洪門" or "Hóng Mén" (Gong Mun) (Petta, 2017, p. 4), literally meaning "Great Gates" or "Secret Society". Another name for these organizations is "黑社会" (Hēi shèhuì)—"Black Society" (Broadhurst, 2012).

A mafia is an organized criminal group with a hierarchical structure, strict rules, and control over certain types of criminal activity (Wing & Kwok, 2012). Any mafia operates based on, and in accordance with, covert rules and customs, often using violence and intimidation to achieve its goals, and typically engages in crimes such as extortion, smuggling, illegal gambling, drug trafficking, and others. The goal of any mafia group is to establish control over certain regions or sectors of the economy to generate profit and exercise power (Albanese, 2011, p. 5).

It is also a characteristic of a mafia to strive to preserve and expand its own influence. To this end, a mafia typically employs any methods, but usually involving corrupt actions, intimidation, direct violence and media capture (Ko-lin Chin, 2003, p. 7).

One "byproduct" of mafia activity in a specific territory is the implementation of some typically law enforcement functions (debt recovery, protection from violence, criminal search, etc.) both for its direct members and for people who directly turn to the mafia for help. This function often gives a kind of carte blanche to the mafia, justifying its existence in the eyes of society. This is particularly relevant in situations where, for various reasons, the completely legal resolution of problems and conflicts is difficult or insufficiently effective (Volkov, 2016).

All the experts we interviewed (businessmen, representatives of law enforcement agencies) unanimously pointed out that environmental crimes such as illegal deforestation, coal mining, gold extraction, jade mining, and poaching are not isolated individual manifestations of criminal activity. Instead, they are carried out under the active influence of, and primarily in the interests of, Chinese organized crime. In turn,

Chinese organized crime (largely represented by Triads) can provide insights into the reasons behind the environmental crimes occurring in the region.

respondents directly connected to China (both Chinese and Russian businessmen) explicitly stated that illegal trade in natural resources is directly or indirectly controlled by Chinese Triads. Next, let's briefly describe the logic behind the criminal activities orchestrated by Triads in the region.

Features of Triads

Today, it can be asserted that the Triads have undergone significant transformations, during which they have spread their influence worldwide (Varese, 2011), maintained an extreme degree of secrecy, and, according to some respondents, collaborate with the authorities in China, although they do not publicly acknowledge such cooperation.[4]

Traditions and strict rules of triads have significantly weakened. It has been very clearly noted is that 'Chinese criminal networks have been branded as secret societies, but recent literature describes them as reliant, flexible, and social networks that are entrepreneurially oriented (Van Uhm & Wong, 2021, p. 2). Triads engage in criminal activities only when the benefits from their illegal operations significantly outweigh the negative consequences. Meanwhile, lower-level members and affiliated structures within the mafia strive to operate as legally as possible.

All triads resemble a social network, involving both triad members and ordinary people. Triad members are familiar only with individual participants at their own level; they have no precise understanding of the entire structure, with the exception of the top tier (Petta, 2017, p. 5).

It is worth comparing the behavior of Chinese organized crime in Eastern Siberia with similar patterns observed in the so-called Golden Triangle of Southeast Asia, as described in a recent comprehensive study (Van Uhm, 2023, pp. 71–108). Let us first address the similarities. Chinese organized crime establishes cooperation with local criminal groups in Eastern Siberia in much the same way. Similarly, interaction with smugglers, poachers, and illegal resource extractors (such as illegal loggers, gold miners, jade extractors, etc.) occurs through outsourcing.

---

[4] Personal communication with Chinese entrepreneurs, March 23, 2024.

At the same time, there are several fundamental differences, the most notable of which are as follows.

First, there is no large-scale investment in the local economy. Moreover, according to statements from Chinese entrepreneurs, the Chinese government has officially advised against investing significant funds in Russia's economy through major projects, deeming this market to be unstable and highly risky. While Chinese capital is present, it typically arrives through private initiatives that are generally not directly linked to the Chinese government.

Secondly, it should be noted that, despite its shortcomings, Russia's law enforcement system generally succeeds in containing the uncontrolled growth of Chinese organized crime. This is evidenced by the structure of the criminal schemes we have identified.[5] Triads in this region are primarily represented by their third and lowest tier, and they are forced to operate in an extremely covert manner through intermediaries (also Chinese), who are not members of the mafia but are rather officially entrepreneurs while simultaneously acting as agents. These intermediaries work in alignment with the triads' direct interests, often unable to refuse for various reasons.

What is particularly noteworthy is the large number of such agents, many of whom operate as double agents (collaborating simultaneously with Russian criminal groups) and even triple agents (interacting with corrupt elements within Russian state institutions), primarily pursuing their own short-term entrepreneurial interests. Surprisingly, triad members themselves prefer, whenever possible, to establish illegal cooperation with regional law enforcement, though this is also done through intermediaries and with an effort to remain discreet.

Thirdly, the current focus of Chinese organized crime is primarily on controlling financial flows related to Chinese contracts, as well as overseeing the logistics and supply chains for the cross-border movement of natural resources such as timber, coal, jade, and gold, alongside the importation of industrial goods from China. The illegal market involving the smuggling of animals and plants is significantly smaller in this region. Additionally, as far as we have been able to determine, Chinese organized

---

[5] Ibid.

crime is not involved in the illegal drug trade; at least, we have been unable to locate any such information.

Fourthly, the current state of the criminal market in the region is spontaneous and far less stable than in the Golden Triangle. An atmosphere of pervasive mistrust prevails, particularly between the local population and foreigners. This is largely a result of significant cultural differences, but also stems from constant competition among Chinese actors themselves.

Fifthly, the majority of individuals involved in illegal activities related to China would gladly forgo the direct "services" of the Chinese triads and form their own independent criminal networks. However, they are currently unable to do so due to their inability to manage certain criminal tasks that the triads can handle, such as financial control, logistics, cross-border movement, and locating buyers. It appears that there are instances in which triad members leave the organization and cease engaging in triad-related criminal activities. However, such cases are typically exceptions rather than a consistent trend. We are aware of only one such instance. Currently, this individual resides in Russia and cooperates with Russian law enforcement authorities.

Not all Chinese and Russian criminals are members of triads, yet almost all of them are directly or indirectly involved with such organizations, often for considerable periods without realizing it. Chinese triads are not a single organization; there are numerous competing groups. Some authors have identified more than 50 different types of southern triads alone (Broadhurst, 2012, p. 2).

The Chinese Triads operating in Russia are part of the global network of Chinese organized crime. In Russia, they operate in the Far East and Siberia.[6] All Chinese Triads can be divided into northern and southern, a split which is historically and culturally determined. Southern triads (with a nominal center in Guangzhou, as well as in Beijing and Shanghai) are interested in the high-tech sector, food products, services, and trade. By contrast, northern triads (with a nominal center in Harbin) are

[6] Sedov, V. (2021, February 19) *Lurking Dragons: Gangsters from the legendary Triad were arrested in Russia. What were Chinese mafiosi doing far from their homeland.* Lenta.ru. https://lenta.ru/articles/2021/02/19/triada/

primarily engaged in the export of natural resources. According to our data, the Irkutsk region is under the influence of northern Chinese triads.[7]

The emergence of such criminal groups in the region began relatively recently, after the collapse of the Soviet Union in 1991. The current situation can be described as a "dynamic equilibrium." In other words, the region simultaneously hosts numerous triad representatives, although only a portion of them significantly influence criminal activity. This state of affairs is linked to economic processes in both China and Russia and could undergo radical changes at any moment.

The name is associated with the fact that members of these groups attribute a mythical significance to the number three. Hence the popular name—TRIads. By the same logic, the structure of such groups is most often built on three levels of hierarchy, with each such level containing three key participants—this is most often a leader and two subordinates (Petta, 2017, p. 9). According to our data, in the Irkutsk region, members of the second- and third-level cells will tend to be almost constantly present. Members of the first level cell may come from time to time, and the direct leader of the highest link (the Dragon) is usually located in China.[8]

Among Chinese citizens, a complex of representations about the indigenous population of Russia living in the Irkutsk region has arisen. The attitude of Chinese citizens, who visit the region towards the local population of the Irkutsk region contains specific representations. Most Chinese believe that all Russians in the region are extremely aggressive and cruel, capable of robbing, raping, and even killing, and that a significant proportion of the Russian population are somehow connected with bandits and the Russian mafia. The key idea is that Chinese visitors should be afraid of the local population and in no case contact them, trust them, let alone befriend them. In fact, among the Chinese coming to Russia, there is an unspoken ban on communication with the native Russian population. Any interaction takes place through senior

[7] Personal communication with Interpol officers, March 30, 2024.

[8] Personal communication with Chinese entrepreneurs, March 23, 2024.

representatives and through Chinese intermediaries, who often become people associated with the Chinese mafia.[9]

One way or another, the described set of representations and myths is directly used by the Chinese mafia to involve all incoming Chinese citizens in their sphere of influence. Moreover, such representations are actively cultivated and propagated in this environment. And most members of the triads also believe in such representations.[10]

The situation has led to a lack of trust among all participants in criminal networks. Mafia representatives distrust each other and their fellow citizens, while the local population in the region also regards all Chinese individuals with considerable suspicion.[11] This dynamic has contributed to the mentioned "dynamic equilibrium" and the decentralization of triads in the area.

There are several main ways for Chinese citizens to arrive in the Irkutsk region. It all depends on the category of the citizen. The first method is as a tourist. One gets a Russian visa and makes a temporary visit to Russia. This method is often used by Chinese mafia leaders, predominantly from the upper echelons of the triads. The second and most popular method is to come to Russia for the purpose of studying.

Chinese Triads in Russia

The goal of the Chinese mafia is to obtain illegal profits, which are ultimately exported to China. To achieve this goal, however, control needs to be established. First and foremost, it's financial control. For example, control over large contracts with China. But they are also interested in any large illegal financial flows. Secondly, triads strive to get control over logistics, supplies, and customs borders. Moreover, control is exercised both from the Russian and Chinese sides. According to some respondents, virtually all Russian–Chinese transit of goods is directly or

[9] Personal communication with Chinese students and entrepreneurs, March 20, 2024.

[10] Personal communication with Chinese students and entrepreneurs, March 20, 2024.

[11] Personal communication with indigenous people, January–March 2024; Results of an official sociological survey 'The state of the migration situation in the Irkutsk region in 2023'.

indirectly controlled by the triads. Thirdly, it is control over the behavior and decision-making of officials, businessmen, and senior regional leaders. This is done, for example, through a system of hostages. In a sense, the Chinese mafia strives to control and exploit Russian corruption.[12]

The primary criminal activity of Chinese triads at present seems to be the organization of smuggling natural resources, primarily timber, coal, jade, gold, and bio-resources. Besides exporting from the Russian Federation, Chinese triads also carry out the supply of counterfeit Chinese industrial products to Russia. According to estimates by some authors, the volumes of such smuggling amount to tens of billions of dollars (Ovchinsky, 2006).

Funds for the extraction of natural resources, timber procurement, and the like are officially issued in the form of interest-free (or low-interest) loans, which are provided by Chinese banks with the support of the Chinese government. In fact, to obtain such a loan, several conditions are necessary:

1. Sign an official written contract for supply with the Russian side.
2. The contract must be concluded for the supply of a product of interest to the Chinese economy.
3. The contract must be bilateral. An enterprise in China must confirm the need for the exported product.
4. The contract should not exceed the allocated quotas. For each specific type of product, the Chinese government allocates certain budgets. The contract at the time of its presentation to the bank should not be excessive, otherwise, they may refuse such a loan.

In fact, these loans are a financial regulator of the natural resources market in Russia. Money for the most priority goods is allocated without interest; subsequently demand is regulated by the volume of allocated money and the interest rate. Sometimes such regulation can bypass market relations. For example, contracts can be concluded for the supply of gold or coal at prices significantly exceeding current market ones.

---

[12] Personal communication with Russian police officers, April 1, 2024.

It is particularly necessary to note that these contracts are concerned with billions of dollars. Chinese triads control the signing of large contracts of interest to them directly in China. Any Chinese businessman who has concluded a contract for the supply of natural resources will, in one way or another, have to deal with the triads and pay them a certain percentage of transactions for patronage. In addition, the triads control the movement of goods under such contracts directly at the customs border between Russia and China. This occurs principally in the city of Blagoveshchensk.

There is a case described by one Chinese entrepreneur in which he and several other Chinese entrepreneurs concluded large contracts for the supply of Siberian timber to China. Officially everything was going well; unofficially, however, everyone was made to understand that they have competitors under the protection of the triads, who wanted to supply timber at an earlier date. As a result, the cargo of entrepreneurs which did not have the approval of the triads was detained for a long time at the Russian border with China in Blagoveshchensk. The timber began to spoil, and their competitors, who did not have such problems, bought up all the detained cargo at a very cheap price.[13]

According to our respondents, the Chinese government is effectively creating and shaping the market for natural resources in the region, including the illicit market.

For instance, the rapid increase in illegal logging in Siberia was almost entirely driven by the sudden demand from Chinese enterprises for large volumes of timber, prompted by subsidies and loans provided by the Chinese government to support its domestic wood-processing industry. A similar situation is currently observed with regard to coal supplies. Jade, for example, was of little interest to illegal traders until a significant demand for certain types of this stone emerged in China, causing its black market price to rise by tens or even hundreds of times.

Meanwhile, the Chinese mafia seeks to control major financial operations in this market through Russian logistics and Chinese customs.

[13] Personal communication with Russian and Chinese entrepreneurs, April 1, 2024.

Interaction with Local Crime, Including Organized Crime

Apparently, the merging of Chinese triads with Russian organized crime began some considerable time ago, in the period immediately after the collapse of the Soviet Union in 1991. Major spheres of natural resource extraction in the region, such as logging, coal mining, gold, and to a lesser extent jade, have been controlled by local organized criminal groups for several decades.

Among all organized forms of crime in the Bratsk region (named after the city of Bratsk in the Irkutsk Oblast), the mafia holds a special place. This well-known criminal organization, both in Russia and abroad, traditionally focused on controlling the extraction of natural resources immediately following the dissolution of the Soviet Union. During this period, the head of the Bratsk mafia, Vladimir Tyutrin, supervised the supply channels for metals, primarily aluminum and timber, in Eastern Siberia.[14]

The Bratsk mafia also engaged in illegal gold trafficking. Although it has lost a significant portion of its influence, it has not been completely eradicated. Currently, it exists as fragmented criminal groups that continue their "traditional" organized criminal activities, including illegal trade in aluminum, gold, and timber.

Russian organized crime groups show interest in triads because the latter effectively control almost all major shipments of timber, gold, steel, and aluminum to China. In practice, any illegal resource shipment to China requires approval from triad members. Another reason Russian criminals turn to triad representatives is their control over the legal and illegal financial flows necessary for export supply chains.

Furthermore, the Chinese mafia's ability to illegally supply specialized industrial goods to the Russian Federation, including items officially prohibited for export from China due to economic sanctions, attracts Russian organized crime. However, their primary motivation lies in the illegal import of industrial machinery essential for natural resource extraction—such as harvesters, draglines, grabs, excavators, and large trucks.

[14] L. Kislynskaya (2012, April 30) *The Gangster Childhood of Russian Business.* Top Secret. https://web.archive.org/web/20121016062735/http://sovsekretno.ru/magazines/article/3120/

Smuggling allows them to acquire such equipment at significantly lower prices than market rates. Additionally, triad members often lend money to local Russian mafia for these acquisitions.

In turn, Chinese triads are interested in swiftly identifying and exploiting Russian corruption schemes and illegal resource supply channels, exerting influence on authorities or industry representatives, and resolving law enforcement and regulatory issues. Russian organized crime provides these "services" to the Chinese mafia, including assistance in hiring local personnel.[15]

Triad's Modus Operandi and Criminal Infrastructure

The Chinese mafia, it appears, has two main approaches to organizing activities: typical and specially disguised (directly related to mafia members). The typical one implies relatively standardly disguised criminal activity, most often carried out by other people's hands. As a rule, this is a Chinese citizen under their control (not a mafia member), who is in Russia for any reason. Usually, it is a young man who came to Russia to study, less often with a private visit or for work. He establishes connections with the local population and organizes a joint venture. The activity can be very diverse; as a rule, however, it is a standard Russian–Chinese business for the region: car wash, car service, shop, cafe, restaurant, supply of Chinese goods, supply of Russian goods to China, extraction of natural resources, etc. In such cases, the business itself is registered to a Russian citizen.

At first, everything happens legally; then, often within a few years, dubious financial operations begin to be carried out inside such a business, such as money laundering, the sale of illegally acquired goods, such as illegally harvested timber, minerals (coal, jade, gold) smuggling of animals and plants, poaching, etc. A specially disguised approach to committing crimes involves the direct participation of triad members. To minimize the risk of being caught and exposed, additional mechanisms

[15] Personal communication with Russian police officers, April 1, 2024.

are used here. Primarily, this is a special type of communication, logistics, financial management, as well as the provision of their own medical services. Thus, in large cities, various large enterprises are officially opened to provide services to the population. Typically, these might include Chinese markets, hotels, recreation centers, cafes, restaurants, and so on. Such enterprises are opened, among other things, for the purpose of illegal activity.

Firstly, large enterprises (recreation centers, hotels, and restaurant complexes) are created with the aim of laundering incomes obtained by criminal means under the guise of their legal activity. According to operational data, it can be asserted that in a number of cases up to 80% of the declared income of such events constitutes "laundered" funds.[16]

Secondly, in such centers, illegal activities are directly carried out by the Chinese for the Chinese. This is primarily in the areas of gambling, prostitution, unlicensed medical and legal services, employment intermediation, and lending. A parallel can also be drawn here with the situation in the previously mentioned Golden Triangle. The patterns of Chinese influence are also similar to those described by other scholars (Gungwu, 2017). The underlying logic is fairly similar: Chinese influence on the economy, including the illegal sector, is evidently present. However, the scale of this influence is somewhat smaller, due both to the lower population and to the reduced levels of investment, as well as the efforts of local law enforcement and cultural differences.

Thirdly, in such places, a system of shadow calculations is carried out between Chinese citizens, including loans. Calculations are made by depositing real money and drawing up corresponding internal receipts, the presentation of which at any other similar place in Russia guarantees the receipt of real money back. However, this method is used for outsiders. Direct commands and calculations between their members are carried out by triad participants via mobile (including Internet) communication, directly and with no paperwork.[17]

Some time ago, we got an unofficial information that in 2010–2015 in the region during the establishment of the mobile network, malefactors

[16] Personal communication with Russian Interpol officers, September 22, 2023.

[17] Personal communication with Russian and Chinese entrepreneurs, 1st April 2024.

(presumably Chinese) introduced a significant volume (several tens of thousands) of fake SIM cards into circulation, thereby bypassing official operators. Initially, it was assumed that they had been obtained in order to bypass official roaming and save on communication services; however, it appears that such SIM cards are predominantly used by Chinese criminals for information exchange. This is primarily concerned with anonymous communication and financial management.[18]

In this way (by the Chinese and for the Chinese), a range of other services are provided in the region. These include medical services, which include full-fledged hospitals with outpatient observation, minor surgical operations, childbirth, and more. In such centers, there is also the basic necessary medical equipment and principal medications. The main purpose of such centers is not commercial, but conspiratorial. In instances of bodily harm, a triad member can receive medical help anonymously. These triads also provide tourist and logistical services: meeting, accommodation, transportation, equipment rental, guide services, treatment, and more with the main goal of controlling logistics and the activities of newly arrived citizens from China.

The activities of the triads related to the commission of environmental crimes are primarily viewed by them as a business endeavor. The region is traditionally rich in natural resources, and China has a significant need for their rapid acquisition. Consequently, activities such as illegal logging, coal smuggling, and the illicit export of jade and gold are assuming colossal proportions.

'Illegal logging' as a Case Study of Chinese Triads in Irkutsk Region

Sooner or later, such criminal schemes, although they can be disguised and quite complex, are nevertheless successfully identified and disclosed by local law enforcement agencies. As a rule, fictitious counterparties, non-existent legal entities, a wide network of intermediaries, front men, a multitude of financial operations, etc. are used. It should be noted that

[18] Personal communication with law enforcement official, 25th August 2023.

these are truly complex financial crimes involving many episodes; the identification and disclosure of such cases require a serious amount of time and intellectual and operational resources from law enforcement agencies. The number of such cases is small; however, each disclosed case is extremely valuable, and even the officially established criminal income is really impressive. Let us give a brief analysis of several successfully disclosed crimes related to illegal logging recently (See Table 1).

The officially documented incomes from criminal activities are impressive, ranging from US$500,000 to almost US$10 million. It is worth noting that the average official salary in the region today barely exceeds US$600 per month.

Unfortunately, it must be stated that the disclosure of the noted crimes, although extremely important, does not significantly damage the positions of the Chinese mafia in the region. Firstly, the majority of the detected criminals are the local population. Secondly, the identified Chinese criminals are merely executors, not members of the triads. Secondly, there are reasons to believe that in this way the Chinese mafia gets rid of competitors. The crimes that have been solved are estimated by law enforcement agencies to represent a small portion of their actual volume, approximately 5%.[19]

Illegal logging not only constitutes a criminal issue but also serves as a vehicle for economic crime within the timber industry. Compounding the problem, the detection and prosecution of related crimes lag behind the actual prevalence due to inadequate practices and evidentiary tools.

Scholars posit that the middleman chain plays a pivotal role in concealing illegal logging activities (Tatarnikov, 2016, p. 42). However, the ultimate motives of criminals are not always straightforward. Timber sales, including exports, are exploited for value-added tax (VAT) reimbursement. Experts estimate that these illicit schemes yield annual reimbursements of up to $350 million (Belokhrebtov, 2015, p. 119).

The complexity of detecting and proving such crimes arises from criminals exploiting legislative gaps across various legal domains (financial, taxation, banking, budgeting, civil, etc.) and at different levels (regional, federal, international). Corruption within ministries and agencies

[19] Personal communication with law enforcement official, April 1, 2024.

Table 1 Some characteristics of economic crimes committed by organized criminal groups involving Chinese citizens disclosed by law enforcement agencies of the Irkutsk region

Period of crime commission	Description/Qualification according to Russian legislation	Number of participants: Total/Chinese citizens	Established criminal income in USD (at the average rate for the period of commission)	Result
2018–2021	Sale of illegally harvested timber in the PRC. Subsequent money laundering through intermediary firms (documented 9 legal entities) on a particularly large scale by an organized group of persons	2/1 (leader is a PRC citizen)	9,730,632 (9.7 million dollars)	Participants were detained; criminal cases are being investigated
2021–2022	Smuggling of timber and natural resources; violation of Russian currency legislation; storage of illegally harvested timber	5/2 (2 leaders are PRC citizens)	2,814,275 (2.8 million dollars)	Participants were detained; criminal cases are being investigated
2019–2020	Sale of illegally harvested timber in the PRC. Subsequent money laundering through intermediary firms on a particularly large scale by an organized group of persons	4/1 (leader is a PRC citizen)	2,542,857 (2.5 million dollars)	Participants were sentenced to imprisonment in a strict regime colony from 6 to 7 years
2021	Storage of illegally harvested timber	2/1 (leader is a PRC citizen)	457,142 (457 thousand dollars)	Participants were detained; criminal cases are being investigated
2023	Full cycle of illegal timber harvesting. Illegal logging, movement, storage, shipment to the PRC; subsequent money laundering	24/2 (were leaders of the criminal group)	Being established	Participants were detained; criminal cases are being investigated

exacerbates the illegal logging epidemic, while inadequate interagency collaboration hampers effective countermeasures.

While a detailed description of all these schemes is impractical, let's highlight key aspects of VAT-related illegal reimbursement. In a remote region such as Vladimirskaya Oblast, a legitimate entity engages in diverse activities, including timber sales and export. This enterprise conducts legal economic operations, paying taxes and maintaining substantial financial turnover. Subsequently, a branch is established in a logging region (e.g., Irkutsk Oblast) to illicitly procure timber.

Typically, the branch doesn't directly acquire timber from loggers but relies on a network of intermediaries—often one-day companies (fly-by-night entities). At this stage, the scheme is straightforward: after purchasing timber, most one-day companies divide the transaction volume, legitimizing the timber purchase. These fly-by-night companies pay cash to physical entities, thereby evading taxes. The branch then sells the timber as an exporter, transferring the proceeds to the parent company. Financial documents are fabricated during this process, often inflating revenue figures. Subsequently, the parent company seeks VAT reimbursement based on the timber's overseas sale. Firms located outside forest regions can easily obscure significant monetary turnovers and illegal actions related to VAT reimbursement. Notably, VAT reimbursement for export operations tied to timber sales in Irkutsk Oblast faces stringent scrutiny from law enforcement bodies and regional regulations.[20]

Illegal Coal Mining

At present, China's demand for coal supply is increasing, as we have previously noted. The market evolves spontaneously in response to emerging needs. However, legal mining operations do not always suffice to provide an acceptable quantity, price, and timeline. Additionally, environmental and sanitary regulations governing coal extraction can slow down the process and significantly raise costs. Consequently, an illegal path is often taken.

[20] Personal communication with law enforcement official, February 3, 2021.

During interviews with law enforcement agencies and local residents, we learned of at least one such case. Specifically, we refer to the Aransakhoy coal deposit in the Irkutsk Oblast. This previously dormant deposit was leased by a local entrepreneur and it is known that the Russian entrepreneur established a joint venture with a Chinese partner. Furthermore, the actual coal deliveries to China by this company far exceed the legitimate capacity of the deposit if coal were extracted legally.[21]

It is highly likely that coal is extracted, including through illegal means, and subsequently legitimized officially through this enterprise's documents. Additionally, cases have been documented in which brown coal was actually delivered to China without official payment by the Chinese side. The Russian entrepreneur involved in these transactions was fined.[22] Presumably, this was done to conceal official revenues and the actual scale of deliveries. Currently, authorities are investigating this case in an attempt to identify signs of criminal activity.[23] This is likely one of many typical examples of the approach used in managing cases and the illicit organization of coal extraction.

Illegal Jade Mining

Approximately 97% of Russia's nephrite reserves are located in the Republic of Buryatia, which shares a border with Irkutsk Oblast. Official nephrite extraction requires obtaining a license through a competitive process. To participate and succeed in such a competition, significant financial resources must be paid, and a comprehensive mining project must be presented, including environmental protection measures and other relevant considerations.[24]

---

[21] Personal communication with indigenous people, January–March 2022.

[22] Kuznetsova, Z. (2020, June 1) *Angarsk company did not receive 6 million rubles for the delivery of 70 thousand tons of coal to China.* Ircity.ru https://ircity.ru/text/economics/2020/06/01/70676528/

[23] Personal communication with law enforcement official, July 2023.

[24] Baikal Daily. (n.d.). Who and how mines jade in Buryatia: Detailed analysis. *Baikal Daily.* https://www.baikal-daily.ru/news/15/420685/

Most of the extracted jade can be sold at relatively low prices; in the meantime, however, a small portion of the jade can be sold at prices several times higher than the market rate. As previously noted, the demand for such jade (specific varieties) is almost entirely driven by the illegal Chinese market. Private buyers in China seek to acquire a small quantity of this special jade as quickly as possible, bypassing legal channels.

The current situation, characterized by the lack of employment and low living standards, has led to the creation of a black labor market among local residents near the jade deposits in Buryatia. These residents engage in illegal jade mining, which involves trespassing on and around official mining territories, digging up riverbeds nearby, as well as directly stealing jade from enterprises. Such illegal activities often result in armed confrontations between illegal miners and law enforcement agencies.[25]

During a survey of local residents, it was established that the jade is directly purchased from these illegal miners on-site by intermediaries using cash transactions. The entrepreneurs we interviewed also confirmed that in the chains of illegal jade supplies to China, in which they participated, it was necessary to engage with representatives of the Chinese mafia to utilize logistics networks under their control.[26]

Illegal Gold Mining

Currently, 125 placer deposits are being exploited in the Bodaybo, Mamsko-Chuysky, Nizhneudinsky, Usolsky, and Kachugsky districts of the Irkutsk region (Shutova, 2008). The largest gold deposits in the region are concentrated in the Bodaybo district of the Irkutsk region. Notably, the Sukhoi Log deposit, located in this area, is one of the largest undeveloped deposits in the world and the largest in Russia..[27]

[25] Ministry of Internal Affairs of Buryatia. (2023, August 2) *Police in Buryatia Detain 30 Illegal Jade Miners.* Vesti Buryatia. https://bgtrk.ru/news/incidents/237131/

[26] Personal communication with Russian and Chinese entrepreneurs, April 1, 2024.

[27] BBC Russian Service. (2017, January 26) *Russia's Largest Gold Deposit Sold at Auction.* https://www.bbc.com/russian/news-38755522

Gold has been industrially mined here since the 1960s. Additionally, illegal gold mining has been prevalent since the discovery of local deposits. Both legal and illegal gold mining are conducted primarily through two methods: lode mining and placer mining. There are also five main methods of gold theft: three of which are directly associated with theft at processing plants, in mining cooperatives, and from decommissioned plants. These methods cause less environmental damage compared to the other two methods: thefts carried out by so-called "black diggers," where local residents independently pan for gold in streams and riverbeds of the taiga, and thefts involving violations of licensing conditions, where licenses issued for geological exploration are misused for actual gold mining.

There are also three primary channels for the distribution of illegal gold, two domestic and one international: within Russia, gold is sold to a jewelry factory in Moscow and a refinery or jewelry factory in Krasnoyarsk; internationally, gold is illegally supplied to China. The latter is the most complex route due to logistical and customs challenges, yet it is also the most profitable, as gold fetches a high price in China (Shutova, 2008).

According to our information, the illegal gold trade with China currently constitutes a significant portion of the black market for gold (over 40%). The Chinese mafia is heavily involved in this trade, primarily by providing financial incentives and solving logistical challenges associated with illegal transport. At the same time, triad members tend to avoid violent crimes, such as murders, armed robberies, extortion, and the intimidation of authorities. These activities are typically carried out by local Russian criminal groups.[28]

Tourism and Hunting

The triads also extract a significant volume of illegal income through the organization of illegal tourism by Chinese citizens. This involves well-established mechanisms for the falsification of documents, illegal

[28] Personal communication with law enforcement official, June 10, 2024.

accommodation, movement, communication, and escorting. There is a well-developed network of their own infrastructure of tourist facilities. These are hotels, guesthouses, restaurants, and their own fleet of vehicles that effectively belong to Chinese citizens. This activity deserves mention in light of the damage it causes to both the environment and the wildlife of the region. In addition to traditional tourism, the illegal format also implies the hunting of wild animals and illegal bio-resources. For example, wild animals caught or killed by poachers, as well as valuable plants. As Professor Repetskaya observes, "We are talking about animals or plants that are used for food (pine nuts, ferns, hogweed, etc.), in medicine (musk deer musk, deer antlers and horns, etc.), for the production of various fur products (skins of valuable fur animals—mink, sable, etc.)" (Repetskaya, 2013).

The volumes of such extraction significantly exceed all established limits necessary for the preservation of the region's natural flora and fauna. Another problem of illegal tourism is the pollution of natural bodies of water, primarily Lake Baikal, due to the violation of sanitary norms for the discharge of sewage from hotels and guesthouses located within the boundaries of specially protected natural territories.

Counteraction Against the Triads by Local Law Enforcement Agencies

The level of interaction between the police of China and Russia is considered to be low. For example, annually, about 1000 requests are sent to China by the police of the Irkutsk region via Interpol. Over the past 10 years (2013–2023), not a single official response has been received from the Chinese police to any such request. Law enforcement agencies of the region, and Russia as a whole, do not have specialized Chinese translators (native speakers of Chinese culture), let alone specialized units. Moreover, such work is almost never carried out due to the circumstances described earlier.[29]

[29] Personal communication with law enforcement official, September 10, 2023.

The Environmental Consequences of Illegal Activities

The most noticeable environmental consequences have resulted from the illegal logging of Siberian forests. Mass illegal logging is directly linked to the illegal smuggling of timber to China. In 2019 and 2021, due to forest fires in the Irkutsk region, a state of emergency was declared.[30] According to representatives of the authorities, the population, and experts, such fires are largely associated with large-scale illegal logging.[31] Overall, this is a complex problem. On the one hand, there are documented facts of intentional forest arson for its subsequent sanitary logging.[32] Secondly, illegal logging does not involve sanitary measures. Only tree trunks are exported from the logging site, while the tops, crowns, and branches are left at the crime scene. Subsequently, the remains of the trees dry out and become potentially fire-prone. Thirdly, deforestation leads to local climate changes and abnormal weather conditions, which also often become the cause of forest fires. It should be noted that climate changes in Siberia are significant, even on a global scale.[33] Especially this problem is recognized by climatologists.[34]

Another, less obvious problem of deforestation and, as a result, the change in the local climate is flooding. Accordingly, the large-scale flood that occurred in 2019 in the Tulun region of the Irkutsk region and which led to human casualties, was, according to Inna Latysheva, an employee of the Department of Meteorology and Physics of Near-Earth Space of

[30] Press Service of the Main Directorate of the Ministry of Emergency Situations of Russia for the Irkutsk Region. (2019, May 8) *A state of emergency has been declared in two districts of the Irkutsk region due to forest fires.* RBK. https://www.rbc.ru/rbcfreenews/5cd28c699a794738db7a9c92

[31] Press Service of the Main Directorate of the Ministry of Emergency Situations of Russia for the Irkutsk Region. (2019, May 16) *Special groups have been created in the Irkutsk region to investigate crimes in the forestry sector.* TASS. https://tass.ru/proisshestviya/6439082

[32] Shamina, O., Kiseleva, M. (2019, August 2) *They set fire to what is profitable to cut down. How 'commercial' forest arson works.* BBC Russian Service. https://www.bbc.com/russian/features-49169767

[33] Dixon, R. (2021, August 11) Siberia's wildfires are bigger than all the world's other blazes combined. *The Washington Post.* https://www.washingtonpost.com/world/2021/08/11/siberia-fires-russia-climate/

[34] Kozlovsky, S. (2020, July 30) *How Siberia is burning. Maps, graphs, and photos of fires in Russia.* BBC Russian Service. https://www.bbc.com/russian/features-53532961

the Geographical Faculty of Irkutsk State University, caused by local climate changes.[35]

Illegal extraction of such natural resources as coal also leads to local air pollution with coal dust, which returns to the earth and pollutes the soil. Both methods of coal extraction, open-pit and underground, disrupt the geological structure of the earth's interior, leading to collapses, landslides, and mudflows. During coal mining, miners either remove the topsoil layers or excavate deep quarries. The reclamation of mined areas is infrequent; more often, the waste rock is simply transported to designated dumps. These dumps consist of low-quality coal and soil, which easily erode, becoming sources of dust pollution and susceptible to spontaneous ignition (Rai & Biswajit, 2010).

Illegal jade mining, although associated with fewer problems, still presents similar issues. The primary concern, akin to illegal gold mining, involves the large-scale washing of rock and the consequent discharge of contaminated water into water bodies, leading to siltation and fish mortality.

Regarding gold mining, the environmental consequences warrant separate discussion. These impacts are linked to extremely harmful methods of extracting gold ore. Several main techniques are employed in illegal gold mining. These include amalgamation, which uses mercury (Yanti et al., 2023); the Planter process, which uses chlorine and iron sulfate; cyanidation, which involves potassium cyanide and sulfuric acid (Bragin & Cherdantsev, 2015); and refining, which uses nitric and hydrochloric acids (Rusalev et al., 2023).

The primary methods of illegal gold ore mining in the area under consideration are refining and cyanidation. Both of these methods are highly polluting, and illegal miners do not adhere to environmental standards, resulting in mercury compounds and other chemicals contaminating rivers and soil, thereby leading to the chemical pollution of the area.

The situation is exacerbated by the fact that such pollution occurs in or near protected and specially conserved natural areas. As a journalist who visited the industrial gold mining region of Bodaibo aptly remarked, the

[35] ISU Meteorologists. (2019, July 2) *An Anomaly Tending Towards Constancy.* KP.ru. https://www.irk.kp.ru/online/news/3525295/

local environment and landscape "resemble photographic images of the Moon or Mars."[36]

Particular attention deserves the pollution of Lake Baikal with sewage due to uncontrolled construction of illegal hotels and guesthouses on its shore. As a result of the pollution of Lake Baikal with untreated sewage, the level of nitrogen and phosphorus content in the coastal waters has increased, which has contributed to the rapid growth of Spirogyra algae. Spirogyra is dangerous for Baikal in that some types of microorganisms are displaced. For example, the indigenous inhabitants of Baikal—gastropod mollusks. After the appearance of Spirogyra in the lake, their number has significantly decreased. Such contamination also destroys the Baikal endemic sponges, thanks to which the water in the lake remains crystal clear. Spirogyra occupies the spawning grounds of the yellow-winged bullhead, which feeds the endemic species of the lake—the Baikal omul. As the cause of Baikal's flowering, specialists name the discharge into the lake of untreated sewage containing detergents with phosphates. This means that the UNESCO World Heritage Site on the brink of an environmental disaster.[37]

Conclusions

Our research indicates that crimes associated with the illegal extraction of natural resources in East Siberia and the Irkutsk region for the local Chinese mafia are merely a means to generate illicit funds for the development of other, more substantial projects.

In fact, all the common conditions for this illegal activity have been established, and local law enforcement has nothing to counter it. These conditions include illegal habitation, medical care, money lending, money laundering, and other systems. Similarly, law enforcement agencies are poorly informed about this illegal activity, and lack the sufficient resources and equipment, or choose to overlook it in exchange for rewards.

[36] Gold of Bodaibo. (2017, November 21) *Oblastnaya* newspaper. https://www.ogirk.ru/2017/11/21/zoloto-bodajbo/

[37] A. Yurkov. (2015, September 22) *How did Baikal end up on the brink of an environmental disaster.* RG.ru https://rg.ru/2015/09/23/baykal.html

The schemes of environmental crimes committed by the Chinese mafia are typical. The field of activity related to the environment is chosen ad hoc by the criminals. It could be logging, coal or gold mining, jade extraction, or any other kind of activity. It all depends on the goals of the criminal groups, existing opportunities, or local authority corruption. Subsequently, they find a Russian citizen partner who directly sets up an entity and prepares the necessary documents. This significantly eases the process of legal body registration, tax payments, and staff employment.

The roles of the partners in such businesses are played by fictitious or real wives, Russian business partners related to crime, including those recruited by the Chinese mafia during visits to China or business dealings with Russian citizens. Sometimes, even homeless people can play this role. This is often initiated for the illegal extraction and legalization of natural resources (timber, coal, gold, jade, etc.) with continuous shipping for sale in China. The immediate illegal activity in the vast majority of cases is carried out by indigenous people. This scheme typically includes the use of a network of counter-agents—'one-day' (fly by night) companies, manipulations with financial and tax reports, cash withdrawal, money laundering in the aforementioned centers, and others.

We need to initiate criminological regional research on the Chinese mafia. We must acknowledge that this is currently a blind spot. It's not just about statistics, but also about anthropological research, similar to the work that Professor Dina Siegel, for example, has been conducting for years in Europe. There is a need to develop programs for the legal arrival and assimilation of Chinese migrants into Russian society. We are capable of developing these programs in partnership with the Chinese government at the federal Russian authority level. We need to create, in partnership with Russia and China, a plan for a law enforcement response to Chinese criminal activity. This includes data exchange with Interpol, and perhaps the establishment of special departments within the structure of Russian law enforcement.

All the aforementioned environmental consequences have ceased to be merely local problems of an individual country, becoming global and significant on a planetary scale. Most of these issues are currently not the subject of proper study or assessment. For example, the true scale of the environmental impact in Bodaibo remains unknown.

It can be stated that Russian criminal law in the area of environmental protection lags significantly behind the actual situation, failing to adequately safeguard public relations in the field of ecology, which are evidently suffering long-term and extensive harm from transnational environmental crime. A reevaluation of criminal policy as a whole is required, taking into account the concepts and approaches proposed by eco-global criminology (White, 2011).

References

Albanese, J. (2011). *Organized crime in our times* (6th ed.). Routledge. https://doi.org/10.4324/9781315721460

Belokhrebtov, V. S. (2015). Problems of detecting and disclosing fraud in the reimbursement of value added tax in the field of the timber industry. *Vestnik Kazanskogo iuridicheskogo instituta MVD Rossii*, (21), 119–123. https://cyberleninka.ru/article/n/problemy-vyyavleniya-i-raskrytiya-moshennichestva-pri-vozmeschenii-naloga-na-dobavlennuyu-stoimost-v-sfere-lesopromyshlennogo/pdf

Bragin, A. A., & Cherdantsev, V. A. (2015). Modern problems and trends in the development of ore gold mining. *Bulletin of NSUEU, 2*, 308–3016. https://cyberleninka.ru/article/n/sovremennye-problemy-i-tendentsii-razvitiya-dobychi-rudnogo-zolota

Broadhurst, R. (2012). *Black societies and triad-like organised crime in China*. Routledge. https://doi.org/10.2139/ssrn.2138650

Bugai, N. F. (2016). Chinese in the USSR and Russia: A policy of double standards (1920–1940s). *ISOM*, 1–2. https://cyberleninka.ru/article/n/kitaytsy-v-sssr-i-rossii-politika-dvuh-standartov-1920-1940-e-gody

Danchevskaya, A. V. (2019). Social characteristics of crime in the Irkutsk and Yenisei provinces at the end of the 19th–beginning of the 20th centuries. *Historical-Economic Research, 4*, 667–687. https://cyberleninka.ru/article/n/sotsialnye-harakteristiki-prestupnosti-v-irkutskoy-i-eniseyskoy-guberniyah-v-kontse-xix-nachale-xx-vv

Gilinsky, Y. I. (2014). *Criminology: Theory, history, empirical base, social control* (3rd ed., revised and expanded). St. Alef-Press.

Greenwald, N. (1872). *Fur trade in Russia and abroad: Its history, statistics, fur processing, and commodity science*. Riga.

Gungwu, W. (2017).*Chinese encounters in Southeast Asia: How people, money, and ideas from China are changing a region* (P. Nyíri & D. Tan, Eds.). University of Washington Press. http://www.jstor.org/stable/j.ctvcwnrns

Institute of History SB RAS. (2009). *Historical Encyclopedia of Siberia* (Vol. 1–3). Siberian Heritage Publishing.

Khamina, N. V. (2017). Dynamics of demographic potential and settlement centers in the regions of Eastern Siberia. *MNIZH*, (4–1(58)). https://cyberleninka.ru/article/n/dinamika-demograficheskogo-potentsiala-i-tsentry-rasseleniya-v-regionah-vostochnoy-sibiri

Ko-lin Chin, H. (2003). *Organized crime, business and politics in Taiwan*. Armonk. https://doi.org/10.4000/chinaperspectives.563

Malenkova, A. A. (2014). The Soviet authorities' policy towards the Chinese diaspora in the Far East of the USSR in the 1920–1930s. *Problems of the Far East, 4*, 120–136.

Nikitin, N. I. (1990). *The development of Siberia in the 17th century: A book for middle and high school students*. Prosveshchenie.

Okladnikov, A. P. (1937). *Essays on the history of Western Buryats–Mongols (17th–18th centuries)*. Leningrad.

Ovchinsky, V. S. (2006). Mafia of the 21st century: Made in China. *Russia in Global Politics, 4*, 65–76.

Petta, D. L. (2017). As Tríades e as Sociedades Secretas na China: Entre o mito e a desmistificação. *Revista Brasileira de Ciências Sociais, 322017*(93), 2–15. https://doi.org/10.17666/329309/2017

Purbrick, M. (2019). Patriotic Chinese triads and secret societies: From the imperial dynasties, to nationalism, and communism. *Asian Affairs, 50*(3), 305–322. https://doi.org/10.1080/03068374.2019.1636515

Rai, A. K., & Biswajit, P. (2010). Degradation of Soil Quality Parameters Due to Coal Mining Operations in Jharia Coalfield, Jharkhand, India. *Journal of Advanced Laboratory Research in Biology, 2*(2), 51–56.

Rednikova, T. V. (2018). Organized Environmental Crime in Russia and Abroad: Criminological Characteristics and Countermeasures. *Lex Russica, 9*(142), 98–111. https://cyberleninka.ru/article/n/organizovannaya-ekologicheskaya-prestupnost-v-rossii-i-za-rubezhom-kriminologicheskaya-harakteristika-i-mery-protivodeystviya

Repetskaya, A. L. (2013). Foreign economic activity of Chinese organized crime groups in Eastern Siberia. *Prologue: Journal of Law, 2*, 53–59. https://cyberleninka.ru/article/n/vneshneekonomicheskaya-deyatelnost-kitayskih-opg-v-vostochnoy-sibiri

Rusalev, R., Rogozhnikov, D., Dizer, O., Golovkin, D., & Karimov, K. (2023). Development of a two-stage hydrometallurgical process for gold-antimony concentrate treatment from the Olimpiadinskoe deposit. *Materials (Basel, Switzerland), 16*(13), 4767. https://doi.org/10.3390/ma16134767

Shakherov, V. P. (2016). Trade entrepreneurship in the cities of Baikal Siberia in the 18th - first half of the 19th centuries. *Bulletin of Irkutsk State University. Series: History*. https://cyberleninka.ru/article/n/torgovoe-predprinimatelstvo-v-gorodah-baykalskoy-sibiri-v-xviii-pervoy-polovine-xix-v

Shutova, V. N. (2008). Criminological characteristics of the mechanism of committing crimes related to the illegal circulation of gold in Eastern Siberia. *All-Russian Criminological Journal, 3*, 25–29. https://cyberleninka.ru/article/n/kriminologicheskaya-harakteristika-mehanizma-soversheniya-prestupleniy-svyazannyh-s-nezakonnym-oborotom-zolota-v-vostochnoy-sibiri

Tatarnikov, V. G. (2016). On improving legislation on criminal liability for crimes in the field of forest protection. *Glagol pravosudiia*, (12), 40–42. https://cyberleninka.ru/article/n/o-sovershenstvovanii-zakonodatelstva-ob-ugolovnoy-otvetstvennosti-za-prestupleniya-v-sfere-ohrany-lesa.pdf

Van Uhm, D. P. (2023). *Organized environmental crime: Black Markets in gold, wildlife, and timber*. Bloomsbury Academic.

Van Uhm, D. P., & Wong, R. W. Y. (2021). Chinese organized crime and the illegal wildlife trade: Diversification and outsourcing in the Golden Triangle. *Trends in Organized Crime, 24*, 486–505. https://doi.org/10.1007/s12117-021-09408-z

Varese, F. (2011). *Mafias on the move: How organized crime conquers new territories*. Princeton University Press.

Volkov, V. (2016). *Violent entrepreneurs: The use of force in the making of Russian capitalism*. Cornell University Press. https://doi.org/10.7591/9781501703294

Weatherford, J. (2004). *Genghis Khan and the making of the modern world*. Crown.

White, R. (2011). *Transnational Environmental Crime: Toward an Eco-global Criminology* (1st ed.). Willan. https://doi.org/10.4324/9780203804001

Wing Lo, T., & Kwok, S. I. (2012). *Traditional organized crime in the modern world: How triad societies respond to socioeconomic change*. Springer US.

Yanti, I., Marliana, T., Anugrahwati, M., Wicaksono, W., & Winata, W. (2023). The comparison of gold extraction methods from the rock using thiourea and thiosulfate. *Open Chemistry, 21*(1), 20230102. https://doi.org/10.1515/chem-2023-0102

The Climate Change–Migrant Smuggling Nexus: Smuggling Services as the Only Way to Reach Europe

Manuel Cabezas Vicente

Climate Change as a Migration Push Factor

Since the advent of the Industrial Revolution, and particularly in the latter half of the twentieth century, human activity has become a primary driver of global change. This is the result of a confluence of factors, including rapid industrialization, accelerated deforestation, population growth, and the disproportionate emission of greenhouse

This research has been developed within the framework of the research project "Diagnóstico y evaluación del cumplimiento por el estado español del Pacto Mundial de Migraciones desde la perspectiva de género (Migration Pact)", funded by the Ministry of Science, Innovation and Universities for the years 2020–2023 (Ref. PID2019-106159RB-100), as well as the grant awarded by the Department of Education of the Junta de Castilla y León under the Order EDU/875/2021, of 13 July, which approves the call for grants to finance the recruitment of pre-doctoral research staff, co-financed by the European Social Fund.

M. Cabezas Vicente (✉)
University of Balearic Islands, Mallorca, Spain

CIDH Diversitas | University of Salamanca, Salamanca, Spain
e-mail: m.cabezas@uib.cat

D. van Uhm, D. Siegel (eds.), *Global Green Crime and Ecojustice*, Palgrave Studies in Green Criminology, https://doi.org/10.1007/978-3-031-91557-4_18

gases. As a result, some scientific disciplines have proposed a new geological epoch called the Anthropocene. In contrast to earlier periods in the planet's history, when human activity played a relatively minor role, humanity has now become a powerful geological force (Chakrabarty, 2009). This is evidenced by the fact that humans are not only embedded in nature but are also fundamentally shaping planetary systems. Anthropogenic climate change provides an illustrative example of this influence. It has been linked to rising atmospheric and oceanic temperatures, which in turn contribute to the melting of ice sheets, rising sea levels, and significant loss of biodiversity and ecosystem health (Clarke, 2014; Steffen et al., 2015). Consequently, the environment is no longer a secure planetary space for humanity; rather, human actions must be acknowledged as a substantial contributor to planetary instability (Holley et al., 2018). In light of these considerations, it becomes evident that humans can be regarded as geological agents, acknowledging their status as biophysical entities engaged in interactions with other biophysical entities. These interactions extend beyond the purely social realm (Shearing, 2015).

As indicated in the scientific basis of the 6th Assessment Report of the Intergovernmental Panel on Climate Change (IPCC), the global temperature has increased by approximately 1.09 °C since the pre-industrial period (1850–1900). This trend is projected to continue in the coming years, with an expected increase of more than 1.5 °C within the next 20 years. Historical data from the last 10,000 years show that temperature changes of 1 °C have rarely occurred (IPCC, 2023). Furthermore, even in an optimistic scenario in which greenhouse gas emissions are halted, the consequences of climate change could persist for centuries. The half-life of these gases in the atmosphere could span hundreds of years, further elevating atmospheric and terrestrial temperatures (Cabezas-Vicente, 2021). For this reason, the IPCC has developed a series of models based on projected future atmospheric GHG emissions and other anthropogenic factors, as measured by radiative forcing. In the most optimistic scenario for the year 2100, temperatures would increase by 1.4 °C compared to 1850–1900, while in the least favorable scenario temperatures would exceed 4.4 °C. However, the IPCC (2021) states that while 1.5 °C is almost certain to be reached, it is virtually impossible to exceed a thermal increase of 2 °C.

The temperature rise has a multitude of physical and biological impacts and consequences (Brisman, 2012, 2013, 2014, 2015), including sea level rise, glacial retreat, and ocean acidification. Additionally, there are indirect impacts resulting from the effects of the temperature rise on the lifestyles and economic conditions of various populations, particularly in rural communities. The consequences of climate change affect millions of people around the world, many of whom choose to leave their homes as a direct or indirect consequence. Others, who refuse or find it impossible to move, are forced to remain in areas with excessive pollution, which implies a violation of their fundamental rights and quality of life (IOM, 2019; Sanz Mulas, 2022a, 2022b).

As detailed in the IPCC 2023 report, the global average sea level rose by 20 centimetres (8 inches) between 1901 and 2018. This rise is primarily due to the thermal expansion of the oceans and the loss of ice from glaciers and ice sheets. Moreover, anthropogenic climate change is contributing to the occurrence of numerous extreme weather and climate events across all regions of the globe (IPCC, 2021, 2023). Such events include heat waves, heavy precipitation, droughts, and tropical cyclones, which are becoming more frequent and intense with each passing year. For example, the rise in sea level is causing the displacement of entire communities in the Pacific Islands, as evidenced by the cases of the States of Tuvalu and Kiribati (Brisman, 2013; Felipe Pérez, 2019; OHCHR, 2018; Schultz, 2017). For instance, some countries have chosen to purchase land in other countries due to the risk of losing their territory, which could lead to statelessness. To illustrate, Kiribati has procured 24 square kilometres of land in Fiji with the objective of establishing a potential resettlement area for its entire population (Bilimoria, 2020; Brisman et al., 2018). Furthermore, the incidence of droughts and floods is increasing, rendering settlements and crops unviable in numerous regions worldwide, including several African states and the Central American Dry Corridor. This may result in heightened competition for increasingly scarce resources, such as water and food, which could potentially contribute to conflict and migration. Additionally, it is anticipated that extreme meteorological events, such as hurricanes and tornadoes, will become more frequent and intense, with considerable implications for social, economic, and security sectors (Cabezas-Vicente, 2021).

The process of establishing a causal relationship between climate change impacts and displacement is inherently complex. As the International Organization for Migration (2019) notes in its World Migration Report 2020, the manner in which environmental impacts contribute to migration is complicated by the fact that the relationship between the environment and migration is rarely linear; rather, "economic, political, cultural and demographic factors interact with environmental drivers to shape intentions of people to move or stay in a given location" (p. 253). A survey conducted by the Mixed Migration Centre (MMC) revealed that environmental factors most frequently interact with economic drivers (59%) and conflict (22%). This illustrates the profound relationship between economic and livelihood losses and the underlying environmental and climate impacts.

In this regard, it is crucial to examine the interconnection between environmental and climate migration and the influence of social and economic factors. It is a verifiable fact that the populations most vulnerable to climate change are generally located in the Global South, a region that has paradoxically contributed the least to climate change, since about 75% of historical greenhouse gas emissions have been produced by the developed countries of the North, where only 20% of the world's population lives. These populations are dependent on the natural environment and have fewer economic resources to cope with its consequences. However, it is essential to consider rural populations and indigenous communities as agents of change and to recognize their traditional knowledge and ways of life as key elements in mitigating and adapting to climate change.

Notwithstanding, authors such as South (2010) and Brisman et al. (2018) have established a correlation between the most significant climate impacts and the most vulnerable populations. Therefore, as long as inequalities remain, the effects of disasters and their impacts will continue to be distributed unevenly. Consequently, wealthier countries possess the adaptive capacity to undertake costly projects aimed at mitigating the effects of climate change, thereby shifting the most adverse impacts to those who have contributed the least to them. As observed by Brisman et al.

> Global warming and climate change are predicted to impact the (im) mobility of those who have contributed the least to anthropogenic climate change. It is a particularly perverse form of climate adaptation that then emerges to enable the elite to enclose themselves in ways that remove them further from the relocating and dislocated masses. (2018, p. 10)

Notable adaptation projects that have been either implemented or funded by wealthier nations include the Clean Air Domes and "Airmageddon" initiatives in Beijing, the "Eko Atlantic" in Nigeria (Brisman et al., 2018), and the Mose Project in Venice. The latter project's objective is to safeguard Venice's coastline through the implementation of an integrated system of barriers with movable gates. These gates can be utilized to isolate the lagoon from the sea during high tide events, while the construction of breakwaters outside the harbor mouths is designed to mitigate the impact of more frequent tides (MOSE, n.d.).

One of the most salient characteristics of environmental and climate migration is its diversity. Thus, there are different forms of environmental migration, which can be classified into three main categories: (1) environmental emergency migrants, who temporarily flee due to a sudden disaster or event; (2) forced environmental migrants, who are forced to leave due to deteriorating environmental conditions or the gradual degradation of their surroundings; and (3) environmentally motivated migrants, who migrate to avoid future problems related to the natural environment, such as the loss of crop productivity due to desertification. Similarly, these migrants can be distinguished as either temporary or permanent, as well as short- or long-distance, or internal (the most prevalent) and transnational migrants (García Ruiz, 2019). In summary, the aforementioned categories of migration can be further distinguished according to their temporal and spatial characteristics.

It is notable that both slow degradation processes and sudden events directly affect populations, with two main trends emerging. Some communities choose to leave their homes, while the vast majority, who refuse or find it impossible to move due to material, psychological, cultural, and preference-based limitations, remain in areas where natural degradation represents a violation of their fundamental rights and quality of life (IOM, 2023). The loss of biodiversity contributes to the extinction of cultural traditions, as the impacts of climate change erode the traditions, cultures, and languages of indigenous groups that are intimately linked to their territories (Brisman et al., 2018).

Similarly, as Felipe Pérez (2019) has observed, slow degradation processes assume greater significance as environmental conditions decline over time. They exert an indirect influence on people through the economic and social transformations they engender, which in turn give rise to permanent migration.

It is important to acknowledge that a significant number of these movements are occurring within the rural communities themselves, driven by the desire of these populations to maintain their traditional way of life. However, other movements are from rural areas to urban centres in search of a superior quality of life, particularly in areas where the effects of climate change make it unfeasible to sustain the agricultural lifestyle characteristic of small communities. To a lesser extent, individuals are also crossing national borders in search of development opportunities, which is likely to accelerate urbanization and increase urban poverty (IOM, 2023; Méndez Barquero, 2017).

Future Prospection on Internal and International Climate Migration

Despite advancements in the collection and analysis of data and information on environmental and climate-related displacement, as well as in the formulation of future projections, the complex and multifaceted nature of displacement presents a significant challenge in discerning the environmental motives that drive migration. As the German Federal Environment Agency (UBA) has observed, "the various studies that have attempted to make quantified predictions about the future scale of environmental migration may be criticized due to methodological flaws" because "slow-onset environmental phenomena in future migration is subject to multiple uncertainties, given the multiple factors at play in decisions to migrate and in long-term development processes" (UBA, 2021, p. 11).

It is therefore not possible to make predictions about the future, and the analysis of such predictions is a challenging task. However, it is possible to quantify environmental displacement due to sudden events and natural disasters when it comes to slow degradation processes. The primary source typically utilized as a reference is the Internal Displacement Monitoring Center (IDMC), which exclusively quantifies internal displacement. In its latest report, published in 2023, the IDMC states that up to 8.7 million people were in this situation at the end of 2022, making it the leading cause of displacement globally, ahead of those displaced by conflict and violence (IDMC, 2023). Furthermore, the report indicates that 2022 represented the highest figure in the last decade, with a 41% increase compared to the annual average of the previous ten years (IDMC, 2023) (Fig. 1).

While a considerable number of individuals have already been displaced by climate change and environmental factors, and this number is likely to increase in the future, the hyperbolic media predictions of millions of climate migrants crossing continents are highly unlikely to materialize. In this regard, the MMC (2022a) asserts that the majority of individuals who relocate due to climate change and other environmental

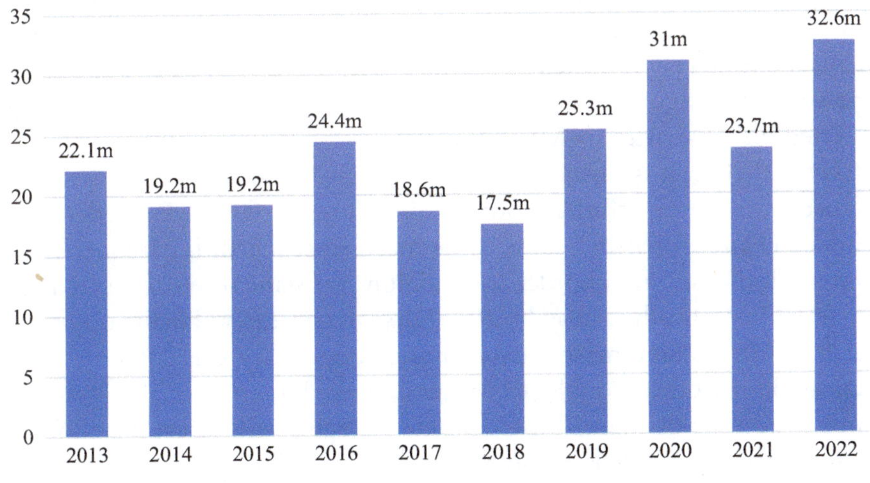

Fig. 1 Disaster displacements in millions (2013–2022). (Source: Own elaboration based on Internal Displacement Monitoring Center (2023))

factors tend to do so within their country of origin. Even if they do cross international borders, they tend to remain within their region. Therefore, the notion of imminent mass transcontinental migration is unsubstantiated (MMC, 2022a, p. 3). This can be attributed to two primary factors. On the one hand, individuals displaced by climate change tend to remain in locations with which they are familiar, where they can draw upon the support of a larger community. Consequently, they tend to utilize their resources as a coping mechanism. On the other hand, transregional and transnational migration requires significant financial resources for relocation, which the majority of individuals do not have, either due to lack of resources or depletion of resources to survive and avoid displacement (MMC, 2022a). This indicates that individuals will likely gravitate towards locations in closer proximity to their place of origin, particularly towards urban centres within their own region. Therefore, alterations in climate patterns will lead to an increase in urbanization and the expansion of secondary cities (Felipe Pérez, 2018; MMC, 2022a, 2022b).

Moreover, identifying the climatological catalysts behind global migrations often eludes displaced people. It is not uncommon for individuals to attribute displacement to immediate triggers such as loss of livelihoods or economic setbacks, thereby overlooking the underlying climatic variables that may be the proximate cause of these economic or livelihood losses. A survey conducted by the MMC revealed that only 1% of respondents identified environmental factors as the primary driver of migration. However, this figure increased to 18% when respondents were directly queried on the matter (MMC, 2022c).

Nevertheless, it is of the utmost importance to acknowledge the continued occurrence of international migration. As the adverse effects of climate change intensify and become more entrenched, the frequency of climate-induced cross-border population movements is expected to increase dramatically (Acevedo, 2014). In this context, as Huang observes, "as disasters become more recurrent and severe, they can combine with economic and other troubles to trigger increased international migration. And as impacts of slower-onset climate change events affect livelihoods, more people might look to move" (2023, para. 34).

In this regard, several reports have established a link between cross-border mobility and climate change. For example, the Platform on

Disaster Displacement and the Office of the United Nations High Commissioner for Human Rights published a report entitled "The slow onset effects of Climate Change and Human Rights Protection for cross-border migrants" (OHCHR, 2018). The report highlights that each region is confronted with distinct challenges to slow the degradation of natural resources, with significant implications for South Asia and the Pacific Islands. In these regions, migration is often a last resort for communities facing the loss of their territory due to sea level rise (Martin, 2010). Additionally, in Central America, both internal and international migration is anticipated to rise in tandem with the intensification of climate change impacts (OHCHR, 2018, p. 44). In the case of the Sahel, climate impacts have resulted in resource scarcity and conflict over their availability, which in turn has contributed to cross-border movements.

As indicated by the OHCHR, international mobility will persist as a prevalent phenomenon among those affected by drought and desertification, despite the various manifestations, as it represents a pervasive mechanism in rural areas. Consequently, international migration will be employed as an adaptation strategy to address the environmental impacts, although the rate of migration will decline in tandem with the deterioration of economic conditions and the depletion of resources. Specifically:

> During a 1968–1973 drought, one million people migrated from Burkina Faso to other countries in the region. In Senegal, migration that began internally expanded to nearby states and then Europe. [...] Cross-border migration may be viewed as a means to secure livelihoods in the face of slow onset events, with Maghreb countries (Egypt, Tunisia, Morocco, and Libya) often serving as destinations or transit points for further movement. (OHCHR, p. 36)

It is thus evident that international climate-related movements will persist as the detrimental impacts of climate change intensify, whether indirectly or, as is increasingly the case, directly. As Enríquez de Salamanca (2022) notes in an article on the influence of climate change, overfishing, and the global pandemic on irregular migration in West Africa, the increasingly adverse effects of climate change will cease to be additional drivers of migration waves and will become sufficient factors to drive international movements.

Consequences for Climate Change Migrants in Spain

Spain's geographical characteristics, including a land border with Morocco in Ceuta and Melilla and a mere 14 kilometres from the African continent by sea through the Strait of Gibraltar and the Canary Islands, have positioned it as a primary gateway to the European continent. As a result, it accounted for 20.4% of the total number of irregular entries into Europe in 2021, after rising from 32.8% in 2020 to 38.6% in 2018. Furthermore, recent years have witnessed a surge in the utilization of the West African route to the Canary Islands, with over 130,000 crossings recorded in the last five years (2020–2024). This represents a 3853% increase compared to the preceding ten years. In addition, the route has seen a significant shift of departure points to the south, with transits from Mauritania, Senegal, Gambia, or Cape Verde, which can last up to two weeks, increasing the potential for tragedy. This is now the most lethal migratory route to Europe, even more dangerous than those in the Mediterranean (Cabezas-Vicente, 2023, 2024).

The origins of migration to Spain are closely linked to the Sahel, a region with significant potential for development and abundant natural resources, yet which faces a multitude of challenges. These include political crises, conflict, violence, forced displacement, and a range of crises related to human needs, the environment, climate change, fishing, food security, and security (Van de Poël, 2024). The effects of climate change, as evidenced by prolonged droughts of up to five years, when combined with the region's valuable natural resources, create considerable tensions and conflicts over resource control, which in turn serve to further exacerbate displacement.

In this region, the prevalence of malnutrition and famine represents a major threat to the lives of more than 4 million children under the age of five, or 35% of the population in this age group. Climate change and conflict together make "the periods between harvests and plantings challenging, compromising essential food supplies and forcing vulnerable families to further restrict their daily meals" (Save the Children, 2022; Van de Poël, 2024, para. 14). The fishing sector is also significantly affected, and irregular migration associated with the sector is prevalent in various ports in Senegal, Gambia, Guinea-Bissau, and Mauritania (Bennett et al., 2020; Cissé, 2020). The combined impact of conflict, industrial overfishing, the absence of effective resource regeneration strategies, and the expansion of the fishmeal and fish oil industries is undermining traditional fishing, which represents the primary source of livelihood for many local communities (Mesa-Pérez et al., 2023).

The conflation of irregular migration, illegality, and criminality, which is predicated on the deployment of labels and myths about migration, gives rise to the implementation of securitizing administrative and penal measures. These can be examined from the theoretical perspective of "crimigration", which contributes to the stigmatization of the irregular migrant (van der Woude et al., 2014, 2017). Consequently, in response to the growth of international migration, receiving states (such as those in Europe or the United States) have increasingly implemented a range of border security measures that have ultimately led to the criminalization of certain practices.

Examples of securitization practices include the strengthening of police controls, the militarization of the management of migration flows, the restrictions around migrants' access to labour markets in destination countries, the lack of regular migration channels, including for work at all skill levels, education, family unity and humanitarian needs (OHCHR, 2018), the construction of large walls and the use of "concertinas"[1], or the strengthening of surveillance and control systems or the privatization of border services (López Sala & Godenau, 2017), In addition, the implementation of restrictive administrative regulations has been employed to regulate access to "regularization" statuses, and the use of criminal law has been expanded to punish virtually any form of collaboration with irregular migration (Arcos Ramírez & Díez Peralta, 2018; Brandariz García & Fernández Bessa, 2017; Daunis Rodríguez, 2017).

[1] Concertina wire is a type of barbed wire or razor wire that is placed on walls for defensive purposes.

Moreover, it is crucial to examine the detention of irregular migrants in alien detention centres (Martínez Escamilla, 2016; Seixas Vicente, 2021) and the extensive system of expulsions—both administrative and criminal—that legitimizes even collective deportations carried out without individualized procedures, commonly referred to as "hot deportations". The Constitutional Court approved Organic Law 4/2015, of March 30, on the Protection of the Security of Citizens. This law allows for the rejection of foreigners detected at the border of the territorial demarcation of Ceuta or Melilla while attempting to circumvent border control measures to gain illegal entry into Spain (tenth additional provision). Furthermore, there is no requirement to identify migrants individually, including those who may be seeking asylum, as the European Court of Human Rights deems expulsion as a consequence of irregular entry without the utilization of existing legal procedures.

It is also noteworthy to mention the Spanish and European policy of border externalization, which entails the management of migration in the countries of origin as a deterrent measure. The externalization of migration control typically manifests as economic agreements with border countries, aimed at strictly regulating migration and preventing it from leaving the state. A case in point is the agreement between the EU and Turkey in 2016, which was reached in the context of the significant influx of Syrian refugees. In exchange for 6 billion euros, the European Union agreed to contain the entry of refugees into its territory, thereby controlling the external borders of the continent (Garcés Mascareñas & Sánchez Montijano, 2017; Garcés Mascareñas, 2020). Similarly, Spain has concluded comparable economic agreements with states such as Morocco, Mauritania, and Senegal (Garver-Affeldt & Mackenzie, 2021). These countries represent the primary sources of migration in the Sahel region, where climate is a pivotal factor.

This practice includes development cooperation agreements with the countries of origin of migrants, in return for which they are obliged to strengthen controls to prevent the departure of people across their borders. Similarly, it is common for receiving countries to conclude agreements with countries of origin for the training of their border control officials, in addition to the intervention of the security forces and agencies of the receiving countries in patrolling and monitoring the territory of the countries of origin. As Anderson (2011) shows, Frontex uses the term deterrence in reference to patrolling and intercepting vessels. The patrols are justified on the basis of the "visibility" they provide to "potential migrants", on the assumption that if they were to withdraw their operations from areas of migration origin, irregular crossings would multiply. A notable example is the three "Africa Plans", a strategic initiative signed by Spain with several African countries to promote trade relations and improve their economies, while also incorporating measures to address climate change. In the III Africa Plan, Spain identifies climate change, desertification and biodiversity loss as major environmental threats. It recognizes the link between irregular migration and the effects of climate change, and commits to providing scientific and technical training to

> promote sustainable practices in fisheries and agriculture. The plan also aims to study and mitigate the effects of climate change and to cooperate in areas such as waste management, which remains a pressing issue, particularly in West African countries. Spain could also offer its experience and scientific and technological expertise to facilitate the sustainable development of emerging economic sectors, including those related to the blue economy. (Ministerio de Asuntos Exteriores, Unión Europea y Cooperación, 2019, p. 38)

This practice encompasses development cooperation agreements with the countries of origin of migrants. In return for these agreements, the countries of origin are obliged to strengthen border control measures to prevent the departure of people across their borders. Similarly, it is common for receiving countries to enter into agreements with countries of origin for the training of their border control officials, and for host countries' security forces and agencies to participate in patrolling and monitoring the territory of countries of origin. As demonstrated by Anderson (2011), the term "deterrence" is employed by Frontex in the context of patrolling and intercepting vessels. The rationale for the patrols is based on the assertion that they enhance the visibility of potential migrants, thereby deterring irregular crossings. This assumption is predicated on the premise that if Frontex were to withdraw its operations from areas of migration origin, the number of irregular crossings would increase. One noteworthy illustration is the three "Africa Plans", a strategic initiative signed by Spain with several African countries. The objective of this initiative is threefold: firstly, to promote trade relations and improve the economies of the aforementioned countries; secondly, to incorporate measures to address climate change; and thirdly, to facilitate the incorporation of environmental concerns into the economic development of the countries in question. In the III Africa Plan, Spain identifies climate change, desertification, and biodiversity loss as significant environmental threats. Furthermore, the document acknowledges the correlation between irregular migration and the consequences of climate change. In this regard, it expresses Spain's commitment to providing scientific and technical training to

> promote sustainable practices in fisheries and agriculture, which is also identified as a key objective. Moreover, the plan aims to examine and address the impacts of climate change and to collaborate on matters such as waste management, which remains a significant concern, particularly in West African countries. Spain could also offer its experience and scientific and technological expertise to facilitate the sustainable development of emerging economic sectors, including those related to the blue economy. (Ministerio de Asuntos Exteriores, Unión Europea y Cooperación, 2019, p. 38)

Notwithstanding the existence of agreements designed to promote the cooperation in development, the strengthening of economic structures, the promotion of sustainable development, and the mitigation of climate change, these measures are often conditioned on the regulation of irregular migration in countries of origin and transit (Dudek & Pestano, 2019; Vives, 2016). This approach has led some authors to characterize it as a new form of colonialism, which serves to consolidate influence over these states. In accordance with this plan, new embassies were established, and substantial financial resources were allocated to economic development initiatives. Concurrently, patrol units of the Spanish Guardia Civil were deployed in Senegal (Dudek & Pestano, 2019; Guillén Pérez, 2011) and Mauritania with the objective of searching for migrant boats (Hernando de Larramendi & Planet, 2009).

Smuggling Services as the Only Way to Reach Europe

The intensifying hostility and border closures have not succeeded in deterring migrants from their need for mobility. In the absence of information, material resources, and technical capabilities, migrants are compelled to seek the services of smugglers (Frontex, 2022), who offer them an alternative to irregular movement in exchange for an economic consideration. In the absence of established legal channels for migration and the reluctance of receiving states to accept irregular migrants, smuggling becomes a necessity. This is corroborated by the findings of Europol and Interpol (2016) and, more recently, Frontex (2022) or the MMC (n.d.), which indicate that over 91% of irregular migrants arriving in the European Union rely on smugglers at some point during their journey. This serves to reinforce the perception of the European Union as a fortress.

In this manner, the escalating number of international migrants, which is exacerbated by climate change, encounters a significant obstacle in Europe's stringent policies, which jeopardize the safeguarding of their fundamental rights and endanger the lives of migrants. This is because, in their pursuit of profit, smugglers frequently prioritize reducing the cost of transfers to the detriment of the means used to achieve it, which can result in significant risks to human health and safety. In this regard, the United Nations Office on Drugs and Crime (UNODC) (2018) has reported that on most smuggling routes, boats and vehicles are poorly maintained, overloaded, defective, and even lack sufficient fuel to reach their destination. Indeed, numerous cases have been identified where vehicles and boats only had the fuel necessary to reach European soil. This is further compounded by the lack of adequate provisions for sustenance, hydration, and hygiene during the transfer process, which has resulted in a considerable number of fatalities due to dehydration (UNODC, 2018; Fargues et al., 2020; Garver-Affeldt & Mackenzie, 2021). Notwithstanding the migrants' consent to the relocation, approximately half of the MCC respondents destined for Europe indicate that they feel partially or entirely misled, particularly with regard to the ultimate conditions of the move, as illustrated in Fig. 2. Nevertheless, it is demonstrated that this business has emerged as a means of compensating for the absence of regular routes due to border restrictions. An analysis of the MMC surveys (n.d.) reveals that the use of smuggling networks as an alternative means of transportation is a prevalent practice. As illustrated in Fig. 3, the majority of respondents who were traveling to Europe indicated that they were helpful and useful in reaching their destination.

In instances of irregular arrivals across the Mediterranean and Atlantic Ocean, it is not uncommon for smugglers to appoint migrants as captain of the vessel in order to circumvent detection on European soil (Frontex, 2020). In exchange for assuming this role, migrants, colloquially referred to as "skippers", are typically exempt from bearing the financial burden of the journey. Ultimately, the "skipper", who is revealed to be a mere migrant, is prosecuted for the crime of migrant smuggling. In contrast, the actual members of the criminal network evade prosecution due to the challenges of pursuing cases in their countries of origin.

Nevertheless, the effort to combat smuggling networks, perceived as a menace to the security of the European Union due to their substantial

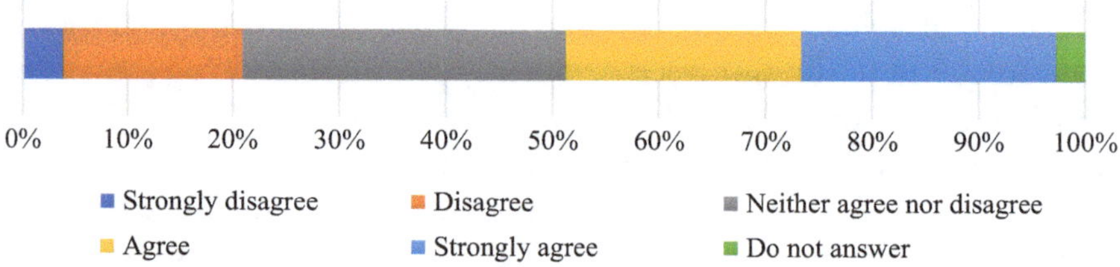

Fig. 2 Degree of agreement with the statement: "I felt cheated by the smuggler". (Source: Own elaboration based on MMC (n.d.))

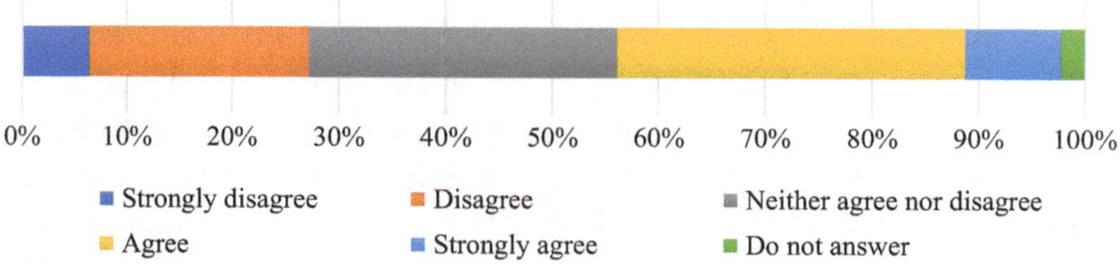

Fig. 3 Degree of agreement with the statement: "The smugglers helped me to achieve my goal". (Source: Own elaboration based on MMC (n.d.))

profits in the illicit market and their purported connections with other criminal organizations, has consistently exhibited a pronounced criminal justice orientation. This approach has resulted, however, in the neglect of the protection of migrants' rights. The criminocentric approach evident in the recently revised EU Action Plan against the Smuggling of Migrants (2021–2025), in the activities of the European Center for the Fight against the Smuggling of Migrants, and in the numerous military operations (such as Indalo or, previously, Irini or Sophia) is inadequate and, in fact, detrimental to the lives of migrants.

In general, as controls on a route become more effective and restrictions increase, movements shift to new routes in order to circumvent the controls. However, as the majority of research has demonstrated, alternative routes frequently prove to be longer and, consequently, more perilous than the original routes, thereby heightening the risks to the safety, health, and lives of migrants (Fargues & Bonfanti, 2014; Godenau, 2012; Grant, 2011; McAuliffe & Lackzo, 2016; Monreal Gainza & Paredes, 2021; UNODC, 2018; Triandafyllidou & McAuliffe, 2018).

In any case, in addition to punishing the migrant smuggling as a business, European legislation also criminalizes any facilitation of illegal entry, whether for economic or altruistic purposes (with the exception of facilitating illegal residence, which is only punishable if it is for profit). Consequently, numerous maritime rescue NGOs and fishermen in the region have elected to refrain from operating in Mediterranean waters, due to concerns that their actions might constitute a criminal offence. The deterrent effect of criminal law has resulted in numerous deaths at sea and has shaped the phenomenon of "solidarity crimes".

In light of the growing criminalization of humanitarian assistance, in accordance with the stipulations of the New Migration and Asylum Pact and in alignment with international conventions on the law of the sea, the European Commission and the European Parliament have collectively determined that criminal penalties for such actions are unwarranted. In this regard, it has been observed that, even though the cause of humanitarian exclusion is not explicitly provided for in the majority of states, such assistance is nevertheless covered by international law of the sea and customary law.

It must be acknowledged that even if international law does indeed provide coverage and exoneration is incorporated, this does not prevent the commencement of criminal proceedings against humanitarian aid actions. Furthermore, when one considers that the burden of proof lies with the accused, it becomes evident that the exclusion on humanitarian grounds does not preclude the deterrent effect of criminal law. This effect has led many non-governmental organizations (NGOs) to refrain from acting, due to the fear of being subjected to prosecution. In this context, the 2013 humanitarian catastrophe off the Italian coast of Lampedusa, in which over 500 individuals perished when a migrant boat was shipwrecked (Laurence, 2013), served as a significant motivating factor for the deployment of Operation Mare Nostrum. In the context of a humanitarian emergency, NGOs and fishermen were reluctant to provide assistance due to concerns about potential criminal liability for migrant smuggling, given the precedent set by previous prosecutions in Italy for similar incidents. Consequently, Italian migration policy and legislation were indirectly responsible for this fateful event (Cabezas-Vicente, 2024).

Accordingly, since May 2017, 38 vessels from different organizations have been operating in the Mediterranean, 23 of which have been the subject of legal action, representing 60.5% of the total (35 proceedings). In addition to these legal proceedings, European authorities have restricted the access of ships to their ports, resulting in delays in disembarkation and leaving individuals at sea for extended periods. The *Sea-Watch 4* ship, carrying 145 individuals, spent up to 14 days at sea before landing in Italy, while the *Aquarius* ship, after 10 days and the rejection of Italy and Malta, ultimately arrived in Spain (Cabezas-Vicente, 2024).

These factors collectively contribute to the criminalization of irregular migration, as Daunis Rodriguez correctly asserts.

> Ultimately, the message conveyed is one of clear criminalization of immigrants. Indeed, by imposing penalties, imposing sanctions, or persecuting those who assist immigrants in entering and/or remaining in Spanish territory illegally, the message is conveyed to society that irregular immigration is a negative and harmful phenomenon, to the extent that criminal law is employed to control it. Accordingly, it can be reasonably inferred that the primary perpetrator of this purported malevolence is the irregular immigrant himself. (2021, p. 785)

Conclusions

The consequences of an uncontrolled economic and production system, which results in a significant release of greenhouse gases into the atmosphere, have contributed to climate change with significant environmental and human consequences. Therefore, in contrast to previous eras in which humans were considered a mere component of the geological system, humanity has now become a powerful geological force with the capacity to influence planetary systems. For this reason, we may designate a new geological era, the Anthropocene, in which human impacts—particularly those occurring since the Industrial Revolution—contribute to planetary instability, such as by inducing anthropogenic climate change.

As a result of climate change, significant numbers of people are being forced to abandon their homes, either within their own countries or to flee to third countries. For example, the rise in sea level is responsible for the loss of territories and entire countries in the Pacific Islands, while the effects of desertification and drought are driving significant population movements in Africa. As the effects of climate change intensify, migratory movements will become more prevalent, occurring both within countries and across international borders.

In recent years, there has been a notable increase in irregular crossings via the West African route to the Canary Islands in Spain. The length of the maritime route and the challenging weather conditions encountered in the Atlantic Ocean render these crossings extremely dangerous. Migrations along this route are closely associated with the various crises confronting countries in the Sahel, including economic inequalities, political crises and coups, conflicts, violence, and, notably, the impacts of climate change, overfishing, and food insecurity. The prolonged droughts currently afflicting countries such as Niger have a significant impact on the economy and livelihoods, while also intensifying conflicts driven by resource scarcity, including those concerning water and food. Moreover, while climate change impacts do not directly cause displacement, they combine with factors such as violence or economic pressures to do so. Scholars emphasize that the increasingly adverse effects of climate change may soon shift from additional contributors to migration to primary drivers of international migration.

However, Western states, which receive only a small fraction of the migrant population and have contributed the most to climate change, view irregular migration as a national security problem. This has led to the fortification of external borders based on a supposed fight against human smuggling at all costs.

A criminocentric approach not only contravenes the overarching duty of solidarity in the reception of European states but also exerts an influence on the structural causes that motivate these migratory movements. It is evident that Spanish and European security policies do not deter migrants from attempting to migrate. Rather, they compel them to seek alternative entry routes. The confluence of the Mediterranean Sea and the Atlantic Ocean constitutes a substantial impediment to land-based migration along Spain's southern border. In addition, the increased restrictions on the land routes through Ceuta and Melilla have led to a reliance on smugglers for those attempting to reach Spain's borders. Smuggling services use small vessels, such as small boats or dinghies, which often lack the necessary safety and hygiene measures. The confluence of multiple factors—the duration of the journey, the severity of the weather conditions, and the lack of adequate safety and hygiene measures at the crossings—has resulted in the Mediterranean and Atlantic becoming some of the largest mass graves on the planet. In this manner, the decisions made by states contribute, to a greater or lesser extent, to the exacerbation of the alarming loss of human life at sea (Cabezas-Vicente, 2024). These policies illustrate how economic considerations supersede the fundamental rights of migrants.

Consequently, the Spanish legislature has chosen to prioritize the enforcement of criminal law over the protection of victims' rights, effectively utilizing the criminal justice system as a migration policy tool for border closure. This approach has resulted in the prosecution of individuals or organizations engaged in any form of assistance or collaboration with those attempting to enter the country illegally. Thus, even humanitarian relief organizations, whose actions are protected by international law, are subject to prosecution. In essence, the imposition of penalties on the migrants themselves and their entire social context makes it challenging, if not impossible, for them to gain entry and become integrated into the country.

In summary, the current approach of viewing irregular migrants, including climate migrants, as a threat to the system and implementing restrictive policies to deter their entry does not guarantee a reduction in irregular migration. Furthermore, it has the unintended consequence of increasing the vulnerability of migrants by fostering a climate conducive to clandestine migration. It is therefore evident that the most prudent course of action would be to address the underlying structural causes that drive migration and the obligation of solidarity in the reception of migrants by the countries of destination. This is particularly pertinent given that many of these migrants originate from developed countries, where social inequality, economic disadvantage, climate change and armed conflict are among the key drivers of migration.

References

Acevedo, P. (2014). *Desplazados ambientales, globalización y cambio climático: mirada desde los derechos humanos y de los pueblos indígenas*. Observatorio Ciudadano.

Anderson, R. (2011). Frontex y la creación de la frontera euroafricana: Golpeando la valla ilusoria. *Revista de derecho migratorio y extranjería, 28*, 177–191.

Arcos Ramírez, F., & Díez Peralta, E. (2018). Control de fronteras y derechos humanos en el Mediterráneo. *Deusto Journal of Human Rights, 3*, 13–47.

Bennett, N. J., Finkbeiner, E. M., Ban, N. C., Belhabib, D., Jupiter, S. D., Kittinger, J. N., Mangubhai, S., Scholtens, J., Gill, D., & Christie, P. (2020). The COVID-19 pandemic, small-scale fisheries and coastal fishing communities. *Coastal Management, 48*(4), 336–347.

Bilimoria, N. N. (2020). *Framework for climate migration readiness for Fiji for relocation of Pacific Islanders to Fiji*. Doctoral dissertation, University of Queensland.

Brandariz García, J., & Fernández Bessa, C. (2017). La crimigración en el contexto español: el creciente protagonismo de lo punitivo en el control migratorio. In A. López Sala & D. Godenau (Eds.), *Estados de contención, estados de detención: El control de la inmigración irregular en España*. Anthropos.

Brisman, A. (2012). The cultural silence of climate change contrarianism. In R. White (Ed.), *Climate change from a criminological perspective*. Springer.

Brisman, A. (2013). Not a bedtime story: Climate change, neoliberalism, and the future of the Arctic. *Michigan State International Law Review, 22*(1), 241–289.

Brisman, A. (2014). The visual acuity of climate change. In P. Davies, P. Francis, & T. Wyatt (Eds.), *Invisible crimes and social harms*. Palgrave Macmillan.

Brisman, A. (2015). 'Multicolored' green criminology and climate change's achromatopsia. *Contemporary Justice Review, 18*(2), 178–196.

Brisman, A., South, N., & Walters, R. (2018). Southernizing green criminology: Human dislocation, environmental injustice and climate apartheid. *Justice, Power and Resistance, 2*(1), 1–21.

Cabezas-Vicente, M. (2021). Cambio climático, conflictos y afección cultural. In A. Yurrebaso Macho (Dir), I. Seixas Vicente (Coord.) & M. Cabezas-Vicente (Coord.), *Políticas Públicas en defensa de la inclusión, la diversidad y el género IV: interculturalidad y Derechos Humanos*. Ediciones Universidad de Salamanca.

Cabezas-Vicente, M. (2023). Criminalizar la solidaridad: la política criminal europea frente al salvamento humanitario. In G. Esteban de la Rosa (Ed.), *Gobernanza multinivel de los movimientos migratorios: retos y perspectivas desde el derecho*. Dykinson.

Cabezas-Vicente, M. (2024). Políticas fronterizas y análisis de la mortalidad en la ruta migratoria de África Occidental. *Revista Vegueta, 24*(2), 1021–1049.

Chakrabarty, D. (2009). The climate of history: Four theses. *Critical Inquiry, 35*(2), 197–222.

Cissé, I. (2020, October 13). *As COVID-19 locks down West Africa, its waters remain open to plunder*. Greenpeace. https://www.greenpeace.org/africa/en/blogs/12395/as-covid-19-locks-down-west-africa-its-waters-remain-open-to-plunder/

Clarke, R. V. (2014). Technology, criminology and crime science. *European Journal on Criminal Policy and Research, 10*, 55–63.

Daunis Rodríguez, A. (2017). La inmigración ante la encrucijada: el tráfico ilegal de personas, la trata de seres humanos y la explotación sexual. In L. Zúñiga Rodríguez (Dir.) & J. Ballesteros Sánchez (Coord.), *Criminalidad organizada transnacional: una amenaza a la seguridad de los Estados democráticos*. Tirant lo Blanch.

Daunis Rodríguez, A. (2021). La criminalización de la inmigración como factor de la construcción del mito inmigrante-delincuente. In N. del Álamo Gómez (Dir.) & E. M. Picado Valverde (Dir.), *Políticas públicas en defensa de la inclusión, la diversidad y el género III: migraciones y derechos humanos*. Tirant lo Blanch.

Dudek, C. M., & Pestano, C. (2019). Canaries in a coal mine: The cayuco migrant crisis and the Europeanization of migration policy. *Revista Española de Ciencia Política, 49*, 85–106.

Enríquez de Salamanca, A. (2022). Influence of climate change, overfishing and COVID19 on irregular migration in West Africa. *Climate and Development*, 15(3), 215–228.

Fargues, P., y Bonfanti, S. (2014). When the best option is a leaky boat : why migrants risk their lives crossing the Mediterranean and what Europe is doing about it. MPC Policy brief, 5, 1–16.

Fargues, P., Rango, M., Börgnas, E., & Schöfberger, I. (Eds.). (2020). *Migration in West and North Africa and across the Mediterranean*. International Organization for Migration.

Felipe Pérez, B. (2018). *Migraciones climáticas: una aproximación al panorama actual*. ECODES.

Felipe Pérez, B. (2019). *Las migraciones climáticas ante el ordenamiento jurídico internacional*. Thomson Reuters Aranzadi.

Frontex. (2020). *Frontex risk analysis 2020*. Risk Analysis Unit.

Frontex. (2022). *Strategic risk analysis 2022*. Risk Analysis Unit.

Garcés Mascareñas, B. (2020). *Cuatro años del acuerdo UE–Turquía*. CIDOB opinión 617.

Garcés Mascareñas, B., & Sánchez Montijano, E. (2017). *El acuerdo UE–Turquía, un año después. Mucho más que externalización*. CIDOB opinión 473.

García Ruiz, A. (2019). Los olvidados de la movilidad humana: migración y desplazamiento de personas frente al desafío climático y medioambiental. *Revista Crítica Penal y Poder, 18*, 134–144.

Garver-Affeldt, J., & Mackenzie, S. (2021). *A gateway re-opens: The growing popularity of the Atlantic route, as told by those who risk it*. Mixed Migration Center.

Godenau, D. (2012). An institutional approach to bordering in Islands: The Canary Islands on the African-European migration routes. *Island Studies Journal, 7*(1), 3–18.

Grant, S. (2011). Recording and identifying European frontier deaths. *European Journal of Migration and Law, 13*, 135–156.

Guillén Pérez, M. I. (2011). Análisis de la migración irregular África-Canarias. *Miradas en Movimiento, 5*, 4–26.

Hernando de Larramendi, M., & Planet, A. I. (2009). *Documentos CIDOB. Mediterráneo y Oriente Medio 16. España y Mauritania: Sáhara, pesca, inmigración y desarrollo en el centro de la agenda bilateral*. CIDOB edicions.

Holley, C., Shearing, C., Harrington, C., et al. (2018). Environmental security and the anthropocene: Law, criminology, and international relations. *Annual Review of Law and Social Science, 14*, 185–203.

Huang, L. (2023). *Climate migration 101: An explainer.* Migration Police Institute.

IDMC (2023). *Global Report on Internal Displacement 2023.* International Displacement Monitoring Center.

IOM. (2019). *World migration report 2020.* International Organization for Migration.

IOM. (2023). *World migration report 2022.* International Organization for Migration.

IPCC. (2021). Summary for policymakers. In V. Masson-Delmotte et al. (Eds.), *Climate change 2021: The physical science basis.* Contribution of Working Group I to the Sixth Assessment Report of the Intergovernmental Panel on Climate Change. Cambridge University Press.

IPCC. (2023). Summary for policymakers. In H. Lee & J. Romero (Eds.), *Climate change 2023: Synthesis report.* Contribution of Working Groups I, II and III to the Sixth Assessment Report of the Intergovernmental Panel on Climate Change. IPCC.

Laurence, P. (2013, October 4). Lampedusa disaster: Europe's migrant dilemma. *BBC.* https://www.bbc.com/news/world-europe-24396020

López Sala, A., & Godenau, D. (2017). *Non-state actors and migration control in Spain. A migration industry perspective.* ISA Forum of Sociology, Vienna, Austria.

Martin, S. F. (2010). *Climate change and international migration.* The German Marshall Fund of the United States.

Martínez Escamilla, M. (2016). Centros de internamiento para extranjeros Estado de la cuestión y perspectivas de futuro. *Revista Electrónica de Ciencia Penal y Criminología*, 18–23, 1-38.

McAuliffe, M., & Lackzo, F. (Eds.). (2016). *Migrant smuggling data and research: A global review of the emerging evidence base.* International Organization for Migration.

Méndez Barquero, J. C. (2017). Cambio climático, movilidad humana y su impacto en las relaciones internacionales del siglo XXI. *Revista Relaciones Internacionales UNA, 90*(2), 1–27.

Mesa-Pérez, C. U., Parreño-Castellano, J. M., & Domínguez-Mujica, J. (2023). Border control and accident rate of irregular immigration in the route to the Canary Islands (Spain) during the COVID-19 pandemic. *Hungarian Geographical Bulletin, 72*(2), 101–117.

Ministerio de Asuntos Exteriores, Unión Europea y Cooperación. (2019). *III Plan África. España y África: desafío y oportunidad*. https://www.exteriores.gob.es/es/PoliticaExterior/Documents/2019_PLAN%20AFRICA.pdf

Mixed Migration Centre. (2022a). *Climate change, environmental stressors, and mixed migration*. Mixed Migration Centre.

Mixed Migration Centre. (2022b). *Climate-related events and environmental stressors' roles in driving migration in West and North Africa*. Mixed Migration Centre.

Mixed Migration Centre. (2022c). *Climate-related drivers of mixed migration in East and the Horn of Africa*. Mixed Migration Centre.

Mixed Migration Centre. (n.d.). *4Mi interactive* [Database]. https://mixedmigration.org/4mi/4mi-interactive/data-on-mixed-migration/

Monreal Gainza, B., & Paredes, C. (2021). *Migración en Canarias, la emergencia previsible*. CEAR.

MOSE. (n.d.). *Il sistema MOSE. Le barriere mobili per la protezione dalle acque alte*. https://mosevenezia.eu/progetto/

OHCHR. (2018). *The slow onset effects of climate change and human rights protection for cross-border migrants*. Office of the United Nations High Commissioner for Human Rights.

Sanz Mulas, N. (2022a). Suicidio ecológico e impunidad La urgencia de una justicia penal efectiva frente al desastre. *Revista Electrónica de Ciencia Penal y Criminología, 24*(15), 1–43.

Sanz Mulas, N. (2022b). Migraciones climáticas como "gendered procces" y política criminal sin perspectiva de género. Delineamientos para generaciones futuras. In Z. Sánchez Sánchez (Dir.) & M. Eiros Bachiller (coord.), *Regulación con prospectiva de futuro y de consenso: gobernanza anticipatoria y prospectiva administrativa*. Thomson Reuters Aranzadi.

Save the Children. (2022, May 12). *More than 1 in 3 children under five at risk of starvation in Central Sahel*. https://www.savethechildren.net/news/more-1-3-children-under-five-risk-starvation-central-sahel

Schultz, K. (2017, March 31). *As tourists flood the Maldives, so does the rising sea*. The New York Times. https://www.nytimes.com/2017/03/30/world/asia/maldives-tourism-beach-erosion.html

Seixas Vicente, I. (2021). Un viaje por los centros de internamiento de extranjeros. In N. del Álamo Gómez (Dir.) & E. M. Picado Valverde (Dir.), *Políticas públicas en defensa de la inclusión, la diversidad y el género III: migraciones y derechos humanos*. Tirant lo Blanch.

Shearing, C. (2015). Criminology and the anthropocene. *Criminology and Criminal Justice, 15*(3), 225–269.

South, N. (2010). The ecocidal tendencies of late modernity: transnational crime, social exclusion, victims and rights. In R. White (ed.), *Global Environmental Harm: Criminological Perspectives*. Willan Publishing.

Steffen, W., Richardson, K., Rockström, J., et al. (2015). Planetary boundaries: Guiding human development on a changing planet. *Science, 347*(6223).

Triandafyllidou, A., & McAuliffe, M. (Eds.). (2018). *Migrant smuggling data and research: A global review of the emerging evidence base* (Vol. 2). International Organization for Migration.

UBA. (2021). *Migration, environment and climate change—Final report*. UBA publikationen.

UNODC. (2018). *Global study on smuggling of migrants*. United Nations publication.

Van de Poël, M. (2024, April 23). *Desbordamiento de la crisis en el Sahel: la atormentada infancia carece de apoyo y protección*. Humanium. https://www.humanium.org/es/desbordamiento-de-la-crisis-en-el-sahel-la-atormentada-infancia-carece-de-apoyo-y-proteccion/

van der Woude, M. A. H., van der Leun, J. P., & Nijland, J. A. A. (2014). Crimmigration in the Netherlands. *Law & Social Inquiry, 39*, 560–579.

van der Woude, M. A. H., Barker, V., & van der Leun, J. P. (2017). Crimmigration in Europe. *European Journal of Criminology, 14*(1), 3-6.

Vives, L. (2016). The European Union—West African sea border: Anti-immigration strategies and territoriality. *European Urban and Regional Studies, 24*(2), 209–224.

Environmental Crimes in Ukraine Under War Conditions

Anna Markovska, Oleksii Serdiuk, Angus Nurse, and Iryna Soldatenko

Introduction

It is impossible to limit the impact of environmental crimes to just one locality. A disruption of the ecological balance in one part of the planet, whether seemingly significant or not, may materially impact the ecology of another.

A. Markovska (✉)
International Policing and Public Protection Research Institute (IPPPRI), Anglia Ruskin University, Cambridge, UK
e-mail: Anna.Markovska@aru.ac.uk

O. Serdiuk
Kharkiv National University of Internal Affairs, Kharkiv, Ukraine

A. Nurse
Anglia Ruskin University, Cambridge, UK

I. Soldatenko
Department of Applied Sociology and Social Communications, V. N. Karazin Kharkiv National University, Kharkiv, Ukraine

D. van Uhm, D. Siegel (eds.), *Global Green Crime and Ecojustice*, Palgrave Studies in Green Criminology, https://doi.org/10.1007/978-3-031-91557-4_19

The environmental impact of the Russian war in Ukraine goes far beyond Ukraine and has a global impact. The Centre for the Environmental Initiative (CEI)[1] in Ukraine conducted three investigations to analyse the consequences of the war on the environment (, De Klerk et al., 2023). The first investigation calculated potential environmental impact based on four factors: "emissions from the movement of refugees, emissions from warfare, uncontrolled fires in forests and cities, and future emissions from the reconstruction of damaged and destroyed buildings, roads, and factories" (De Klerk et al., 2023, p. 5). The report noted that "Green House Gas (GHG) emissions attributable to 18 months (or 555 days) of the Russian full-scale war in Ukraine amounted to 150 million tCO_2e. This is more than the annual GHG emissions of a highly industrialized country such as Belgium" (ibid., p. 5). The second report considered, among other issues, the impact of the war on the European and Ukrainian energy sector. While the third suggested a methodology for "mapping carbon emissions during an ongoing conflict" (ibid., p. 5), thus considering GHG emissions connected with military activity in various countries and specifically in Russia and Ukraine.

It is increasingly acknowledged that armed conflicts and management of natural resources are closely interlinked. The natural environment may serve as a cause, a means, or as a casualty of war. Brewster (2022) reports that the timing of the Russian invasions of 2014 and 2022 should not be considered as coincidental. Identified as a potential 'critical minerals superpower', in 2014 Ukraine engaged in talks with the "Shell Oil Co. and Chevron Corp. to develop its Black Sea reserves—plans that were scuppered because of Russia's actions" (Brewster, 2022, p. 1). In effect, the Russian war in Ukraine has not only caused catastrophic damage to atmosphere, water and soil, but also facilitated Russian state control of the natural resources of Ukraine. For example, by February 2024, occupying Russian forces had destroyed over 66,000 hectares of forest (EcoZagroza, 2024). According to Smal' (2023), about 30% of Ukrainian forest has been badly damaged, either completely destroyed or mined (Verbovska, 2024). In some cases, entire ecosystems have been destroyed

[1] A civil society organisation uniting experts and activists 'in a joint struggle to protect the environment' https://en.ecoaction.org.ua/.

(EcoZagroza, 2024). In such areas rich in natural resources, Russia is either using these resources itself or is employing a tactic of "occupy and deny" (Davies, 2023)—by occupying these Ukrainian territories Russia is denying Ukraine the use of its natural resources. By the end of 2023 Russia has seized "27% of Ukraine's iron ore deposits, 50% of its manganese ore deposits, 100% of its strontium supplies, and roughly 20% of its precious metal deposits" (ibid.). SecDev[2] calculates "the total worth of all Ukrainian mineral wealth now under Russian control to be around $12 trillion" (Muggah & Dryganov, 2022). "Even if Russia were to do nothing and sit on Ukraine's mineral resources and infrastructure, it benefits by controlling supplies and rising prices" (Muggah quoted in Davies, 2023).

The damage to the environment caused by ongoing warfare is vast, and the restorative action that will be required post-conflict far exceeds the domestic capacity of Ukraine. Consequently, both a national and an international response to the environmental crimes committed as a result of the Russian war in Ukraine will be required. The question arises as to how the evidence collected can be used to make the parties accountable for environmental catastrophe. A further controversial point concerns the contribution of modern warfare to climate change in general (Depledge, 2023). In 2022, as a result of Russia's war in Ukraine, SIPRI estimated global military expenditure to be $2.24 trillion, the highest-level recorded since the end of the cold war (Tian et al., 2023). Emissions resulting from military action reached a total of 37.0 million tCO2e (De Klerk et al., 2023, p. 7) with war-related fuel consumption by Russian forces contributing 24.9 million tCO2e to this total (ibid.).

Prior to the full-scale invasion, Ukrainian environmental activists campaigned to stop illegal natural resource exploitation in the country, such as logging and the extraction of amber in the West Ukraine (Earthsight, 2018; van Duyne et al., 2022) and informal mining in the East (Zabyelina & Kalczynski, 2020). These researchers argued that the presence of organised criminal networks in Ukraine (and in wider Europe), as well as endemic corruption among Ukrainian state officials and the judiciary, had prevented the effective control and punishment of those involved.

[2] SecDev is a global digital risk and resilience.

Research publications relied on evidence collected by a handful of investigative journalists working with civic organisations (Maksimov, 2018; Earthsight, 2018; Malko, 2018; Besser, 2020). Official police data, though available, was difficult to contextualise (Van Duyne et al., 2022). The fragile regulatory framework and corrupt local administrations complicated any realistic assessment of the situation on the ground. Consequently, powerful open-source tools enabled effective data gathering and analysis. Since the beginning of the war in Syria, and the start of the 2014 conflict in the Donbas region of Ukraine, various technological approaches have been used to document environmental crimes (Roca, 2017). Open-source information-gathering and investigations (OSINT) by organisations such as Bellingcat, have demonstrated their effectiveness in gathering evidence (Müller & Wiik, 2023), including data on the different ways in which warfare damages the environment (Zwijnenburg, 2017).

Considering the consequences of Russia's full-scale invasion and resulting damage to the Ukrainian environment, this chapter critically examines the existing constraints in fighting some forms of environmental crime and specifically protecting the environment of Ukraine. It further investigates global challenges in dealing with the consequences of such warfare.

Established in 2023, the International Working Group on the Environmental Consequences of War considers the following three factors: "Assessing the environmental consequences of war; recommendations for finding mechanisms to bring an aggressor to justice in this case so that Russia pays in full for the destruction it causes; and ecological restoration" (Office of the President of Ukraine, 2023). In this chapter we are concerned with the capacity to deliver the third of these objectives, namely ecological restoration. More specifically, we consider the capability and capacity of central and local Ukrainian authorities, as well as the role of the international community, in delivering environmental justice for all.

We review the growing body of literature on war and conflict studies within the filed of green criminology in order to critically address the following two issues: (i) the use of anti-corruption methodologies to analyse and investigate environmental crimes in a conflict situation; and (ii) the employment of international law to punish the Russian state for the environmental damage caused by its full-scale war.

Environmental Criminology, Conflicts and Wars

Over the years, the field of green criminology has afforded opportunities for multidisciplinary studies that are both critical and sensitive (South et al., 2013; Brisman & South, 2014; White, 2015; van Uhm, 2018; van Uhm & Nijman, 2022; Robinson, 2022). The investigation of environmental harm and crimes are increasingly considered through the lens of comparative politics to understand the ways in which wars and conflict affect the environment (Sowers & Weinthal, 2023). Brisman et al. (2016) propose a typology of conflict–environment relationships: namely, conflicts over natural resource possession, conflicts over declining resources, conflicts "over natural resource extraction process", and "conflicts that destroy the environment" (p. 2). Generally, these four types of conflict involve different motivations for the control of the use and exploitation of natural resources, in some cases escalating to conflicts that lead to the environmental destruction, often several or all the above factors can be observed within one location, such as that of the Russian war in Ukraine.

Conflicts that result in environmental destruction can be situated along a spectrum, with state crime against the environment at one end and the enemy action at the other. In the Russo-Ukrainian context, we are thus able to consider the crimes against environment committed by corrupt state authority of an invaded country on one side, and the crimes of an invading country on the other side. The role of the military in this continuum is both crucial and controversial.

Sowers and Weinthal (2023) analyse the impact of military operations on the natural environment and identify "the many ways in which the environment and natural resources are not only affected by war, but also are instrumental in conflicts" (p. 789). For example, the pollution of land due to military activity, the intensive consumption of natural resources by military-related industry, and the weaponisation of critical infrastructure. Drawing on a recent example, Sowers and Weinthal (2023) continue Ruggiero's (2010) discussion of war as corporate crime and discuss the privileged position of national security that results in secrecy and systemic lack of transparency. Sowers and Weinthal (2023) hold that the impact of war on the environment remains "understudied and rarely figures in public policy debates about the decision to make war", thus the "natural environment continues to be a casualty of war" and is used as a weapon (ibid., p. 776). Developing the concept of green criminology, Fichtelberg (2016) argues that "international war crimes law is particularly ill-suited for the realities of modern war" and suggests that the identification of "a new international crime ... that of the criminal exploitation of natural resources in wartime" is now very important (p. 177). Moreover, Fichtelberg (2016) suggests that war crime can, in effect, incorporate not only the environmental crimes in times of war, but also the related environmental crimes of the military industry.

Recently, the US Department of Defense (2021) published a Climate Risk Analysis, outlining the major challenges that climate change poses for US national security. However, it also contains an acknowledgement that "the environmental footprints of military installations and operations are difficult to understand". This is so for various reasons, including a lack of transparency (Sowers & Weinthal, 2023, p. 789). "The close relationships between militaries and associated industrial complexities that privilege secrecy and national security create a systemic lack of transparency that often results in environmental contamination and public health crises for marginalized regions and populations" (ibid., p. 789). Also to be considered in this context are the environmental cost of running the war (De Klerk et al., 2023) and the "indirect effects of the disposal of military equipment and ammunition once peace is secured" (Lawrence et al., 2015). The long-term impact of war on the environment is well illustrated by Hupy (2006), who discusses that in some areas of France even decades after World War II, one can still see deep craters that were produced by heavy artillery fire and make any vegetative growth and restoration impossible. Lawrence et al. (2015) argue that, due to new and emerging military technologies, it is important to continue the study of "war's impact on ecosystem structure and to assist with developing potential mitigation strategies to minimize negative consequences and implement effective rehabilitation and restoration approaches" (p. 445).

Ukrainian Forest: 'The Monopoly Management', Corruption and Wartime Emergencies

While the East of Ukraine is heavily industrialised, the north and the west of the country are predominantly richly forested. The total forest area of Ukraine is 10.4 million hectares (SAFR, 2022). About 87% of these forests are state-owned, and the remaining 13% are owned communally. Private forests amount to only 0.1% of the total (SAFR, 2022). Two state institutions manage most of the forests in Ukraine: the Ministry of Ecology and Natural Resources of Ukraine (SAFR, 2022), and the State Agency of Forest Resources of Ukraine. The Ministry of Ecology and Natural Resources of Ukraine (SAFR, 2022) is the principal government body overseeing environmental policy, including forestry, and is responsible for the development and implementation of state policy regarding the protection, use, and restoration of forests. The State Agency of Forest Resources of Ukraine (SAFR, 2022) is a specialised agency that is directly responsible for the management and exploitation of the country's forest resources. Shumilo et al. (2023) report on some positive developments in connection with the Emerald Network,[3] stating that "despite the war, territories remaining under Ukraine control after 2014 continued reforestation (+9%), while sites under Russian control experienced dramatic forest loss (-25%)" (p. 2). However, most researchers discuss corruption, lack of transparency, and accountability in discussing these strategies. For example, Earthsight (2018) infamously reported that "field investigations indicate that 40 per cent of the timber being produced by the country's state-owned enterprises is illegally cut through the abuse of a loophole allowing trees to be harvested to prevent the spread of disease" (p. 47). This revelation prompted a number of European Union (EU) countries and Ukrainian officials to promise an urgent investigation into the management and control of forest enterprises in Ukraine. Official data from the State Inventory of Greenhouse Gases (MEEP, 2018)

[3] Emerald Network is founded by the EU as a conservation network. Emerald Network of Areas of Special Conservation Interest—Convention on the Conservation of European Wildlife and Natural Habitats (coe.int).

suggest that in 2018, "for the first time since independence, emissions from agriculture exceeded the amount of carbon stored by forests". Earthsight (2018, 2019) further reported the systemic nature of corruption and abuse of power amongst those who are responsible for the protection of the Ukrainian forestry. While Hrynyk et al. (2023) support the Earthsight findings and argue that the biggest threat to forests comes from corrupt documentation and management in Ukraine which sabotages most anti-corruption initiatives. Matsala et al. (2024) study the effect of war on forests in three areas of Ukraine, observing that "the state of these forests prior to the war and the legacy of their recent management might have increased their vulnerability to war-related disturbances" (p. 5). For example, forest fires in the Luhansk region in 2020 resulted in extensive damage and revealed significant flaws in the emergency response system. The State Bureau of Investigation concluded that employees of the State Emergency Service failed to properly carry out their duties during the firefighting efforts, and had falsified documents regarding the fire suppression leadership, the forces and resources allegedly involved (SBI, 2021). Key violations included a lack of coordination between various units, failure to establish a firefighting headquarters and a fire-fighting strategy, and the neglect of essential fire prevention measures such as creating firebreaks and initiating counterfires (ibid.). The consequences of the fires were catastrophic, with two forestry departments in the Luhansk region suffering damages amounting to over 5 billion hryvnias (RBC-Ukraine, 2022).

Boreiko (2023) writes about the loss of qualified specialists from the Ministry of Ecology. While the reasons are complex, the author singles out the Cabinet of Ministers of Ukraine decision to merge two completely distinct entities—the Ministry of Ecology and the Ministry of Energy and Environmental Protection of Ukraine (2019–2020)—into a single ministry, despite their fundamentally different mandates. This restructuring occurred without any public discussion. As a result, the Ministry of Ecology was virtually inactive for nearly a year, leading to the collapse of several state environmental programmes and initiatives (Boreiko, 2023).

The full-scale Russian invasion of Ukraine has presented unparalleled challenges to the environment: on one side, the forestry management system is already compromised by systemic corruption and on another enemy forces are destroying the forests, turning them into the battlefields. While in the East the occupying forces have been destroying the forests in order to build military infrastructure and increase revenues generated by illegal timber sales (Tril, 2024), the Ukrainian territories in the West of the country continue to experience the corrupt management of natural resources (for example, in March 2024 the State Bureau of Investigation [SBI] reported the arrest of a criminal group including the directors and chief engineers of five state-owned forestry enterprises in the Lviv region who illegally cut down trees worth more than UAH 64 million). This was reported by State Bureau of Investigation (2024). Moreover, following the full-scale invasion, the Ukrainian authorities held on a moratorium on inspections by the SEI (Decree of the Cabinet of Ministers of Ukraine No 303, 2022). The moratorium made a controversial decision "to restrict public access to previously publicly available documents and data on logging (forest maps, logging plans, permits, results of SEI inspections etc.)" (Hrynyk et al., 2023, p. 37). Russian attacks on Ukrainian critical infrastructure have "significantly increased the country's need for firewood" (ibid., p. 37). In 2022, the Head of the State Agency for Forest Resources admitted the need to increase the volume of logging in those territories still under Ukrainian control (Epravda, 2022). For example, Kurkov (2024) explores why rural residents need more firewood and wartime wood controversies around. Moreover, the military requirements for timber have grown significantly (from construction of fortification to heating and cooking) (Tkachuk, 2023). A combination of the above factors has created a very fragile and dangerous situation where the abuse of natural resources could go undetected under cover of emergency legislation. Solving these problems by a "collective collaborative approach" might prove a way forward. For example, Myroniuk et al. (2020) discuss ways to assess the damage to state forests and thus control associated illegal activities (for example, they write that "uncontrolled amber mining causes huge environmental damage and has contributed to the vast degradation of forest stands" [ibid.]). They employed an innovative approach to calculate forest damage. "We used virtual reality (VR) 360 interactive panoramic images taken from the sites to attribute four levels of forest disturbance associated with the delta normalized burn ratio (dNBR) and then calculated the carbon loss. We revealed that illegal amber extraction in Ukraine has been occurring since the middle of the 1990s, yielding 3260 ha of total disturbed area up to 2019. Consequently, were able to calculate that the area of forest disturbance increased dramatically during 2013–2014, and illegal amber operations persist" (pp. 6–7). The new ways of measuring damage represent an important step toward greater transparency and better understanding of how natural resources are managed and impacted.

Anti-corruption Methodology and Open-Source Investigation/Intelligence (OSINT)

Le Billon (2013) explores the pivotal role of the political geography of natural resources, asserting that "the ability of governments to exercise legitimate and effective control over resource revenue" is a key determinant in shaping the dynamics of conflict (p. 35). Le Billon further asserts that a comprehensive understanding of the resource environment and its management can provide valuable insights into the underlying causes of conflict. Concurrently, green criminologists have drawn attention to issues of transparency and accountability, particularly in relation to the military-industrial complex and its environmental impact (Sowers & Weinthal, 2023), as well as state-owned corporations and their accountability practices (Ruggiero, 2010). Lemaître (2023) acknowledges that significant progress in addressing the issue of corruption in relation to environmental crime was reached globally within the last five years.[4] This includes a Statistical Framework by which to measure corruption pub-

[4] Lemaître (2023, p. 9) singles out the following three resolutions: Resolution 8/12 adopted by the United Nations Convention against Corruption in 2019, "underlies the importance of addressing corruption linked to crimes that have an impact on the environment. It urges States parties to prevent, investigate and prosecute corruption offences where they may be linked to crimes that have an impact on the environment" (p. 9); Resolution 10/6 on Preventing and combating crimes that affect the environment falling within the scope of the United Nations Convention against Transnational Organized Crime, adopted in December 2020; Resolution 76/185—Preventing and combating crimes that affect the environment adopted by the United Nations General Assembly in 2021.

lished in 2023 by the United Nations Office on Drugs and Crime (UNODC, 2023). Despite these advancements, a considerable obstacle remains in the challenging of bad governance and corrupt organisational cultures, "strengthening integrity and ethical practices within governmental institutions and agencies tasked with preventing crimes that have an impact on the environment is essential for building strong institutions that can effectively tackle corruption linked to the environment" (Lemaître, 2023, p. 14).

The anti-corruption methodology developed by Klitgaard (2000) and the recent rethinking on anti-corruption strategy by Mungiu-Pippidi (2023) offer measures that can be applied within green criminology. There are intersections between the fast-moving OSINT, and its potential to improve evidence collection, accountability, and subsequent prosecution. Klitgaard et al. (2000) analysed the contexts which give rise to corruption with the formula $K = M+D-P/P$, where K is corruption, M is monopoly of execution, D is discretion (broad decision-making power), and P/P is accountability/transparency. Klitgaard (2015) considers progress in the control of corruption over a broad perspective, ranging from a change in attitude towards corrupt practices, to the development of a new anti-corruption culture (p. 33). Over the two years from 2018 to 2020, Klitgaard's methodology (the Islands of Integrity™ Anti-Corruption Methodology, United National Development Programme, 2019) was successfully tested in Ukraine within several local administrations in the Luhansk and Donetsk regions. Both Klitgaard et al. (2000) and Mungiu-Pippidi (2023) defend such a contextual explanation of corruption. The critical aspect of this approach lies in finding mechanisms to challenge the power of established monopolies and entrenched corrupt practice. For example, as discussed earlier, the Ukrainian state has a monopoly in the management of forests in Ukraine, a situation that can create sufficient conditions for corrupt practices at the level of both state and local government. Such centralised power results in increased bureaucracy and a consequent heightened risk of corruption.

Zwijnenburg and Ballinger (2023) note that the use of open-source data in assessing the damage generated by war has democratised environmental research. Investigative journalism has been transformed, contributing to the development of new methodologies by which to study the impact of environmental crime using unobtrusive methods (Müller & Wiik, 2023). Posthuma et al. (2014) consider the Flash Environmental Assessment Tool (FEAT) used to "support disaster assessment and coordination actions by United Nations Disaster Assessment and Coordination (UNDAC)" (p. 1). In 2017, FEAT was used to study the impact of war on highly industrialised areas of Donbas in order to "highlight the likely environmental threats and hazards associated with damage to these locations" (Zwijnenburg, 2017, p. 2). The environmental consequences of war were well described in a World Bank report, specifically detailing the high concentration of the heavy industry in the East: "The Donbas area was estimated to host about 900 large industrial plants, including 140 collieries, 40 metallurgical plants, 7 thermal power stations, and 177 chemically dangerous operations, including 113 using radioactive materials. Environmentally, the most harmful industry of these is mining. There are 248 mines, many of which are run-down or nonfunctioning. In addition, the region is also traversed by 1230 kilometres of oil, gas, and ammonia pipelines. By 2002, an estimated 10 billion tons of industrial waste had accumulated in Donbas, equivalent to a total of 320,000 tons per square kilometre" (RPA, 2015, p. 12). Zwijnenburg (2017) used open-source data such as the Wikimap online data base Evard Toxic Sites and Identification program, to identify and map "potential environmental hotspots" in Donbas and argued that the region is a "ticking toxic timebomb". RPA (2015) stressed an urgent need to reinstate environmental monitoring in its affected areas. However, even in those regions unaffected by the war since 2014, environmental monitoring was rather weak. Since independence in 1991, environment protection in Ukraine has been very slow to develop and official tolerance of the abuse of natural resources by individuals and organised criminal groups remains rather high (Earthsight, 2018). Unfortunately, many cases of corruption and complacency towards corruption by state officials were reported but not many were successfully prosecuted (Hrynyk et al., 2023; Earthsight, 2018, 2019; Lemaître, 2023). Recently, the full-scale invasion has somewhat changed this attitude and has transformed the ways in which data on war related environmental damage is collected.

Mungiu-Pippidi (2023) develops seven arguments in favour of rethinking strategies to challenge corruption. One of these concerns tackling corruption by means of a "collective action problem". Mungiu-Pippidi (ibid.) observes that "human anticorruption agency has to be conceived of in broader terms", than only as coalitions of interested parties, of which some are sufficiently altruistic to make initial investment and might thus "be able to trigger substantial change if they manage to outnumber those who profit from the status quo" (pp. 300–304). Any international agencies involved "should be supporting broader national anti-corruption coalitions besides simply governmental interest, as the ownership of anti-corruption actions should rest with citizens, rather than kleptocrats" (Mungiu-Pippidi, 2023, p. 304). "Collective action problem solving" ideally employs a collaborative framework and accessible modern investigative journalism, with its freely available data, international networks and use of technological developments. Collective action is thus important not only nationally but internationally, where joint efforts are made to develop workable international instruments by which to tackle the environmental harm caused by war. The question arises as to whether such an approach could challenge systemic corruption within the forestry industry of Ukraine.

The Russian War in Ukraine and the Need for a Collective Action

In 2022 President Volodymyr Zelenskyy presented A Ukrainian 10-point peace formula as the basis for achieving a just and lasting peace in Ukraine. Environmental protection is a key element of this formula, with the notion of 'ecocide' and its restoration presented in the following terms "Countering ecocide: creation of a platform for assessing the environmental damage in Ukraine as a result of the Russian war of aggression; strengthening of international assistance for demining of the de-occupied territories of Ukraine; provision of financing and technologies for the restoration of treatment facilities in Ukraine; countering consequences of the Russian Federation's detonation of the dam of the Kakhovka Hydroelectric Power Plant (KHPP), which is located near the town of Nova Kakhovka in the temporarily occupied territory of the Kherson region; ensuring the international legal responsibility of Russia for ecocide in Ukraine, preventing repetition of such crimes in the world" (Zelenskyy, 2022, p. 1). According to many, the delivery of this formula in Ukraine largely depends on reducing the risk of internal corruption, such as breaking state monopoly, reducing managerial discretion, and increasing transparency (Roshchina, 2022).

While many efforts to curb corruption and crimes against the environment have previously failed, since the full-scale invasion of February 2022, the political will to record the environmental damage caused by warfare has been established. As a result, since 2022, various methodologies for assessing such damage have been developed (Resolution No. 326 of the Cabinet of Ministers of Ukraine dated March 20, 2022 approved A Procedure for determining damage and losses caused to Ukraine as a result of the armed aggression of the Russian Federation). The State Environmental Unit (SEU) has established an operational headquarters (Order of the State Environmental Inspection dated March 1, 2022 No. 73) and a Working Group (Order of the State Environmental Inspection dated April 15, 2022 No. 78/1) for the production of methodological documents and procedures for determining damage caused to natural resources and the surrounding natural environment. The SEU has also collaborated with the Ministry of the Environment to develop quantitative methods for calculating the financial cost of war on the environment in addition to assessing the level of structural damage (Ministry of Environment, 2022a, 2022b, 2022c, 2022d; Ministry of Justice, 2022a, 2022b).

In accordance with the above, the Resolution of the Cabinet of Ministers of Ukraine dated September 30, 2022 No. 1111 has amended Regulations on the State Environmental Inspection of the country, according to which specialist State Environmental Inspectors are authorised to undertake laboratory analyses of soil. These include radiological studies, in zones directly influenced by pollutants, in the event of accidents, emergency situations, military aggression, and terrorist or other criminal acts. Such representatives conduct inspections without obstruction in order to calculate the extent of damage and loss caused to the natural environment and natural resources.

Before the full-scale war, the Ministry of the Environment had created an online platform to improve the reporting of environmental issues named "EkoSistema" (https://eco.gov.ua/). The purpose of this platform is to provide up-to-date and reliable information about the state of the environment and natural resources. It was intended to interact with citizens' concerns in a transparent and convenient way. The platform holds monitoring data on the state of air, water, and soils in populated areas (by means of an interactive map) and its intended to be maintained by the Ministry of the Environment. Access of the platform is facilitated by means of an electronic signature, replacing the need for a physical visit to the relevant office.

In addition to these initiatives, the Ministry of the Environment has launched the EkoZagroza app—an official web resource that provides immediate and reliable information about the state of air, water, soil and other environmental data (EcoZagroza, 2024) in specific locations. By means of an interactive map of Ukraine, it provides data from monitoring systems on air quality and levels of radioactivity throughout the country and also specifies current environmental threats caused by the Russian invasion.

By means of the app, Ukrainian citizens can also report environmental crimes that they have witnessed. Such reports can include the following: the burning of military equipment; forest fires; the spillage of petroleum products or poisonous substances into the soil or water; and the emission of poisonous substances (e.g., chlorine, ammonia, hydrogen sulphide, hydrocyanic acid, nitric acid) into the air. The app user can also contact the emergency services directly (Ecozagroza, 2024).

Overall, the following stakeholders are involved in collecting data on the environmental impact of war: The Ministry of Environmental Protection and Natural Resources with its specialised bodies and agencies; the Office of the Prosecutor General, and various intergovernmental organisations and NGOs (Weir & Denisov, 2024). Weir and Denisov (2024) argue that awareness of war-related environmental damage therefore "enjoys a far higher profile because of energetic advocacy by Ukraine's government". However, it is observed that, historically, government concern regarding war-related environmental damage "has tended to have a low priority" (ibid., p. 2). Despite the above, a number of governmental agencies have experienced "human and technical capacity constraints around the volume of data it can collect, and how such collection and data analysis can be standardised" (Weir & Denisov, 2024, p. 4). Nevertheless, various national NGOs as well as international organisations have very quickly adapted to the situation and contributed to effective monitoring. While support in data collection was and is needed, some experts criticised the "spontaneous bottom-up data collection and lack of coordination among international actors, as well as the duplication of efforts at the national level" (ibid., p. 5).

One challenge is that the data that is currently collected as evidence of war-related environmental harm should be believed to be accurate, but it is not clear what "precise international avenue" might be used in the future to seek justice on its basis. Weir and Denisov (2024) argue that it is important to ensure that data collection, analysis, and storage are efficient. It is also important to move beyond simply recording incidents to a broader understanding of trends (ibid.). However, the Ukrainian parliament has decided to prioritise domestic methodologies to "place monetary value on war damage, rather than on evidence collection, analysis and storage *per se*" (ibid., p. 6). For example, it has been estimated that the damage caused by enemy forces in the first ten months of war following the February invasion amounted to almost 1.7 trillion UAH (more than US$25 billion). In 2023 alone, the losses amounted to 490,460,811,500 UAH (SEI, 2023) according to the following criteria:

1. 374.3 Bn UAH for soil contaminated with fuel, lubricants and other petroleum products plus the destruction of land resources due to missile and bomb damage;
2. 82.9 Bn UAH for atmospheric pollution caused by combustion products;
3. 25.9 Bn UAH for water pollution, and the arbitrary use of water resources (of which damage to water resources from the Kakhovka Reservoir breached as a result of Russian aggression amounted to 9.4 Bn UAH). (SEI, 2023)

Nevertheless, according to Weir and Denisov (2024), although assigning such monetary value to the damage has "some use in publicly communicating environmental narratives around the degree of harm, it is unlikely that they will greatly assist a future compensation case unless aligned with internationally best practice" (p. 6).

Since the full-scale invasion, Ukraine has made significant steps towards improving transparency and accountability in the field of ecology, the beginning of what will hopefully be an ongoing process. Despite Russian aggression, a comprehensive anti-corruption framework has been established for interaction between citizens and the state on questions of environmental justice. Whereas earlier environmental investigations were due to investigative journalism and the concerns of eco-activists, since the invasion, as related above, the state has developed a platform for citizens to access environmental data and report issues. Thereby "developing a comprehensive strategy for data collection and its preservation for accountability purposes" (Weir & Denisov, 2024, p. 8) by such means, the authorities might seek environmental justice on the international stage.

International Law, Warfare and Environmental Harm

As the preceding discussion suggests, Russian aggression in Ukraine has had a significant impact on the environment. Accordingly, there is now scope to consider what remedies may exist to seek redress for damage caused (according to monetary calculation) or to prosecute specifically for the environmental harm Russia has caused.

International law provides a potential avenue for such action. The Rome Statute which set up the International Criminal Court (ICC) is one potential mechanism, given the ICC's jurisdiction over war crimes, genocide, crimes against humanity, and crimes of aggression (Arsanjani, 1999). Though the ICC does not explicitly recognise environmental harm as a specific crime a model definition of ecocide has been produced by an independent expert panel and model rules for ecocide were also produced by the European Law Institute (ELI) in 2023. The ELI Report proposed to extend the powers of the European Public Prosecutor's Office to include offences of ecocide affecting more than one Member State or one or more Member States and one or more third countries. Potentially, this could allow for environmental harm amounting to ecocide to be linked to action under the Rome Statute where this affects an EU state and a third country (ELI, 2023). The definitions contained in the Rome Statute allow for prosecution where a state adopts a systematic approach to damage or aggression during an armed conflict although thus far, the ICC has failed to recognise ecocide as the fifth crime against peace (Mehta & Merz, 2015). However, a theoretical case has been proposed whereby Russia's aggression in Ukraine has far-reaching environmental consequences and thus could meet the threshold for a breach of Article 8(2)(b) of the Rome Statute (Palarczyk, 2023). Article 8 concerns war crimes committed intentionally as a planned part of large-scale activity (e.g. the large-scale commission of environmental damage as systematic policy on the part of Russian forces). While the text of the Rome Statute does not specifically refer to general environmental harm as a war crime, Article 8(2)(a)(iv) refers to the "extensive destruction and appropriation of property, not justified by military necessity and carried out unlawfully and wantonly". A high threshold for environmental damage is arguably set by Article 8(2)(b)(iv) which refers to:

> Intentionally launching an attack in the knowledge that such attack will cause incidental loss of life or injury to civilians or damage to civilian objects or widespread, long-term and severe damage to the natural environment which would be clearly excessive in relation to the concrete and direct overall military advantage anticipated.

Thus, there is a need to consider the nature and extent of the environmental damage caused during an attack and to contextualise this within the military objectives being pursued. Wartime treatment of the environment can be categorised as either active or passive. Active means 'using the environment as a weapon' whereas passive means "acts that do not 'use' the environment but have deleterious effects on the environment" (Jensen, 2005, p. 149). The scale of the damage being caused by Russian forces is certainly extensive, but it is nevertheless difficult to conclude that it thereby meets the threshold of 'clearly excessive'. Should further investigation conclude that a systematic policy of environmental destruction is being deployed as a punitive measure (i.e. to discourage and punish Ukrainian resistance) this may assist in bringing a case against Russia before the ICC.

While in reality the ICC may be of limited practical use in bringing a case against Russia for the environmental harm caused during its acts military campaign, the International Court of Justice (ICJ) offers a possible alternative. The ICJ is the UN's primary court and is available either for one state to pursue a case against another for breaches of international law that impact it negatively or harm its interests. A state can also obtain a ruling on the lawfulness of certain state actions. International law provides that a breach of an international obligation involves an obligation to make 'adequate' reparation for that breach. Reparations made to individuals for damage caused by gross violations of international human rights and humanitarian laws are now established features of the international law framework. In addition, reparation to states where breaches of international law have a negative impact on a state's interests are also features of ICJ jurisdiction. The key principle is that, so far as is possible, reparations should provide appropriate redress for the consequences of an illegal act and, in principle, should re-establish the position that would have existed had the illegal act not taken place. This is clearly difficult to enforce where extensive environmental harm is caused during an armed conflict. However, the ICJ has previously confirmed this principle in the case of Armed Activities on the Territory of the Congo (Democratic Republic of the Congo v Uganda) (2005) where the Democratic Republic of the Congo brought a case against Uganda before the ICJ seeking reparation for acts of intentional destruction, the looting of national property, and for resources that had been appropriated during the conflict. The ICJ's judgment confirmed that a State that has violated a rule of international law, thereby causing damage has the obligation to make reparation for that damage. The ICJ ruled that, in accordance with customary international law, Uganda was required to make reparation for all types of damage caused by wrongful acts attributable to its activities, no matter how remote the causal link between those acts and the damage concerned. The ICJ also ordered Uganda to make reparation in kind where possible, for Congolese resources, assets, or wealth now in its possession, or, alternatively, make financial compensation for the damage suffered. Finally, the ICJ ruled that in any event Uganda, as the aggressor, should render satisfaction for the injuries inflicted upon the Democratic Republic of the Congo in the form of an official apology, and the prosecution of those responsible for the damage.

By the logic of the ICG's Uganda judgement, it would appear to be irrelevant whether environmental damage caused in Ukraine was a policy sanctioned by the Russian state, or by local military commanders as part of their operational jurisdiction. The key question rests as to whether the actions of the Russian army, which prosecuted the war on behalf of the Russian state, caused damage to their Ukrainian neighbours. The evidence, as outlined earlier in this chapter, we believe clearly established a pattern of environmental harm employed intentionally as a warfare 'tactic', and in respect of the extent to which this harm contravenes customary international law we would point to the principles enshrined in the World Charter for Nature 1982, Principle 5 which states that "Nature shall be secured against degradation caused by warfare or other hostile activities" (United Nations, 1983; Wood, 1985). This Charter was adopted by UN Member States which arguably brings these states within the remit of customary international law which is typically defined as a "general and consistent practice of states and followed by them from a sense of legal obligation" (Goldsmith & Posner, 1999, p. 1113). In this case the legal obligation is heavily implied by the UN's prohibition of wanton environmental destruction during warfare, strengthened by the Rome Statute's stipulation that excessive environmental destruction and destruction of a state's property, which is arguably not an essential component of any military objective, cannot be justified.

Remedies Under International Law

There are several potential options to consider when seeking to prosecute or seek compensation for Russian environmental destruction in Ukraine. At a domestic level, individual actors might be prosecuted for breaches of Ukrainian environmental protection law. Similarly, violations of international human rights law (e.g. the right to peaceful enjoyment of property and the surrounding environment) or international human rights law (including environmental damage that impacts negatively on citizens' private lives) might be pursued in the domestic courts where perpetrators can be identified.[5] Prosecution via the ICC remains a possibility under war crimes provisions which could require criminal penalties against the perpetrators, even though our initial analysis has identified potential difficulties with prosecuting such environmental damage. Indeed, in 2024 the ICC has announced a further consultation on incorporating a crime of ecocide into its jurisdiction more directly which indicates at least a tacit acknowledgement of the current problems of doing so, and of the need to increase clarity on a prosecutor's understanding of environmental harm (ICC, 2024).

The ICJ would appear to be a more obvious route to achieve some form of redress.[6] In addition to considering a full case, the ICJ has the option to issue an advisory opinion concerning the lawfulness of any environmental harm caused or to consider whether such amounts to a breach of international or customary law. As the ICJ's Uganda decision indicates, where reparations are sought this should address all of the various harms arising from a conflict. If successful, this could result in compensation that may go some way to establishing a programme of environmental restoration once hostilities have ceased.

Conclusion

The environmental damage caused by the Russian invasion of Ukraine is severe and extensive, impacting ecosystems that extend beyond Ukraine's borders. The scale of this damage demands an urgent need for action to combat such environmental crimes. Although Ukraine itself has taken significant steps to create a comprehensive legal and institutional framework for identifying and quantifying environmental damage, the sheer

[5] For example, an inquiry into war crimes allegedly committed by British soldiers in Afghanistan has the potential to result in prosecutions (Sabbagh, 2023).

[6] It should be noted that Ukraine has already raised a case against Russia with the ICJ, relating to the 2014 conflict in the Donbas region and Russia's claims of genocide as a justification for its 'special military operation'. This case is ongoing with a preliminary ruling issued in February 2024, in which the ICJ concluded it had jurisdiction to consider Ukraine's request for a declaration that it did not breach its obligations under the Convention on the Prevention and Punishment of the Crime of Genocide.

scale of the problem and the continuing challenge posed by the ongoing conflict must require international cooperation and support.

Our analysis critically examines the possibility of employing international law, in particular the provisions of the Rome Statute, the jurisdiction of the International Criminal Court (ICC) and the International Court of Justice (ICJ), to hold Russia accountable for its environmental crimes committed during the war. However, while the ICC's jurisdiction over environmental crimes is currently limited, the International Court of Justice offers the potential to seek eventual redress for environmental damage.

In light of the current situation, we highlight the importance of collective action and cooperation among various stakeholders, including government agencies, civil society organisations, investigative journalists, and the general public, in documenting and responding to environmental crimes. The use of open-source intelligence (OSINT) and new technologies has proven effective in gathering evidence and raising awareness of these issues. Moreover, the international community now recognises the need for a comprehensive and united approach in addressing the long-term consequences of environmental damage caused by war. To do so will require significant resources, and a strong commitment to environmental justice and accountability to be evidenced by ensuring reparations on behalf of an aggressor.

Our analysis underpins a call to action for the international community to prioritise environmental protection in armed conflicts. We emphasise the urgent need to develop robust mechanisms and legal frameworks to hold the perpetrators of environmental crime accountable and ensure that the costs of environmental restoration are borne by the responsible parties.

We conclude our chapter with a pertinent question, namely: How correct might it be to consider among those responsible for environmental damage not only an aggressor, but also the policy makers and manufacturers of weapons that will inevitably cause significant damage to the environment? Perhaps it is worth considering some international legal instruments not only for obtaining for the actual damage, which, as we show, is very difficult to achieve, but also to assign responsibility for environmental crimes to those who allow the use of destructive weaponry as a matter of policy.

Acknowledgement Iryna Soldatenko would like to acknowledge the support of the British Academy grant (LTRSF\100497).

References

Arsanjani, M. H. (1999). The Rome Statute of the International Criminal Court. *American Journal of International Law, 93*(1), 22–43. https://doi.org/10.2307/2997954

Besser, L. (2020). Ukraine's illegal amber mining boom is scarring the earth and making criminal gangs rich. [Online]. https://www.abc.net.au/news/2020-01-21/illegal-amber-mining-creating-environmental-disaster-in-ukraine/11745470

Boreiko, V. (2023). *Лісова реформа"—шлях нікуди* (in Ukrainian) [Forest reform—A path to nowhere]. https://interfax.com.ua/news/blog/931888.html

Brewster, M. (2022). Natural gas, rare earth minerals: What's at stake for Ukraine in the territory Russia is trying to conquer. The invasion may be about Russian imperialism, but it's also about securing natural resources. *CBC News*. [Online]. https://www.cbc.ca/news/politics/natural-resources-ukraine-war-1.6467039

Brisman, A., & South, N. (2014). *Green cultural criminology: Constructions of environmental harm, consumerism, and resistance to ecocide*. Routledge.

Brisman, A., South, N., & White, R. (2016). Toward a criminology of environment-conflict relationships. In A. Brisman, N. South, & R. White (Eds.), *Environmental crime and social conflict: Contemporary and emerging issues*. Routledge.

Davies, K. M. (2023). As sanctions bite, Russia eyes Ukraine's mineral resources to fund its invasion. *Kyiv Independent*. [Online]. https://kyivindependent.com/as-sanctions-bite-russia-eyes-ukraines-mineral-resources-to-fund-its-invasion/

De Klerk, L., et al. (2023). Climate damage caused by Russia's war in Ukraine. [Online]. https://en.ecoaction.org.ua/wp-content/uploads/2023/06/clim-damage-by-russia-war-12months.pdf

Decree of the Cabinet of Ministers of Ukraine No 303. (2022, March 13). [Online]. https://zakon.rada.gov.ua/laws/show/303-2022-%D0%BF?lang=en#Text

Department of Defence, Office of the Undersecretary for Policy (Strategy, Plans, and Capabilities). (2021). *Department of Defense climate risk analysis*. Report

Submitted to National Security Council. [Online]. https://media.defense.gov/2021/Oct/21/2002877353/-1/-1/0/DOD-CLIMATE-RISK-ANALYSIS-FINAL.PDF

Depledge, D. (2023). Low-carbon warfare: Climate change, net zero and military operations. *International Affairs, 99*(2), 667–685. https://doi.org/10.1093/ia/iiad001

Earthsight. (2018). Complicit in corruption: How billion-dollar firms and EU governments are failing Ukraine's forests. [Online]. https://fe8a03e2-1131-44e7-a06a-fb468c2a30d4.filesusr.com/ugd/624187_673e3aa69ed84129bdfeb91b6aa9ec17.pdf

Earthsight. (2019). EU firms buying timber from wanted suspect at centre of $30m Ukrainian bribery case. [Online]. https://www.earthsight.org.uk/news/timberleaks/eu-firms-buying-timber-wanted-suspect-30million-ukraine-bribery-case

EcoZagroza. (2024). Official resource of the Ministry of Environmental Protection and Natural Resources of Ukraine. [Online]. www.ecozagroza.gov.ua

ELI. (2023). ELI report on ecocide. https://www.europeanlawinstitute.eu/fileadmin/user_upload/p_eli/Publications/ELI_Report_on_Ecocide.pdf

Epravda. (2022). *Потреба в альтернативному паливі під час війни зросла в рази. Як збільшити заготівлю дров без шкоди для лісів?* [The demand for the fuel has risen significantly during the war. How do we increase fire wood production without damaging forests?]. https://www.epravda.com.ua/columns/2022/07/25/689559/

Fichtelberg, A. (2016). Resource wars, environmental crime, and the laws of war. Updating war crimes in a resource scarce world. In A. Brisman, N. South, & R. White (Eds.), *Environmental crime and social conflict: Contemporary and emerging issues*. Routledge.

Goldsmith, J. L., & Posner, E. A. (1999). Theory of customary international law. *University of Chicago Law Review, 66*(4), 1113–1178.

Hrynyk, Y., Biletskyi, A., & le Roux, A. C. (2023). *How corruption threatens the forests of Ukraine: Typology and case studies on corruption and illegal logging.* Working Paper 43 / Environmental Corruption Deep Dive Series, Basel Institute on Governance. [Online]. https://baselgovernance.org/publications/deepdive1-ukraine

Hupy, J. P. (2006). The long-term effects of explosive munitions on the WWI battlefield surface of Verdun, France. *Scottish Geographical Journal, 122*(3), 167–184.

ICC. (2024). The Office of the Prosecutor launches public consultation on a new policy initiative to advance accountability for environmental crimes under the Rome Statute. https://www.icc-cpi.int/news/office-prosecutor-launches-public-consultation-new-policy-initiative-advance-accountability-0

Jensen, E. T. (2005). The international law of environmental warfare: Active and passive damage during armed conflict. *Vanderbilt Journal of Transnational Law, 38*(1), 145–186.

Klitgaard, R. (2015). Final addressing corruption together. [Online]. https://www.oecd.org/dac/conflict-fragility-resilience/publications/FINAL%20Addressing%20corruption%20together.pdf

Klitgaard, R., MacLean-Abaroa, R., & Parris, H. L. (2000). *Corrupt cities: A practical guide to cure and prevention.* ICS Press & World Bank Institute, RUR 162.

Kurkov, A. (2024). Opinion: Adding fuel to the fire. [Online]. Retrieved December 12, 2024, from https://www.kyivpost.com/opinion/41961

Lawrence, M., Stemberger, H. L. J., Zolderdo, A. J., Daniel, P., et al. (2015). The effects of modern war and military activities on biodiversity and the environment. *Environmental Reviews, 23*(4), 443–460. https://doi.org/10.1139/er-2015-0039

Lemaître, S. (2023). *Seedlings of hope: Addressing corruption linked to crimes that impact the environment in line with UNCAC Resolution 8/12.* Working Paper 50. Basel Institute on Governance. [Online]. https://baselgovernance.org/publications/wp-50

Maksimov, M. (2018). Volyn's armed conflict. [Online]. https://ukraina.ru/amp/20181101/1021605707.html

Malko, R. (2018). Ukraine to save biodiversity. *Ukrainian Week.* [Online]. https://euneighbourseast.eu/news/stories/ukraine-to-save-biodiversity/

Matsala, M., Odruzhenko, A., Hinchuk, T., et al. (2024). War drives forest fire risks and highlights the need for more ecologically-sound forest management in post-war Ukraine. *Scientific Reports, 14*, 4131. https://doi.org/10.1038/s41598-024-54811-5

Mehta, S., & Merz, P. (2015). Ecocide—A new crime against peace? *Environmental Law Review, 17*(1), 3–7. https://doi.org/10.1177/1461452914564730

Ministry of Energy and Environmental Protection of Ukraine (MEEP). (2018). *Ukraine's greenhouse gas inventory 1990–2018.* Annual National Inventory Report for Submission under the United Nations Framework Convention on Climate Change and the Kyoto Protocol. [Online]. https://mepr.gov.ua/wp-content/uploads/2023/07/5_Natsionalnyj-kadastr-antropogennyh-vykydiv-iz-dzherel-ta-absorbtsiyi-poglynachamy-parnykovyh-gaziv-v-Ukrayini-za-1990-2018-roky.pdf

Ministry of Environment. (2022a). *Methodology for determining the amount of damage caused to land and soil as a result of emergency situations and/or armed aggression and hostilities during martial law*. Dated 04/04/2022 No. 167, registered with the Ministry of Justice of Ukraine on 04/11/2022 under No. 406/377. [Online]. https://zakon.rada.gov.ua/laws/show/z0433-22#Text

Ministry of Environment. (2022b). *Methodology for calculating unorganized emissions of polluting substances or a mixture of such substances into atmospheric air as a result of emergency situations and/or during martial law and determining the amount of damage caused*. Dated 04/13/2022 No. 175, registered with the Ministry of Justice of Ukraine on 04/16/2022 under No. 433/37769. [Online]. https://zakon.rada.gov.ua/laws/show/z0433-22#Text

Ministry of Environment. (2022c). *Methodology for determining damage caused to the surrounding natural environment within the territorial sea, the exclusive maritime (economic) zone and internal sea waters of Ukraine in the Azov and Black seas*. Approved by the order of the Ministry of Environment dated 19/08/2022 No. 309, registered with the Ministry of Justice of Ukraine on 17/10/2022 under No. 1253/38589. [Online]. https://ips.ligazakon.net/document/RE38589?an=1

Ministry of Environment. (2022d). *The method of determining damage and losses caused to the forest fund as a result of the armed aggression of the Russian Federation*. Approved by the order of the Ministry of Environment dated 05/10/2022 No. 414, registered in the Ministry of Justice of Ukraine 10/24/2022 under No. 1308/38644. [Online]. https://ips.ligazakon.net/document/MN026061

Ministry of Justice. (2022a). *Methodology for determining the amount of compensation for damages caused to the state as a result of arbitrary use of subsoil*. Approved by the order of the Ministry of Environment of September 15, 2022 No. 366, registered in the Ministry of Justice of Ukraine on October 28, 2022 under No. 1337/38673. [Online]. https://ips.ligazakon.net/document/RE38673

Ministry of Justice. (2022b). *Methodology for determining damages caused by water pollution and/or clogging, arbitrary use of water resources*. Approved by the order of the Ministry of Environment dated 07/21/2022 No. 252, registered in the Ministry of Justice of Ukraine on 08/09/2022 under No. 900/38236. [Online]. https://www.reestrnpa.gov.ua/

Muggah, R., & Dryganov, V. (2022). *Ukraine's extraordinary riches in energy, minerals, and agriculture are a prize for the Kremlin*. Russia's Resource Grab in Ukraine, Foreign Policy. [Online]. https://foreignpolicy.com/2022/04/28/ukraine-war-russia-resources-energy-oil-gas-commodities-agriculture/

Müller, N. C., & Wiik, J. (2023). From gatekeeper to gate-opener: Open-source spaces in investigative journalism. *Journalism Practice, 17*(2), 189–208. https://doi.org/10.1080/17512786.2021.1919543

Mungiu-Pippidi, A. (2023). Seven arguments in favour of rethinking corruption. *Eucrim, Issue, 3*, 300–304. https://doi.org/10.30709/eucrim-2023-028

Myroniuk, V., Bilous, A., Khan, Y., Terentiev, A., Kravets, P., Kovalevskyi, S., & See, L. (2020). Tracking rates of forest disturbance and associated carbon loss in areas of illegal amber mining in Ukraine using Landsat time series. *Remote Sens, 12*, 2235. https://doi.org/10.3390/rs12142235

Office of the President of Ukraine. (2023, June 30). *First meeting of the international working group on the preparation of the Special Tribunal held in Kyiv* [Internet]. Retrieved November 18, 2023, from https://www.president.gov.ua/en/news/u-kiyevi-vidbulosya-pershe-zasidannya-mizhnarodnoyi-robochoy-83949

Palarczyk, D. (2023). Ecocide before the International Criminal Court: Simplicity is better than an elaborate embellishment. *Criminal Law Forum, 34*, 147–207. https://doi.org/10.1007/s10609-023-09453-z

Le Billon, P. (2013). *Fuelling war: Natural resources and armed conflicts*. https://doi.org/10.4324/9781315019529

Posthuma, L., Wahlstrom, E., Nijenhuis, R., Dijkens, C., Zwart, D., Meent, D., Hollander, A., Brand, E., den Hollander, H. A., van Middelaar, J., van Dijk, S., Hall, E. F., & Hoffer, S. (2014). The flash environmental assessment tool: Worldwide first aid for chemical accidents response, pro action, prevention and preparedness. *Environment International, 72*, 140–156. https://doi.org/10.1016/j.envint.2014.04.018

RBC-Ukraine. (2022, January 31). Large-scale fires in Luhansk region: Officials notified of suspicion. Retrieved September 3, 2024, from https://www.rbc.ua/ukr/news/masshtabnye-pozhary-luganskoy-oblasti-dolzhnostnym-1643635378.html

Robinson, D. (2022). Ecocide—Puzzles and possibilities. *Journal of International Criminal Justice, 20*(2), 313–347. https://doi.org/10.1093/jicj/mqac021

Roca, C. (2017). Long read: How the Syrian war changed how war crimes are documented. *The New Humanitarian*. [Online]. https://deeply.thenewhumanitarian.org/syria/articles/2017/06/01/long-read-how-the-syrian-war-changed-how-war-crimes-are-documented

Roshchina, O. (2022). Zelenskyy outlines Ukraine's 10-point "formula for peace" at G20 summit. *Ukrainska Pravda*. [Online]. Retrieved March 20, 2024, from https://www.pravda.com.ua/eng/news/2022/11/15/7376378/

RPA. (2015). Recovery and peacebuilding assessment: Analysis of crisis impacts and needs in Eastern Ukraine. [Online]. https://reliefweb.int/report/ukraine/recovery-and-peacebuilding-assessment-analysis-crisis-impacts-and-needs-eastern

Ruggiero, V. (2010). War as state crime. In W. Chambliss, R. Michalowski, & R. Kramer (Eds.), *State crime in the global age* (1st ed.). Willan). https://doi.org/10.4324/9781843927051

Sabbagh, D. (2023). SAS shot dead two sleeping Afghans in war crime, public inquiry told. *The Guardian*. https://www.theguardian.com/uk-news/2023/oct/11/two-sleeping-afghan-adults-shot-dead-by-sas-was-war-inquiry-hears

SBI. (2021, February 16). *SBI: Information about firefighting in Luhansk region in July 2020 was distorted*. LB.ua.

SEI Iconografika. (2023). *State Environmental Inspectorate of Ukraine*, SEI statistics.

Shumilo, L., Skakun, S., Gore, M. L., et al. (2023). Conservation policies and management in the Ukrainian Emerald Network have maintained reforestation rate despite the war. *Communications Earth and Environment, 4*, 443. https://doi.org/10.1038/s43247-023-01099-4

Smal', V. (2023). *Публічний звіт голови Державного агентства лісових ресурсів за 2023 рік* [Public reporting by the head of the State Agency of Forest Resources]. [Online]. https://www.ukrinform.ua/rubric-presshall/3833078-publicnij-zvit-golovi-derzavnogo-agentstva-lisovih-resursiv-za-2023-rik.html

South, N., Brisman, A., & Beirne, P. (2013). A guide to a green criminology. In N. South & A. Brisman (Eds.), *Routledge international handbook of green criminology* (pp. 27–43). Routledge.

Sowers, J., & Weinthal, E. (2023). War and environmental politics: A comparative perspective. In J. Sowers, S. D. Vandeveer, & E. Weinthal (Eds.), *The Oxford handbook of comparative environmental politics* (pp. 775–796). OUP.

State Agency of Forest Resources of Ukraine (SAFR). (2022). *Публічний звіт Голови Державного агентства лісових ресурсів України за 2021 рік* (Public report of the Head of the State Agency of Forest Resources of Ukraine for 2021). https://forest.gov.ua/storage/app/sites/8/publich_zvit/publichnii-zvit-za-2022.pdf

State Bureau of Investigation. (2024). In Lviv region, the SBI exposed officials who illegally cut down trees worth more than UAH 64 million. [Online]. https://dbr.gov.ua/news/na-lvivshhini-dbr-vikrilo-posadovciv-yaki-nezakonno-virubali-derev-na-ponad-64-mln-grn

Tian, N., Silva, L. D., Liang, X., Scarazzato, L., Béraud-Sudreau, L., & Assis, A. (2023). Trends in military expenditure, 2022. *SIPRI*. https://doi.org/10.55163/PNVP2622

Tkachuk, O. (2023). Pine and Spruce logs are being prepared for shipment. [Online]. https://www.5.ua/amp/ru/rehyoni/55-vahonov-lesa-dlia-vsu-kakye-derevia-zakazaly-v-mynoboroni-y-dlia-cheho-318673.html

Tril, M. (2024). Russia destroys over 60,000 hectares of Ukrainian forests worth over $360 mn—investigation. *Euromaidan Press*. [Online]. https://euromaidanpress.com/2024/04/09/russia-destroys-over-60000-hectares-of-ukrainian-forests-worth-over-360-mn-investigation/

United Nations. (1983). *World charter for nature*. UN General Assembly 37th Session: 1982-1983. https://digitallibrary.un.org/record/39295?ln=en&v=pdf

UNODC. (2023). Statistical framework to measure corruption. [Online]. https://unstats.un.org/UNSDWebsite/statcom/session_54/documents/BG-4d-Framework-E.pdf

Van Duyne, P., Markovska, A., & Serduyk, A. (2022). The organisation of eco-crime and corruption in green Ukraine. In P. Van Duyne et al. (Eds.), *Narrating organised crime stories and Aristotelian principles of drama* (pp. 233–269). Eleven Publishing.

Van Uhm, D. P. (2018). *The diversification of organized crime into the illegal trade in natural resources*. Utrecht University.

Van Uhm, D. P., & Nijman, R. C. C. (2022). The convergence of environmental crime with other serious crimes: Subtypes within the environmental crime continuum. *European Journal of Criminology, 19*(4), 542–561. https://doi.org/10.1177/1477370820904585

Verbovska, M. (2024). To cut it all. *NGL Media Report*. [Online]. https://ngl.media/en/2024/04/08/to-cut-it-all/

Weir, D., & Denisov, N. (2024). Assessing environmental damage in Ukraine. Conflict and environment observatory. Input to An Environmental Compact for Ukraine. [Online].

White, R. (2015). Climate change, ecocide and the crimes of the powerful. In G. Barak (Ed.), *The Routledge international handbook of the crimes of the powerful* (pp. 211–222). Routledge.

Wood, H. W. (1985). The United Nations World Charter for Nature: The Developing Nations' Initiative to Establish Protections for the Environment. *Ecology Law Quarterly, 977*, 977–996.

Zabyelina, Y., & Kalczynski, N. (2020). Shadowy deals with "Sunny Stone": Organized crime, informal mining, and the illicit trade of amber in Ukraine. In Y. Zabyelina & D. Van Uhm (Eds.), *Illegal mining: Organized crime, corruption, and ecocide in a resource-scarce world*. Palgrave Macmillan.

Zelenskyy, V. (2022). Peace formula of the President of Ukraine. [Online]. Ukraine has always been a leader in peacemaking efforts; if Russia wants to end this war, let it prove it with actions—Speech by the President of Ukraine at the G20 Summit—Official website of the President of Ukraine.

Zwijnenburg, W. (2017, March 30). Donbas: A ticking toxic time bomb. *Bellingcat*. http://www.bellingcat.com/news/uk-and-europe/2017/03/30/donbas-ticking-toxic-time-bomb/

Zwijnenburg, W., & Ballinger, O. (2023). Protecting the environment in armed conflict. *International Review of the Red Cross, 105*(924), 1497–1521. https://doi.org/10.1017/S1816383123000383

The Hidden Toll: Women and Girls in the Intersection of Climate Change and Domestic Servitude

Nieves Sanz Mulas

Climate Emergency and Human Displacement

Human activity has been so crucial in determining the climate of the last two centuries that we can speak of a new geological interval: the Anthropocene. This new epoch of the Earth is characterized by the dominant influence of human activity on climate and the environment (Felipe, 2019). The advance of neoliberal economic policies and the unstoppable state deregulation have increased inequalities and caused unprecedented global environmental degradation. According to the Intergovernmental

Paper result of the I+D+i project "Diagnosis and evaluation of compliance by the Spanish State with the Global Compact on Migration from a gender perspective (Migration Pact)", funded by the Ministry of Science, Innovation and Universities for the years 2020–2023 (Ref.: PID2019-106159RB-100/AEI/10.13039/501100011033). Principal Investigator: Nieves Sanz Mulas.

N. Sanz Mulas (✉)
Centre for Human Rights and Public Policies Research, University of Salamanca, Salamanca, Spain
e-mail: ixmucane@usal.es

D. van Uhm, D. Siegel (eds.), *Global Green Crime and Ecojustice*, Palgrave Studies in Green Criminology, https://doi.org/10.1007/978-3-031-91557-4_20

Panel on Climate Change (IPCC, 2021), temperatures have already risen by an average of 1.1 °C, and this, together with unprecedented population growth, has had serious negative effects on health and the availability of food and water[1] (UN, 2019). Shortages of basic resources have led to the displacement of large groups of people known as climate migrants (Felipe, 2019). This chapter reviews the literature to demonstrate the relationship between climate migrations, gender, and domestic servitude, a necessary feminist approach that has not been adopted in consequent measures.

According to Haley and Arrigo, "climate change is a primary driver of migration, including both migration within countries of residency and immigration to other countries" (2022). The reference names in the study of this phenomenon are Lester Brown (1976) and Essam EL-Hinnawi (1985). However, it was not until 2019 that the International Organization for Migration (IOM) included *climate migration* for the first time in its glossary on human mobility, which it defines as: "the movement of a person or groups of persons who, predominantly for reasons of sudden or progressive change in the environment due to climate change, are obliged to leave their habitual place of residence, or choose to do so, either temporarily or permanently, within a State or across an international border" (IOM, 2014). As Agnew describes, "drought will force farmers and herders off the land; sea-level rise will force coastal inhabitants to move; social conflict will force many to flee to safer areas (more below); and extreme weather events and forest fires will destroy homes and livelihoods—again prompting migration. … While a portion of this migration will cross borders, much of it will be internal. Many of the migrants will move to megacities in developing countries, which are often polluted, overcrowded, and lacking in basic infrastructure such as clean water and sewerage" (2012). A good (and very sad) example is Bangladesh.

[1] The number of inhabitants worldwide will increase from the current 7.7 billion to 9.7 billion in 2050. This figure will continue to rise to 11 billion by 2100. India will become the most populous country from 2027 (1.45 billion inhabitants) and China will move into second place (1.065) followed by Nigeria (733), which will overtake the United States in fourth place (434 million inhabitants). Sub-Saharan Africa will see the largest population increases. Countries such as Niger, Angola, Tanzania, Somalia, Somalia, Zambia, Burundi, Republic of Congo, Burkina Faso, Mali, and Mozambique will see growth of more than 300%. At the other extreme, Europe will lose population, with countries such as Poland and Romania registering declines of more than 38%.

In an average year, about 20% of the land area of the country is flooded in one or other part of the country. Floods, therefore, are an important climatological factor pushing migrants from coastal areas to the cities, especially the country's capital. According to World Bank approximately 300,000 to 400,000 people migrate to Dhaka each year, a lot of them living in slums and squatter settlements (CEAR, 2021). Overcrowding and lack of planning is creating traffic jams, pollution, land grabbing, imbalance in the climate, and impossible daily demands for energy and water. That is, these people face problems of sanitation facility, lack of safe water, shortage of water and poor congested shelter, which makes them even more vulnerable (Islam et al., 2018).

The challenge now is to quantify to this phenomenon. The most reliable data come from statistics related to the impact of disasters on internal migratory movements (Global Report on Internal Displacement) of the Internal Displacement Monitoring Centre (IDCM). In this regard, while 18.8 million were counted in 2017 (1.3 million associated specifically with droughts) (IDCM, 2018), the figure fell to 17.2 million in 2018 (IDCM, 2019), rising again to 24.9 million in 2019, which is three times the number of displacements caused by conflict and violence (8.5 million). In 2023 alone, disaster displacement was detected in 148 countries and territories, with significant events across six continents. Some high-income countries, such as Canada and New Zealand, reported their highest figures ever (IDCM, 2024).

Since the 1990s (Lynch, 1990; South, 1998), green criminology describes some of the ways in which global warming and climate change may impact the (im)mobility[2] of those who have contributed the least to anthropogenic climate change (Brisman et al., 2018). How they are the ones who suffer most from its extreme events: drought, floods, and disease.

As an inherently unjust phenomenon, climate change takes its toll on impoverished countries. Although climate change is a global phenomenon, the responses to it are grossly inequitable. While rich nations protect

[2] Having the need to migrate does not mean being able to do so. Those who are forced to leave their families and friends, home, livelihoods and land, may become "climate refugees". Those who remain involuntarily, because they don't have the necessary means to migrate (money, contacts, skills, etc.) to migrate, may become the "climate abandoned".

their cities with expensive seawalls and storm barriers, areas inhabited by the poor and indigenous people remain at the mercy of storms and rising seas.

In short, we face a complex, heterogeneous, and multifactorial issue that affects, above all, the neediest and socially isolated people in cities and rural areas. This is due to their close relationship with their environment, the lack of resources, and the remoteness of the institutions and political power that allow for a rapid and adequate response. "It's always the poor who are hurt the most because they are less able to absorb the shocks and adapt because they have fewer assets" (Vidal, 2011). And as every vulnerable group has within it a group that is always more fragile, the situation for women and girls becomes considerably more complicated. Approximately 80% of climate refugees are women (Haigh & Vallely, 2010). "As climate disasters increase, women and single-mother households displaced bay such events are forced to resettle elsewhere" (Henrici et al., 2015).

Female Climate Migration and Human Exploitation

The Feminization of Migration

This new era of global conflicts and economic (under)development has led to millions of women migrating, a clear effect of the "global feminization of poverty". These disparities have more to do with social positions within families and communities than genetics. Among the complex and multiple reasons behind female migration, factors related to social constraints stand out. Constraints influenced by models of "sexual selectivity" perpetuate women's submissive roles: domestic violence, marital break-ups, the impossibility of getting a divorce, "desires for self-improvement", racism, or sexism.

And what happens if environmental degradation is added to all this? Generally, women face various vulnerabilities in the context of climate change (intersectionality). Green criminology analyzes the impact of sexism, racism, and capitalism at the intersection of climate justice and the literature shows that poor women and women of color are impacted the most by climate change. But gendered vulnerability to climate change is not due to characteristics intrinsic to women as a group, "the same cultural, historical, political, and economic factors that influence the cycle of poverty disproportionately impacting women are also responsible for the inequitable distribution of climate change impacts" (Haley & Arrigo, 2022).

To begin with, in times of crisis, men flee earlier while women stay longer in their homes to protect them. In many countries, women are supposed to care for and protect children and the elderly, as well as the family's household property, which hampers their own rescue efforts in almost any type of natural disaster (Schwoebel & Menon, 2004; Alston, 2013). As a result, the harm suffered by women in relation to climate impacts is both direct (death, injury) and indirect (less access to education, fewer opportunities for economic development). More women than men usually die in natural disasters due to social inhibitions, dress, and a lack of survival skills (swimming, climbing trees, etc.) (Neumayer & Plümper, 2007; MacDonald, 2005).[3] Even food insecurity caused by climate change affects the female population more: in much of the world, women are systematically discriminated against and do not have equal access to resources such as land, water, seeds, fertilizers, or credit for production (Dominelli, 2013). Although women are responsible for approximately 50–80% of the global food-producing workforce, they only own around 20% of the land (Pachauri & Meyer, 2014). In the end, it is also women who are the first to see their food consumption reduced in the face of extreme weather events.

In most cases, disasters place an additional burden on women and girls, as they are responsible for unpaid work (providing care, water, and food for households, etc.). In Kenya, 85% of women's energy is used to fetch water, which—in times of drought—can take up to eight hours a

[3] A 2005 study on the aftermath of the tsunami in Indonesia, Sri Lanka, and India found that the rate of men who survived the storm outnumbered surviving women by four to one. Another study conducted in 2007, which looked at natural disasters in 141 countries over a 20-year period, found that women represent the majority of those most impacted by climate disasters and are 14 times more likely to die in a disaster than men. These disparities are reduced in countries where women have increased socio-economic power.

day (Duncan, 2007). This work overload is not compensated by an increase in caloric intake, since some cultures have food hierarchies and women are relegated to last place despite of their increased needs during menstruation, childbirth, and breastfeeding (Sorensen et al., 2018). This is the case, for example, in sub-Saharan Africa, where cultural rules mean that women consume fewer calories despite lifting much heavier loads than men (FAO, 2000).

In short, being poor and being women means that they are doubly affected by climate change (and it is even worse if they are indigenous) with climate change affecting their basic livelihoods. Food, water, and energy (e.g., firewood) are usually provided by women; therefore, it is women who are most affected by their depletion. In Africa, for example, women are often involved in agriculture, so water scarcity puts them and their families at risk, increasing their workload as they must spend much of their time fetching water. Women and girls are the ones who must collect water in 8 out of 10 households. Of course, this work is neither recognized nor remunerated and ends up meaning they are unable to attend schools (Molinares & Echeverría, 2011; Lane & McNaught, 2009). Climate migration is thus another gendered process that does not receive adequate attention and protection (IOM, 2014).

Human Exploitation of Migrant Women and Girls

Millions of women, expelled from their homes, find themselves ensnared by harsh immigration laws, exacerbating their vulnerability and rendering them perfect targets for trafficking and both sexual and labour exploitation. If work and employment opportunities are limited and if resources are scarce in the receiving region, this could create conditions ripe for human trafficking of migrant women and girls (Brisman et al., 2018). As displacements, migration, and service disruptions increase due to climate disasters, vulnerability to human trafficking also increases. Research shows the relationship between unstable housing and unemployment with involuntary and voluntary entry into sex work. Once a woman enters sex work, the cycle of poverty becomes more pronounced as an increased dependency on others emerges and socio-economic power declines (Gerrard, 2017).

However, this escalating human exploitation often remains concealed for several reasons. Firstly, the crime is frequently mistaken for offences against workers or, more commonly, conflated with migrant smuggling. Additionally, the criminal justice system's inadequate investigative capacity, compounded by prevalent corruption, further obscures these injustices. Moreover, the circumstances of female workers whose prevailing cultural norms—such as the perception that caregiving is exclusively women's responsibility—often renders reporting impossible. Furthermore, the victim's imperative to sustain employment leads many women to be reluctant to cooperate with authorities and leads many to opt for collaboration with traffickers over repatriation. In some cases, victims even seek to advance within the labour hierarchy, transitioning from being exploited to becoming exploiters (Villacampa, 2022a).

Ironically, human trafficking and exploitation serves a precarious labour market that exists as a complement to the capitalist economic system (Hurtado & Pereira-Villa, 2018). While capitalism has fostered efficiency in certain production processes and reduced poverty rates in some regions, its emphasis on economic accumulation has also facilitated the exploitation of individuals within labour markets (Villacampa, 2019). Referred to as the "modern Indies Companies", these entities profit from the influx of desperate migrants, exemplified vividly in Spain through industries such as the cultivation and harvesting of "red gold" (tomatoes and strawberries), with revenues from such activities even included in the GDP. This paradoxical scenario underscores the failure to incorporate this economic dimension into policies aimed at preventing and prosecuting these crimes (García-Sedano, 2020).

Slavery and Domestic Servitude of Climate Migrant Women and Girls

The Feminization of "Dirty Work"

A quintessential example of labour exploitation affecting women and girls is found in domestic service, encompassing tasks such as cleaning and caregiving for the elderly, sick, disabled individuals, children, and pets. The crucial question arises: who undertakes this vital work, essential not only for sustaining human life but, paradoxically, also for maintaining the functioning of the liberal capitalist system? This labour, often labelled as "dirty", involves tasks and roles perceived as repugnant and degrading. The term "dirty" here stems not from the nature of the work itself but rather from its devaluation and pejorative perception, both individually and socially constructed (Lerussi, 2020, p. 409). This labelling, however, significantly impacts the working conditions associated with such tasks, resulting in diminished rights and guarantees, precarious employment contracts often in temporary or part-time formats, heightened levels of informality, and even instances of labour exploitation. Workspaces are characterized by substandard conditions, with limited or no organizational or union representation, and frequently lack legal recognition as legitimate employment (Lerussi, 2020).

In essence, dirty work is structured along social hierarchies based on various factors such as class, race, ethnicity, migratory status, sexuality, and gender. Predominantly, individuals engaged in such tasks are Black, Latino, Caribbean, Asian, indigenous, migrant, and predominantly female. The feminization of poverty exacerbates this dynamic, pushing women and girls into domestic labour and facilitates their exploitation. Given that women often shoulder the responsibility of supporting children, they bear a disproportionate burden when social welfare programmes are cut or entire sectors, such as agriculture, collapse. Furthermore, as families frequently prioritize the education of their sons over their daughters, this leads to girls being compelled to abandoning schooling to assist with family obligations. These tasks, undervalued and often unseen within the occupational structure, are influenced by societal expectations of mothers, wives, and housewives, as dictated by the heterosexual framework in which we are born, live, and die (Lerussi, 2014; Butler, 1990). Despite being essential, such work is consistently devalued, with a prevailing notion that the more it involves helping others, the less it warrants adequate compensation. This perpetuates a narrative of sacrifice expected from those employed in feminized sectors (Lerussi, 2020). It's noteworthy that migrant women and girls are not exempt from these sectors, further emphasizing their vulnerability within these dynamics.

Female Migration and Domestic Exploitation

Outlines and Characteristics of the Phenomenon

It is certainly difficult to link migration to climate change, especially when it comes to slow degradation processes such as drought. But the data do not lie: the climate is forcing more and more people to migrate (CEAR, 2021). And this is a reality that affects women the most, as they are responsible for supplying their families with increasingly scarce resources (water, firewood, agricultural products). Women and girls who, culturally relegated to domestic and care tasks, seek jobs in the destination countries in line with these skills. Tasks carried out in the private sphere, and which give rise to all kinds of abuses (labour, sexual, etc.).

Situations of slavery and domestic servitude are notably pervasive within the context of migration, owing to the heightened vulnerability experienced by migrant workers. Particularly, women and girls in irregular situations often gravitate towards clandestine employment to avoid migration authorities. Payment in cash and the occasional opportunity to spend nights at home provide a semblance of accommodation and security. Paradoxically, it is this very situation that renders them more vulnerable, especially when their regularization depends on their employment status.[4]

Human trafficking can indeed serve as a conduit for domestic slavery or servitude, yet detecting this form of trafficking proves exceedingly challenging, often occurring under the guise of seemingly legal activities and sometimes even socially sanctioned. As such, exploitation can go unnoticed, aided by the complicity of parents or guardians.[5] The United

---

[4] For instance, in numerous Asian and Middle Eastern nations, such as those operating under the kafalah system, migrant worker visas are contingent upon sponsorship by a specific family. Domestic workers are typically unable to change their visa sponsorship without their employer's consent, a feat that is practically unattainable in most cases. Should a domestic worker be terminated from their employment, they may find themselves abruptly without income, legal residency, or familial support. With no provision for a return flight ticket or the right to seek alternative employment, they risk being left destitute and vulnerable on the streets.

[5] As another example, the Committee on the Rights of the Child expressed concern at the growing number of children trafficked internally in Pakistan, often sold by their parents, or forced into marriage, sexual exploitation, or domestic servitude.

Nations Office on Drugs and Crime (UNODC) has cautioned that statistics on this form of trafficking may be biased, as it frequently goes unreported (UNODC, 2022).

Likewise, the Organization for Security and Co-operation in Europe (OSCE) has highlighted how courts may misapply criminal offences, opting for charges such as harbouring a foreigner with irregular documents or exploiting their vulnerable position, rather than recognizing it as human trafficking (OSCE, 2014a). This underscores the need for heightened awareness and improved mechanisms for identifying and addressing instances of domestic slavery and servitude within the context of human trafficking.

Slave-like practices are not confined to what we may perceive as "less developed" regions, serving as a stark reminder of the abuses present even within Europe—an uncomfortable reality often overlooked to evade acknowledging our own transgressions. Despite the voluntary migration of many domestic workers to Europe, some fall victim to deception by their employers, agencies, or intermediaries, finding themselves ensnared in debt bondage or even trafficked, thereby enabling abusive employers or spouses (such as in cases of mail-order brides) to subject them to enslavement. These deplorable situations are further compounded when they occur within the residences of diplomats. In addition to the inherent challenges in detecting domestic servitude due to its clandestine nature within households, the sanctity of diplomatic residences under international law poses an additional barrier to intervention. This combination of factors underscores the urgent need for heightened vigilance and concerted efforts to combat domestic slavery and servitude, even within seemingly privileged contexts.

It is indeed startling, and indeed deeply disconcerting, to discover that a significant portion—approximately one-third—of domestic servitude cases involve diplomats. In these instances, officials exploit the privileges and immunities afforded by their positions to engage in corrupt practices that facilitate the trafficking of domestic workers. These practices may include falsifying visas, confiscating passports, issuing threats of denunciation, and, in some egregious cases, resorting to physical and sexual violence (IBA, 2016). The isolation and invisibility experienced by such victims are particularly pronounced. Beyond language and cultural barriers and the physical distance from their families and support networks, the tasks performed by these workers often occur behind closed doors—cleaning, cooking, childcare, and the like (Rodríguez, 2022).[6] This phenomenon is both unexpected and multifaceted, presenting a formidable initial barrier: diplomatic immunity, which affords perpetrators impunity for their actions. Addressing this issue requires concerted efforts to dismantle the systemic mechanisms that perpetuate such exploitation and to hold accountable those who abuse their positions of power and privilege.

Ultimately, the recognition of this reality in several Member States prompted the publication of an ad hoc handbook by the Organization for Security and Co-operation in Europe (OSCE) in 2014 (OSCE, 2014b). This acknowledgment also spurred further study of a phenomenon that is more prevalent than previously imagined, revealing common characteristics shared across different contexts (Lousada, 2020):

- *The context in which the services are provided.* Unlike other forms of human exploitation, this one take place in the private sphere, which makes it difficult to eradicate, both because of its invisibility and because it is often condoned by the society in which it takes place.
- *The versatility of the services* provided extends beyond traditional domestic chores such as cleaning, cooking, and childcare, often encompassing additional responsibilities like gardening or work in agricultural, industrial, or commercial settings. Unfortunately, exploitative situations may also involve instances of sexual abuse. In some cases, performing tasks becomes secondary to enduring conditions of labour slavery, forced begging, or coerced involvement in prostitution.

---

[6] According to the US State Department's Trafficking in Persons Report (TIP), between 2009 and 2018, cases have been reported in Austria, whose government reported that several foreign diplomats left the country in 2011 and 2012 after being accused of subjecting their domestic workers to forced labour. Similarly, between 2004 and 2005, several diplomats were sanctioned for identical acts in Belgium. In August 2011, a German diplomat in Cyprus was investigated for subjecting a Nepalese domestic worker to forced labour. In April 2013, an official of the Italian consulate in California was found guilty of exploiting a Brazilian domestic worker whose documents he had previously seized. Also in 2013, the Irish government reported two investigations into domestic servitude in diplomats' residences. Similarly, in 2011, a federal court in the canton of Geneva, Switzerland, convicted the Saudi Arabian consul general of violating the rights of two domestic workers from Indonesia, who were required to work 14-hour days for around 300 euros per month.

- *The gender impacts are profound.* Between 80 and 90% of the victims are adult women, adolescents, and girls, a statistic attributed to the pervasive discrimination faced by women globally (ILO, 2013).[7] As observed, sexist stereotypes dictate that domestic chores fall within the purview of women within the family, where they are expected to fulfil these duties without reward, recognition, or room for dissent. Regrettably, these stereotypes permeate the professional realm as well, rendering domestic workers vulnerable to exploitation and objectification, often as objects of sexual gratification. Furthermore, their reproductive rights are frequently denied, with dismissal being a common consequence of pregnancy.
- *The vulnerability of victims is a significant concern*, as underscored by the Report of the Special Rapporteur on Contemporary Forms of Slavery (UN, 2010). This report particularly emphasizes the heightened vulnerability experienced by migrants, exacerbated by intersecting forms of discrimination based on ethnicity, race, colour, or caste. In essence, arbitrary distinctions regarding an individual's worth and dignity, often stemming from insignificant differences in appearance or lineage, serve as the foundation for ruthless exploitation.[8]

In summary, these intricacies are interconnected and mutually reinforcing, shaping domestic work as a personal or quasi-family relationship between the household head, often occupying a higher social status, and the servant, perceived as inferior due to socio-cultural attributes such as sex, age, ethnicity, race, colour, caste, or nationality. Within this dynamic, the household head exerts control over the servant's actions in exchange

[7] The percentages fluctuate between regions: from two-thirds in the Middle East, Eastern Europe, and the former Soviet Union, to over 90% in Latin America and the Caribbean, with female participation at both extremes in Western Europe.

[8] As examples, she cites Brazil, where there is a greater tendency for girls of African origin to be subjected to domestic servitude than lighter-skinned Brazilian girls. In Middle Eastern countries, meanwhile, Filipino migrant domestic workers are paid more than dark-skinned migrants from South Asia or Africa. And in India, the so-called untouchables (Dalits) and some indigenous communities make up most people in captive domestic work, which is a form of domestic servitude. In addition, specific types of degrading domestic work are specifically associated with lower castes and exploitative conditions. In some states in India, despite the government's campaign to end the practice, many Dalit women are still engaged in manual scavenging: because of their caste, they are expected to remove excreta from dry toilets in private homes or public places.

for a level of protection typically limited to basic survival needs, shielded from public scrutiny (Lousada, 2020). This framework echoes medieval attitudes, reminiscent of the authoritarian context of Franco's Spain. Here, domestic servants were regarded as extended family members requiring paternalistic guidance and moral instruction to protect them from perceived societal risks, with the housewife assuming the role of "mother and teacher". This discourse reflects "late-feudal reminiscences", as evidenced by terminology such as servant, servitude, lord, and master. Tragically, this discourse facilitated the exploitation of thousands of young, unmarried women from impoverished provinces by the emerging middle classes (Cañabate, 2014).

The Impunity of Human Exploitation in Spain

However, despite the severity of this reality, the truth remains that convictions for the use and abuse of the labour force in our country are virtually non-existent. The rate of prosecution and punishment for these offences is alarmingly low, contributing to the proliferation of such cases, particularly in sectors like agriculture, domestic work, and cannabis cultivation in Catalonia. To prove this point, between 2015 and 2021, there were only 14 convictions for trafficking and exploitation in the workplace, with just 3 resulting in prison sentences. The scarcity of sentences is especially striking in cases involving exploitation[9] of vulnerable individuals and participation in criminal activities.[10]

Indeed, as the number of female victims of workplace exploitation continues to rise, convictions are typically only attainable in cases where extreme violence can be proved, resulting in the victims suffering physical or psychological trauma. Conviction[11] becomes practically unattainable when exploitation occurs through the abuse threats. This is largely due to the fact that victims, driven by fear, necessity, or other insurmountable

[9] Understood as forced labour since Directive 2011/36/EU.

[10] Up to 2016, there had been only one conviction for exploitation of begging. Between 2016 and 2021, there were only two convictions for trafficking for the purpose of criminal activity.

[11] According to the reports of the Spanish State Prosecutor's Office, between 2013 and 2018, only 9 prosecutions were opened to follow up on the crime of domestic employment.

factors (often including their irregular administrative status), rarely confront their traffickers. Furthermore, victims are often perceived as having consented to their exploitation, as they are believed to have had, or to have had, a "real or acceptable choice" to submit to the abuse.

One of the primary challenges hindering the prosecution and punishment of human exploitation in Spain stems from its distinct differentiation from labour trafficking. Unlike the approach outlined in Article 3 of the Palermo Protocol (UN, 2000), which encompasses exploitation as a separate offense, the Spanish Criminal Code only penalizes exploitation for trafficking (Ramón, 2022). The absence of human exploitation being considered a crime which encompasses both sexual and non-sexual forms, coupled with the complexities surrounding the identification of trafficking for labour exploitation, often results in courts dismissing trafficking charges when victims claim to have engaged in work voluntarily, irrespective of the conditions under which the work was conducted.

The primary recourse, albeit used sparingly, has been to categorize such conduct as crimes against workers' rights (Articles 311 and 312 PC). As labour trafficking and crimes against workers' rights cannot be equated, it makes punishment more difficult. This is because in such cases, "not only is there a violation of the labour or social rights of the worker as such, but there is also a violation of their freedom to decide whether or not to engage in work, as they are subjected to conditions akin to slavery, serfdom, or similar forms of exploitation" (Pomares, 2011; Villacampa, 2022b).

Towards Ending Impunity: Prevention and Response to Human Exploitation

Criminalization of Modern Forms of Slavery

Since the end of 2022, Spain has introduced the *Draft Comprehensive Law against Trafficking and Exploitation of Human Beings* (Ministerio de Justicia, 2022). While pending approval, this legislation aims to combat what are commonly referred to as modern forms of slavery. Aligned with measures undertaken in neighbouring countries such as Germany, France, Italy, and the United Kingdom,[12] Article 177 criminalizes forced labour or services, servitude, and slavery. It's important to note that these concepts are not synonymous; rather, the severity of the offense depends on the level of coercion and the extent to which the victim's freedoms are restricted or denied.

While the underlying characteristic of all these behaviours is the absence of voluntarism in the provision of work or services, distinctions arise based on the extent to which areas of freedom are restricted or eliminated. In instances of *forced labour or services* (Art. 177 ter.1 PC), victim subjugation typically involves the exertion of control, use of violence, intimidation, deception, or exploitation of a position of superiority or the victim's vulnerability (such as threatening to report their irregular situation to authorities). Conversely, *servitude* (Art. 177 ter.2 PC) occurs when the victim is compelled to reside where they provide services or activities, or when their freedom of movement is otherwise restricted, leading them to perceive their condition as permanent with no possibility of change (for example, debt bondage forcing acceptance of subjugated worker status until the debt is repaid) (UN, 1956).[13] Finally, *slavery* (Art. 177 ter.3 PC) manifests itself when the victim's availability is absolute, resulting in a clear objectification of the individual. While this does not equate to ownership of the person, it does involve *de facto* possession, wherein the perpetrator appropriates the value of the victim's labour by stripping them of their inherent status, effectively assuming control as though they were the owner.

[12] 19 European countries have criminalized slavery, servitude and similar practices, and even forced labour, including Austria, France, Germany, Italy, Norway and the United Kingdom.

[13] *Supplementary Convention on the Abolition of Slavery, the Slave Trade, and Institutions and Practices Similar to Slavery*, identified as practices analogous or equivalent to slavery: debt bondage; serfdom of the glebe (a slave attached to an estate and who does not disengage from it upon change of ownership); marriage by promise, assignment, or inheritance; and the assignment of children for labour. Specifically, debt bondage, the most common in sectors such as domestic service, garment workshops and agriculture, is the situation where the debtor undertakes to provide his personal services as security for a debt, provided that the services provided "fairly valued" are not applied to the payment of the debt (it is an undue debt or subject to unfair conditions), or where the duration is not limited and the nature of such services is not defined (Art. 1.1).

Globalization of Human Rights and Decrease in Demand: Another Challenge to Feminism

Make no mistake: modern slavery is not merely a malady to be cured, but rather an integral component of a system characterized by domination, exploitation, and a structure perpetuated by capitalism over centuries. It represents a profound violation of the fundamental aspects of human existence. But the solution is not easy "with most policymakers being white affluent men who are not substantially impacted by climate change, those making policy decisions do not understand the unique needs of those most impacted by a warming planet nor appreciate the urgency to act" (Haley & Arrigo, 2022). Consequently, the reversal of this economic paradigm necessitates the global promotion of human rights in a climate change context. Achieving this objective demands a reordering of priorities and the adoption of a comprehensive, multi-tiered approach. Addressing this issue effectively requires concerted action from all stakeholders: governments, the private sector, workers, and citizens alike.

Particularly crucial are the roles of citizens, both as consumers and as members of society (Carballo, 2021). It is imperative to address demand, especially focusing on aspects that violate human rights. Discrimination based on race, gender, ethnicity, or age often manifests itself in the demand for services which result in trafficking and exploitation, including domestic work. Consequently, international guidance also aims to penalize demand for these services. Both the Warsaw Convention (Art. 19)(Council of Europe, 2005), Directive 2011/36/EU of European Parliament and Council of European Union on preventing and combating trafficking in human beings and protecting its victims (Art. 18.4) and the EU Strategy on Combatting Trafficking of Human Beings (2021–2025) (COM (2021) 171 final) mention the (optional) possibility of implementing measures that criminalize the use of exploitative services, when it is known that the individual is a victim of trafficking. From an economic perspective on trafficking, the law of supply and demand underpins this illicit industry; therefore, reducing demand leads to a decrease in supply.

Aligned with this objective, the Spanish Draft Bill penalizes the solicitation, even though gross negligence (Art. 177 quater2), of anyone who, without having directly participated as a perpetrator or accomplice, benefits from the services, advantages, or activities of a victim subjected to forced labour or services, servitude, or slavery (paragraph 1). This measure is necessary but insufficient, as demand persists, inevitably fuelling supply. Moreover, demand not only fails to diminish but is also often encouraged within society using new technologies or the consumption of pornography.

Recognizing this reality and driven by a preventive approach, the *Draft Bill* also emphasizes efforts on awareness-raising initiatives across various sectors, including education, training, health, advertising, and the media (Art. 16). Specifically, within the realm of higher education, the legislation anticipates the incorporation of comprehensive knowledge on the phenomenon of trafficking and exploitation of human beings, across diverse academic disciplines, underscoring the grave violations of human rights inherent in these practices and fostering teaching, research, and knowledge dissemination on these issues (Art. 7). Naturally, achieving this objective necessitates both initial and continuous training for educators in human rights, with a specific focus on combating trafficking and exploitation (Art. 9).

Education and awareness-raising serve as crucial initial steps. If we genuinely aspire to eradicate the appalling reality of human exploitation, taking consumer actions against products or services made or offered by victims of trafficking and exploitation becomes imperative. It is incumbent upon all of us to adopt a more proactive stance in combating these practices by deepening our understanding of the ramifications of our decisions to purchase and utilize certain services (Carballo, 2021). If the pursuit of cheap goods or services outweighs concern for the individuals who produce or provide them, little progress will be made (ICAT, 2014; ITUC, 2011). Unfortunately, this seems to be the trajectory of our consumerist, sexist, and apathetic society. A society that, in its relentless pursuit of consumerist comfort, even denies undeniable truths such as climate change and its myriad consequences, including the forced migration of countless vulnerable women and girls. These individuals, infinitely vulnerable, may eventually gain access to necessities like water and energy, but at the cost of continued deprivation of education and essential opportunities. These women and girls, who undertake the often-denigrated "dirty work" traditionally associated with femininity, in a way, sacrifice their own aspirations so that others may pursue theirs. This poses yet another profound contradiction to be addressed by feminism.

In any case, "if societies continue to tolerate the environmental and social harm of industrial processes in the name of economic growth, capitalism and consumerism will continue to be two of the biggest drivers of social injustice" (Haley & Arrigo, 2022). Women and girls in the countries most vulnerable to climate change have the most to lose. The lack of options, both in the country of origin and in the host country, turns them into objects of marketing and exploitation. They will continue to be treated as things in the service of those who were fortunate enough to be born in the right time and place.

Human exploitation is, therefore, another painful consequence of a climate change that affects us all, but for which we are not all equally responsible. Those of us who inhabit the Global North are mainly to blame for this climate disaster. Economic growth and limitless consumerism cannot be the eternal excuse for slowing down measures that are extremely urgent. Time is really running out while we remain obsessed with consuming and accumulating a series of products that, in their case, would never quench our thirst or hunger. And without food and water, we simply do not exist.

Bibliography

Agnew, R. (2012). Dire forecast: A theoretical model of the impact of climate change on crime. *Theoretical Criminology, 16*(1), 21–42.

Alston, M. (2013). Introducing gender and climate change: Research, policy, and action. In *Research, action, and policy: Addressing the gendered impacts of climate change* (pp. 3–14). Springer.

Brisman, A., South, N., & Walters, R. (2018). Southernizing green criminology: Human dislocation, environmental injustice and climate apartheid. *Justice, Power & Resistance*, 1–22. https://repository.essex.ac.uk/23195/1/Brisman_South%20and%20Walters%202018_Southernizing%20in%20European%20Group.pdf

Brown, L. R. (1976). *Twenty-two dimensions of the population problem.* Worldwatch Institute.

Butler, J. (1990). *Cuerpos que importan. Sobre los límites materiales y discursivos del "sexo".* Paidós.

Cañabate, J. (2014). "Servir es distinto a trabajar": Potestad del cabeza de familia vs Protección social del servicio doméstico en la dictadura franquista. In G. Garcia & M. Espugny I Tomas (Coord.), *Relaciones laborales y empleados del hogar: reflexiones jurídicas* (pp. 61–84). Dykinson.

Carballo, M. (2021). *Explotación, esclavitud y trata de seres humanos. Historia, debates y limitaciones jurídicas.* Tirant lo Blanch.

CEAR-Greenpeace. (2021). *Huir del clima. Cómo influye la crisis climática en las migraciones humanas.* https://www.cear.es/wp-content/uploads/2021/10/informe-huir-del-clima.pdf

Council of Europe. (2005). *Convention on action against trafficking in human beings.* Council of Europe Treaty Series, no. 197. https://rm.coe.int/168008371d

Dominelli, L. (2013). Gendering climate change: Implications for debates, policies, and practices. In *Research, action, and policy: Addressing the gendered impacts of climate change* (pp. 77–93). Springer.

Duncan, K. (2007). Global climate change and women's health. Women and Environments International Magazine, 74/75, 10–12. https://www.yorku.ca/weimag/BACKISSUES/images/WEI%20scan%202007%20iss%2074-75%20part%201_2.pdf

El-Hinnawi, E. (1985). *Environmental refugees.* United Nations Environment Programme.

FAO. (2000). *The state of food and agriculture.* United Nations.

Felipe, B. (2019). *Las migraciones climáticas ante el ordenamiento jurídico internacional.* Aranzadi-Thomson Reuters.

García-Sedano, T. (2020). *La detección, identificación y protección de las víctimas de trata de seres humanos.* Editorial Reus.

Gerrard, M. B. (2017). Climate change and human trafficking after the Paris agreement. *University of Miami Law Review, 72*, 345.

Haigh, C., & Vallely, B. (2010). *Gender and the climate change agenda: The impacts of climate change on women and public policy.* Women's Environmental Network.

Haley, S., & Arrigo, B. (2022). Ethical considerations at the intersection of climate change and reproductive justice: Directions from green criminology.

Critical Criminology, 30, 1001–1018. https://doi.org/10.1007/s10612-022-09661-w

Henrici, J., Childers, R., & Shaw, E. (2015). *Get to the bricks: The experiences of Black women from New Orleans public housing after Hurricane Katrina*. Institute For Women's Policy Research. https://iwpr.org/wp-content/uploads/2020/08/D506_GetToTheBricks.pdf

Hurtado, M., & Pereira-Villa, C. (2018). Inserción laboral adversa: otra cara del "demonio amorfo" de la trata de seres humanos. *Cadernos pagu, 53*, 1–24.

IDCM. (2018). *Global report on internal displacement*. Internal Displacement Monitoring Centre.

IDCM. (2019). *Global report on internal displacement*. Internal Displacement Monitoring Centre.

IDCM. (2024). *Global report on internal displacement*. Internal Displacement Monitoring Centre.

Inter-Agency Coordination Group Against Trafficking in Persons (ICAT). (2014). *Preventing trafficking in persons by addressing demand*. ICAT Paper Series, Issue 2. https://www.unodc.org/documents/human-trafficking/2014/ICAT_Demand_paper_FINAL.pdf

International Bar Association (IBA). (2016). *Human trafficking and corruption. A report by the IBA's presidential task force against human trafficking*. https://www.ibanet.org/MediaHandler?id=E34FFA1D-8038-4AEC-A631-E0E2A7E0AD86

International Labour Organization (ILO). (2013). *Domestic workers across the world: Global and regional statistics and the extent of legal protection*. https://www.ilo.org/wcmsp5/groups/public/%2D%2D-dgreports/%2D%2D-dcomm/%2D%2D-publ/documents/publication/wcms_173363.pdf

International Organization for Migration (IOM). (2014). *IOM outlook on migration, environment and climate change*. https://publications.iom.int/system/files/pdf/mecc_outlook.pdf

International Trade Unión Organization (ITUC). (2011). *Never work alone. Trade Unions and NGOs joining forces to combat Forced Labour and Trafficking in Europe*. https://www.antislavery.org/wp-content/uploads/2017/01/forced_labour_en_final.pdf

IPCC. (2021). *Climate change 2021: The physical science basis*. Cambridge University Press. https://report.ipcc.ch/ar6/wg1/IPCC_AR6_WGI_FullReport.pdf

Islam, N., Ehsan, S. A., & Khanam, B. (2018). Factors of population movement towards Dhaka: An empirical study. *Bangladesh Journal of Administration and Management, 30*(2), 82–98.

Lane, R., & McNaught, R. (2009). Building gendered approaches to adaptation in the Pacific. *Gender & Development, 17*(1), 67–80.

Lerussi, R. (2014). De vuelta al debate sobre la domesticidad. *Mora (Buenos Aires), 20*(2), 93-104

Lerussi, R. (2020). Trabajo sucio y feminización. Lecturas jurídicas feministas. In S. Olarte & E. Pérez (Dir.), *Formas contemporáneas de esclavitud y derechos humanos en clave de globalización, género y trata de personas* (pp. 415–430). Tirant Lo Blanch.

Lousada, J. (2020). La servidumbre doméstica y su tratamiento en el derecho español. In E. Pérez & S. Olarte (Dir.), *Formas contemporáneas de esclavitud y derechos humanos en clave de globalización, género y trata de personas* (pp. 421–460). Tirant lo Blanch.

Lynch, M. J. (1990). The greening of criminology: A perspective on the 1990s. *Critical Criminologist, 2*(3–4), 11–12.

MacDonald, R. (2005). How women were affected by the tsunami: A perspective from Oxfam. *PLoS Medicine, 2*(6), e178.

Ministerio de Justicia. (2022). *Anteproyecto de Ley Integral contra la Trata y la Explotación de Seres Humanos*. https://www.mjusticia.gob.es/es/AreaTematica/ActividadLegislativa/Documents/Anteproyecto%20de%20Ley%20Org%C3%A1nica%20Trata%20TAIP.pdf

Molinares, V., & Echeverría, J. (2011). El derecho humano al agua: posibilidades desde una perspectiva de género. *International Law: Revista Colombiana de Derecho Internacional, 19*, 269–301.

Neumayer, E., & Plümper, T. (2007). The gendered nature of natural disasters: The impact of catastrophic events on the gender gap in life expectancy, 1981–2002. *Annals of the Association of American Geographers, 97*(3), 551–566. https://doi.org/10.1111/j.1467-8306.2007.00563.x

Organization for Security and Co-Operation in Europe (OSCE). (2014a). *Unprotected work, invisible exploitation: Trafficking for the purpose of domestic servitude*. https://www.osce.org/files/f/documents/2/a/75804.pdf

Organization for Security and Co-Operation in Europe (OSCE). (2014b). *How to prevent human trafficking for domestic servitude in diplomatic households and protect private domestic workers*. https://www.osce.org/files/f/documents/2/c/126303.pdf

Pachauri, R. K., & Meyer, L. A. (2014). *Fifth assessment report of the intergovernmental panel on climate change* (R. K. Pachauri & L. A. Meyer, Ed.). IPCC, 1–151.

Pomares, E. (2011). El delito de trata de seres humanos con finalidad de explotación laboral. *Revista Electrónica de Ciencia Penal y Criminología, 13–15*, 15:1–15:31. http://criminet.ugr.es/recpc/13/recpc13-15.pdf

Ramón, E. (2022). La explotación laboral como finalidad propia del delito de trata de personas. In C. Villacampa (Dir.), *La trata de seres humanos tras un decenio de incriminación ¿Es necesaria una ley integral para luchar contra la trata y la explotación de seres humanos?* (pp. 423–458). Tirant lo Blanch.

Rodríguez, S. (2022). *Trata de seres humanos y corrupción*. Tirant lo Blanch.

Schwoebel, M. H., & Menon, G. (2004). *Mainstreaming gender in disaster management support project*. CEDPA and Chemonics International Inc.

Sorensen, C., Murray, V., Lemery, J., & Balbus, J. (2018). Climate change and women's health: Impacts and policy directions. *PLoS Medicine, 15*(7), e1002603.

South, N. (1998). A green field for criminology? A proposal for a perspective. *Theoretical Criminology, 2*(2), 211–233.

UN. (1956). *Supplementary convention on the abolition of slavery, the slave trade, and institutions and practices similar to slavery*. https://www.ohchr.org/en/instruments-mechanisms/instruments/supplementary-convention-abolition-slavery-slave-trade-and

UN. (2010). *Report of the Special Rapporteur on contemporary forms of slavery, including its causes and consequences*. https://digitallibrary.un.org/record/847436/files/A_HRC_33_46-EN.pdf?ln=es

UN. (2019). *World population prospects 2019: Highlights*. Department of Economic and Social Affairs Population Division. https://population.un.org/wpp/publications/files/wpp2019_highlights.pdf

United Nations Office on Drugs and Crime (UNODC). (2022). *Global report on trafficking in persons 2022*. https://www.unodc.org/documents/data-and-analysis/glotip/2022/GLOTiP_2022_web.pdf

Vidal, J. (2011). "Drier, hotter: can Egypt escape its climate future?", The Guardian.

Villacampa, C. (2019). La trata de seres humanos: concepto y caracterización. In E. Pérez & E. Pomares Cintas (Coord.), *La trata de seres humanos en el contexto penal iberoamericano* (pp. 21–62). Tirant lo Blanch.

Villacampa, C. (2022a). Dificultades en la persecución penal de la trata de seres humanos para explotación laboral. *InDret, 2*. https://doi.org/10.31009/InDret.2022.i2.06

Villacampa, C. (2022b). Introducción: acerca de la conveniencia de una ley integral para afrontar la trata y la explotación severa de seres humanos. In C. Villacampa (Dir.), *La trata de seres humanos tras un decenio de incriminación ¿Es necesaria una ley integral para luchar contra la trata y la explotación de seres humanos?* (pp. 29–50). Tirant lo Blanch.

UN. (2000). Protocol to Prevent, Suppress and Punish Trafficking in Persons Especially Women and Children. General Assambly resolution 55/25, 15 November 2000. https://www.ohchr.org/en/instruments-mechanisms/instruments/protocol-prevent-suppress-and-punish-trafficking-persons

Correction To: Conceptualizing Ecocide: Interdisciplinary Perspectives

Daan van Uhm, Cedric Ryngaert, Susanne Knittel, and Marijke van Kuijk

Correction to: Chapter 5 in : D. van Uhm, D. Siegel (eds.), *Global Green Crime and Ecojustice*, Palgrave Studies in Green Criminology, https://doi.org/10.1007/978-3-031-91557-4_5

The Chapter "Conceptualizing Ecocide: Interdisciplinary Perspectives", by Daan van Uhm, Cedric Ryngaert, Susanne Knittel, and Marijke van Kuijk, was published without open access. Following the authors' decision to opt for open access, the copyright of the chapter changed on 1 February 2026 to © The Author(s) 2026 and the chapter is now distributed under the terms of the **Creative Commons AttributionNonCommercial-NoDerivatives 4.0 International License** (https://creativecommons.org/licenses/by-nc-nd/4.0/), which permits use, sharing, adaptation, distribution and reproduction in any medium or format, as long as you

The updated version of this chapter can be found at https://doi.org/10.1007/978-3-031-91557-4_5

D. van Uhm, D. Siegel (eds.), *Global Green Crime and Ecojustice*, Palgrave Studies in Green Criminology, https://doi.org/10.1007/978-3-031-91557-4_21

Index[1]

[1] Note: Page numbers followed by 'n' refer to notes.

D. van Uhm, D. Siegel (eds.), *Global Green Crime and Ecojustice*, Palgrave Studies in Green Criminology, https://doi.org/10.1007/978-3-031-91557-4

FSC
www.fsc.org
MIX
Papier | Fördert
gute Waldnutzung
FSC® C083411

Kolibri 360 GmbH
Ferdinand-Jühlke-Straße 7
99095 Erfurt, Deutschland
produktsicherheit@kolibri360.de